Springer Proceedings in Mathematics & Statistics

Volume 27

T0205635

For further volumes:
http://www.springer.com/series/10533

Springer Proceedings in Mathematics & Statistics

This book series features volumes composed of select contributions from workshops and conferences in all areas of current research in mathematics and statistics, including OR and optimization. In addition to an overall evaluation of the interest, scientific quality, and timeliness of each proposal at the hands of the publisher, individual contributions are all refereed to the high quality standards of leading journals in the field. Thus, this series provides the research community with well-edited, authoritative reports on developments in the most exciting areas of mathematical and statistical research today.

Guy Latouche • Vaidyanathan Ramaswami
Jay Sethuraman • Karl Sigman
Mark S. Squillante • David D. Yao
Editors

Matrix-Analytic Methods in Stochastic Models

Springer

Editors
Guy Latouche
Blvd. du Triomphe, CP 212
Bruxelles
Belgium

Vaidyanathan Ramaswami
AT & T Labs Research
Florham Park
New Jersey, USA

Jay Sethuraman
IEOR
Columbia University
New York City
New York, USA

Karl Sigman
Department of Industrial Engineering
and Operation Research
New York City
New York, USA

Mark S. Squillante
IBM Thomas J. Watson Research Center
Yorktown Heights
New York, USA

David D. Yao
IEOR
Columbia University
New York City
New York, USA

ISSN 2194-1009 ISSN 2194-1017 (electronic)
ISBN 978-1-4899-9424-0 ISBN 978-1-4614-4909-6 (eBook)
DOI 10.1007/978-1-4614-4909-6
Springer New York Heidelberg Dordrecht London

Mathematics Subject Classification (2010): 60-06, 60G, 60H, 60J, 60K, 15B05, 65F99

Printed on acid-free paper

Springer is part of Springer Science+Business Media (www.springer.com)

Preface

Matrix-analytic and related methods have become recognized as an important and fundamental approach to the mathematical analysis of general classes of complex stochastic models. Research in the area of matrix-analytic and related methods seeks to discover underlying probabilistic structures intrinsic in such stochastic models, develop numerical algorithms for computing functionals (e.g., performance measures) of the underlying stochastic processes, and apply these probabilistic structures or computational algorithms within a wide variety of fields including computer science and engineering, telephony and communication networks, electrical and industrial engineering, operations research, management science, financial and risk analysis, and biostatistics. These research studies provide deep insights into and understanding of the stochastic models of interest from a mathematical or applications perspective.

From 13 through 16 June 2011, the Seventh International Conference on Matrix-Analytic Methods in Stochastic Models – MAM7 – was held at Columbia University in New York, NY, USA continuing the rich tradition of previous successful MAM conferences in Flint (1995), Winnipeg (1998), Leuven (2000), Adelaide (2002), Pisa (2005), and Beijing (2008). The MAM7 conference was sponsored by the Center for Applied Probability (CAP) at Columbia University and IBM Research, as well as the Applied Probability Society of INFORMS; MAM7 also thanks ACM SIGMETRICS for financial support.

The conference brought together researchers working on the theoretical, algorithmic, and methodological aspects of matrix-analytic and related methods in stochastic models, as well as the applications of such mathematical research across a broad spectrum of fields. In particular, the conference provided an international forum for presenting recent research results on the theory, algorithms, and methodologies concerning matrix-analytic and related methods in stochastic models; presenting recent research results on the application of matrix-analytic and related methods to address problems arising within a wide variety of fields; reviewing and discussing methodologies and related algorithmic analysis; improving collaborations among researchers in applied probability, operations research, computer science, engineering, and numerical analysis; and identifying directions for future research.

All submitted chapters were reviewed by at least 4 members of the scientific advisory committee, resulting in a total of 37 submissions being selected for presentation at the MAM7 conference and inclusion in an informal proceedings distributed at the conference. In addition, plenary talks were given by Edward Coffman, Steven Kou, Marcel Neuts, and Devavrat Shah. This book, the formal proceedings of MAM7, contains a selection of papers from the conference program, covering various aspects of matrix-analytic and related methods in stochastic models and their applications across many different fields.

In Chap. 1, Baek et al. establish the factorization properties of a MAP-modulated fluid flow model under generalized server vacations and two types of increasing fluid patterns during idle periods. Chapter 2, by Bini et al., considers quasi-birth-and-death processes with low-rank downward and upward transitions and show how this structure can be exploited to reduce the computational cost of the cyclic reduction iteration. In Chap. 3, Bladt et al. define and study the classes of bilateral and multivariate bilateral matrix-exponential distributions that have support on the entire real space and have rational moment-generating functions. In Chap. 4, Casale and Harrison propose algorithms to automatically generate exact and approximate product-form solutions for large Markov processes that cannot be solved by direct numerical methods. In Chap. 5, Hautphenne et al. consider multitype Markovian branching processes subject to catastrophes that kill random numbers of living individuals at random epochs, providing characterizations of certain cases. He et al. present in Chap. 6 majorization results for phase-type generators on the basis of which bounds for the moments and Laplace–Stieltjes transforms of phase-type distributions are obtained. In Chap. 7, Horváth and Telek propose efficient random variate generation methods to support simulation evaluation of matrix exponential stochastic models based on appropriate representations of the models. The chapter by Kobayashi and Miyazawa, Chap. 8, considers a two-dimensional skip-free reflecting random walk on a nonnegative integer quadrant and derives exact tail asymptotics for the stationary probabilities on the coordinate axis, assuming it exists. In Chap. 9, Latouche et al. consider a two-dimensional stochastic fluid model with multiple inputs and temporary assistance and derive the marginal distribution of the first buffer and bounds for that of the second buffer. Chapter 10, by Ramaswami, provides an introduction to Brownian motion and stochastic integrals using linear fluid flows on finite-state Markov chains, which can facilitate the development of algorithms for stochastic integration. In Chap. 11, Van Houdt and Pérez study a supply chain consisting of one manufacturer and two retailers, develop a GI/M/1-type Markov chain to analyze this supply chain, and exploit fast numerical methods to solve the chain.

Many people deserve thanks for the important roles they played in making the MAM7 conference a great success. We thank the plenary and regular speakers and coauthors for their presentations at and participation in the conference and express our gratitude to all other conference attendees as well. We also thank our fellow scientific advisory committee members, listed in the next section. Special thanks go to Parijat Dube and Risa Cho, webmasters for the conference Web site, and to Jessie Gray, Adina Brooks, and other staff members from the IEOR Department at Columbia University for their many efforts and assistance. Without all the work and support from these groups of people, the MAM7 conference would not have been possible. We thank Columbia University for hosting the conference and our sponsors for their financial support. Finally, we thank Donna Chernyk and the editorial staff at Springer for all of their assistance with this book as part of the Proceedings in Mathematics series.

Brussels, Belgium — Guy Latouche
Florham Park, NJ — Vaidyanathan Ramaswami
New York City, NY — Jay Sethuraman
New York City, NY — Karl Sigman
Yorktown Heights, NY — Mark S. Squillante
New York City, NY — David D. Yao

Scientific Advisory Committee

Ivo J.B.F. Adan, Eindhoven University of Technology, the Netherlands
Attahiru S. Alfa, University of Manitoba, Canada
Jesus R. Artalejo, Complutense University of Madrid, Spain
Konstantin Avrachenkov, INRIA, France
Florin Avram, University of Pau, France
Nigel Bean, University of Adelaide, Australia
Dario Bini, University of Pisa, Italy
Mogens Bladt, National University of Mexico, Mexico
Srinivas R. Chakravarthy, Kettering University, USA
Tugrul Dayar, Bilkent University, Turkey
Erik van Doorn, University of Twente, the Netherlands
David Gamarnik, Massachusetts Institute of Technology, USA
Ayalvadi Ganesh, University of Bristol, UK
Peter Harrison, Imperial College, UK
Qi-Ming He, University of Waterloo, Canada
Andras Horvath, University of Turin, Italy
William J. Knottenbelt, Imperial College, UK
Udo R. Krieger, University of Bamberg, Germany
Dirk Kroese, University of Queensland, Australia
Guy Latouche, Université Libre de Bruxelles, Belgium
Ho Woo Lee, Sungkyunkwan University, Korea
Johan S.H. van Leeuwaarden, Eindhoven University of Technology, the Netherlands
Yingdong Lu, IBM T.J. Watson Research Center, USA
Beatrice Meini, University of Pisa, Italy
Masakiyo Miyazawa, Tokyo University of Science, Japan
Marcel F. Neuts, University of Arizona, USA
Bo Friis Nielsen, Technical University of Denmark, Denmark
Rudesindo Nunez Queija, University of Amsterdam & CWI, the Netherlands
Takayuki Osogami, IBM Tokyo Research Laboratory, Japan
Antonio Pacheco, Instituto Superior Técnico, Portugal
Harry Perros, North Carolina State University, USA

Contents

Contributors

Soohan Ahn Department of Statistics, University of Seoul, Seoul, Korea

Jung Woo Baek Department of Electrical and Computer Engineering, University of Manitoba, Winnipeg, Canada

Dario A. Bini Dipartimento di Matematica, Università di Pisa, Italy

Mogens Bladt Instituto de Investigaciones en Matematicas Aplicadas y en Sistemas, UNAM, Mexico, DF

Giuliano Casale Imperial College London, Department of Computing, London, UK

Luz Judith R. Esparza Department of Informatics and Mathematical Modelling, DTU, Lyngby, Denmark

Paola Favati Istituto di Informatica e Telematica, CNR, Pisa, Italy

Peter G. Harrison Imperial College London, Department of Computing, London, UK

Sophie Hautphenne Department of Mathematics and Statistics, University of Melbourne, Melbourne, Victoria, Australia

Qi-Ming He University of Waterloo, Waterloo, ON, Canada

Gábor Horváth Department of Telecommunications Technical University of Budapest, Budapest, Hungary

Masahiro Kobayashi Tokyo University of Science, Noda, Chiba, Japan

Guy Latouche Départment d'informatique, Université Libre de Bruxelles, Blvd du Triomphe, Bruxelles, Belgium

Ho Woo Lee Department of Systems Management Engineering, Sungkyunkwan University, Suwon, Korea

Se Won Lee Department of Business Administration, Dongguk University-Seoul, Seoul, Korea

Beatrice Meini Dipartimento di Matematica, Università di Pisa, Italy

Masakiyo Miyazawa Tokyo University of Science, Noda, Chiba, Japan

Giang T. Nguyen Département d'informatique, Université Libre de Bruxelles, Blvd du Triomphe, Brussels, Belgium

Bo Friis Nielsen Department of Informatics and Mathematical Modelling, DTU, Lyngby, Denmark

Zbigniew Palmowski University of Wroclaw, Mathematical Institute, Wroclaw, Poland

J.F. Pérez Department of Electrical and Electronics Engineering, Universidad de los Andes, Bogotá, Colombia

V. Ramaswami AT&T Labs, Research, Florham Park, NJ, USA

Miklós Telek Department of Telecommunications Technical University of Budapest, Budapest, Hungary

B. Van Houdt Department of Mathematics and Computer Science, University of Antwerp – IBBT, Antwerp, Belgium

Juan C. Vera Tilburg University, Tilburg, The Netherlands

Hanqin Zhang University of Singapore, Singapore

Chapter 1
Factorization Properties for a MAP-Modulated Fluid Flow Model Under Server Vacation Policies

Jung Woo Baek, Ho Woo Lee, Se Won Lee, and Soohan Ahn

Introduction

The classic Markov-modulated fluid flow (MMFF) model is a stochastic model in which the rate of change of the fluid level is modulated by an underlying Markov chain (UMC). It was introduced by Anick et al. to analyze a data-handling system with multi-input sources [6]. More details about the conventional MMFF model can be found in Aggarwal et al. [1], Ahn [2], Ahn and Ramaswami [3–5], Asmussen [7], Mitra [22], and references therein.

In the conventional MMFF model, the idle server begins to process the fluid as soon as the zero fluid level becomes positive. This simple behavior of the MMFF model has limited wider applications to real-world systems. In an effort to overcome this drawback, a feedback fluid flow model was introduced recently by Malhotra et al. [20]. They considered buffer thresholds to improve the system performance. We refer the interested reader to Da Silva and Latouche [11], Mandjes et al. [21] and Van Foreest et al. [23] for more details about the feedback queue.

J.W. Baek
Department of Electrical and Computer Engineering, University of Manitoba, Winnipeg, Canada
e-mail: rainbeak@gmail.com

H.W. Lee (✉)
Department of Systems Management Engineering, Sungkyunkwan University, Suwon, Korea
e-mail: hwlee@skku.edu

S.W. Lee
Department of Business Administration, Dongguk University-Seoul, Seoul, Korea
e-mail: swlee94@dongguk.edu

S. Ahn
Department of Statistics, University of Seoul, Seoul, Korea
e-mail: sahn@uos.ac.kr

G. Latouche et al. (eds.), *Matrix-Analytic Methods in Stochastic Models*, Springer Proceedings in Mathematics & Statistics 27, DOI 10.1007/978-1-4614-4909-6_1,

In this chapter, we consider a new modification of the conventional MMFF model such that the system has a vacation period whenever the fluid level reaches zero. During the vacation period, no service is rendered by the server and the fluid level only increases by the inflow of the fluid. We consider two patterns of fluid increase during the vacation period: the vertical increase (Type V) and the linear increase (Type L). For Type V systems, we assume that the fluid arrives from outside according to the Markovian arrival process (MAP) [19]. For type L systems, we assume that the fluid level increases linearly or stays unchanged depending on the phase of the UMC. When the vacation period ends, the server starts to process the fluid immediately. We will call this model a MAP-modulated fluid flow model.

Similar, but not equivalent, models can be found in the literature. Kulkarni and Yan [15] studied a fluid inventory model with instant stock replenishment. They assumed that buffer content increased or decreased according to the flow rates governed by a UMC. The buffer is replenished to a predetermined level instantaneously whenever it becomes empty. Their model is similar to ours if lead times are considered. For vacation policies and their applications in queueing systems, readers are referred to Doshi [12], Fuhrmann and Cooper [13], Heyman [14], Baek et al. [8,9], Chang et al. [10], Lee and Baek [16], and Lee et al. [17,18].

For the MAP(BMAP)/G/1 queue under generalized vacation policies, it is known that the vector Laplace–Stieltjes transform (LST) $\mathbf{u}^*(\theta)$ of workload distribution at an arbitrary time point is factored into the following two parts [10]:

$$\mathbf{u}^*(\theta) = \mathbf{u}^*_{\text{idle}}(\theta) \cdot \mathbf{W}^*(\theta), \tag{1.1}$$

where

$$\mathbf{W}^*(\theta) = \theta\left[\theta\mathbf{I} + \mathbf{D}[S^*(\theta)]\right]^{-1}. \tag{1.2}$$

In (1.1) and (1.2), $\mathbf{u}^*_{\text{idle}}(\theta)$ is the vector LST of workload level at an arbitrary idle time point and $\mathbf{W}^*(\theta)$ is the matrix LST common to all generalized vacation systems. $\mathbf{D}(z)$ is the matrix-generating function of the parameter matrices $\{\mathbf{D}_1, \mathbf{D}_2, \ldots\}$ of BMAP, and $S^*(\theta)$ is the LST of the service time distribution. The importance of factorization (1.1) is that the analysis of any BMAP/G/1 queue with vacations is reduced to obtaining $\mathbf{u}^*_{\text{idle}}(\theta)$.

The objective of this chapter is to propose a unified fluid level formula like (1.1) for a MAP-modulated fluid flow model under generalized server vacations.

The System and the Model

In this section, we describe our model. We also review some known results of the conventional MMFF model for later use.

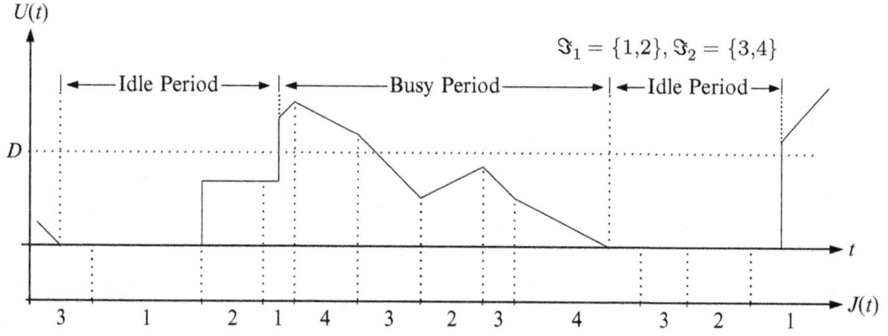

Fig. 1.1 Type V MAP-modulated fluid flow model under the D-policy

The System

We consider a class of stochastic fluid flow systems with the following specifications:

1. The fluid level increases with arriving customers and decreases with the server's processing (service). The arrival process and the processing rates are governed by a MAP with parameter matrices $\mathbf{D}_0$ and $\mathbf{D}_1$. Thus, the governing background process is a continuous-time Markov chain with an infinitesimal generator $\mathbf{Q} = \mathbf{D}_0 + \mathbf{D}_1$. We will call this background process $\{J(t), t \geq 0\}$ the UMC.
2. During the busy period (processing period), the rate of change in fluid level is r_i if the UMC is in phase i. If $r_i > 0$, then the fluid level increases linearly. If $r_i < 0$, then the fluid level decreases linearly. We have two sets $\{\Im_1, \Im_2\}$ of phases, where $\Im_1$ is the set of UMC phases with increasing rates and $\Im_2$ is the set of UMC phases with decreasing rates.
3. As soon as the fluid level becomes zero, the system becomes idle and the server leaves for a vacation until a predetermined reactivation condition is satisfied. During the idle period, the server does not process the fluid. Examples of vacation policies are given in the section "Example systems."

Types of Level Increases During Idle Period

We assume that each system may have either one of the following two increase patterns during the idle period:

(a) Type V: During the idle period, arriving customers bring in a random amount S of fluid and the system level jumps up vertically (Fig. 1.1).
(b) Type L: During the idle period, the fluid level increases linearly at a rate v_i, v_i being either zero or positive depending upon the UMC phase. We have two sets $\{\Im_1^{\text{idle}}, \Im_2^{\text{idle}}\}$, where $\Im_1^{\text{idle}}$ is the set of UMC phases with increasing rates and $\Im_2^{\text{idle}}$ is the set of UMC phases without any level change (Fig. 1.2).

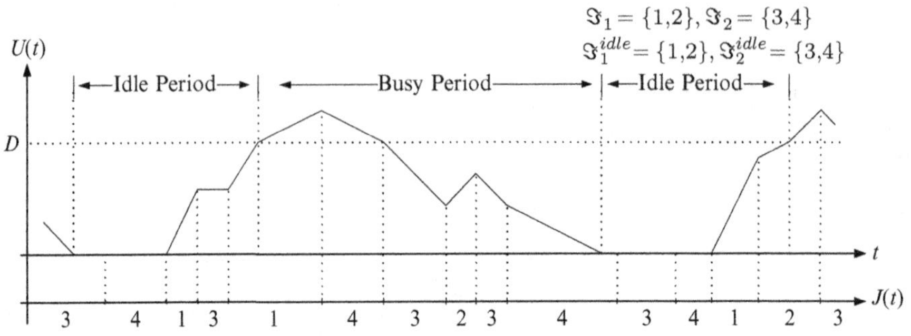

Fig. 1.2 Type L MAP-modulated fluid flow model under the D-policy

Assumptions

We define the MAP-modulated fluid flow system under generalized vacations as the fluid flow system that satisfies the foregoing specifications and the following assumptions:

Assumptions 1.1 The fluid model we are studying satisfies the following assumptions:

1. The jump sizes in the Type V system are independent and identically distributed (i.i.d.) and independent of the arrival process, the vacation process, and the phases of the UMC.
2. The system is work-conserving and stable.
3. The input and output rates of fluid are independent of the fluid level and depend only on the UMC phase.
4. As soon as the system becomes busy, the server begins to process the fluid until the system becomes empty (exhaustive service).

Example Systems

What follows are some descriptive examples of a MAP-modulated fluid flow system under generalized server vacations.

Example 1.1 (Multiple vacations). In this system, the server leaves for repeated vacations of i.i.d. random length $\{V_1, V_2, \ldots\}$ as soon as the fluid level reaches zero. If there exists any fluid at the end of a vacation, then the server begins to process the fluid immediately. If not, then the server leaves for another vacation.

Example 1.2 (Single vacation). In this system, the server leaves for a vacation of random length V as soon as the fluid level becomes zero. If there exists any fluid at the end of the vacation, then the busy period starts immediately. If not, then either the server stays dormant in the system waiting for an influx of fluid (for the Type V case) or a $\mathfrak{I}_2 \to \mathfrak{I}_1$ transition of the UMC occurs (for the Type L case) and the busy period starts.

Example 1.3 (D-policy). In this system, as soon as the fluid level becomes zero, the server becomes idle until the accumulated fluid level exceeds D.

Figures 1.1 and 1.2 show the Type V and Type L sample paths of the MAP-modulated fluid flow system under the D-policy.

Preliminaries

In this section, we review the important theoretical results of the conventional MMFF model. This section is based on the results of Ahn and Ramaswami [3–5].

The conventional MMFF system is a fluid input–output system in which all the rates of change in fluid level are linear and governed by a UMC with infinitesimal generator $\mathbf{Q}$. Let us divide the UMC phases into two sets $\{\mathfrak{I}_1, \mathfrak{I}_2\}$, where $\mathfrak{I}_1$ is the set of UMC phases with increasing rates and $\mathfrak{I}_2$ is the set of UMC phases with decreasing rates. Conforming to $\mathfrak{I}_1$ and $\mathfrak{I}_2$, the infinitesimal generator can be partitioned as $\mathbf{Q} = \begin{pmatrix} \mathbf{Q}_{11} & \mathbf{Q}_{12} \\ \mathbf{Q}_{21} & \mathbf{Q}_{22} \end{pmatrix}$. The fluid level increases with slope $r_i > 0$, $i \in \mathfrak{I}_1$, and decreases with slope $r_i < 0$, $i \in \mathfrak{I}_2$. We define $\mathbf{R} = \{r_i\}$.

Let $U(t)$ be the fluid level at time t and $J(t)$ be the phase of UMC at time t. Then the two-dimensional stochastic process $\{U(t), J(t), t \geq 0\}$ is called an MMFF process.

Let us define $\mathbf{\Gamma}$ as the diagonal matrix of $\gamma_i = |r_i|$ and $\mathbf{P}$ as

$$\mathbf{P} = \frac{\mathbf{\Gamma}^{-1}\mathbf{Q}}{\lambda} + \mathbf{I}, \tag{1.3}$$

where λ is a positive number with

$$\lambda \geq \max_{i \in \mathfrak{I}} [-\mathbf{\Gamma}^{-1}\mathbf{Q}]_{ii}.$$

We partition $\mathbf{P}$, $\mathbf{R}$, and $\mathbf{\Gamma}$ according to $\mathfrak{I}_1$ and $\mathfrak{I}_2$ as

$$\mathbf{P} = \begin{pmatrix} \mathbf{P}_{11} & \mathbf{P}_{12} \\ \mathbf{P}_{21} & \mathbf{P}_{22} \end{pmatrix}, \quad \mathbf{R} = \begin{pmatrix} \mathbf{R}_1 & \mathbf{0} \\ \mathbf{0} & \mathbf{R}_2 \end{pmatrix}, \quad \mathbf{\Gamma} = \begin{pmatrix} \mathbf{\Gamma}_1 & \mathbf{0} \\ \mathbf{0} & \mathbf{\Gamma}_2 \end{pmatrix}.$$

In the analysis of MMFF-related systems, the first passage times play important roles. Let $\tau(x) = \inf\{t > 0, U(t) = x\}$ be the first passage time to level x. Let us define the following LSTs:

$$[\boldsymbol{\Psi}^*(\theta)]_{ij} = E[\mathrm{e}^{-\theta\tau(0)}, J(\tau(0)) = j | U(0) = 0, J(0) = i], (i \in \mathfrak{I}_1, j \in \mathfrak{I}_2),$$

$$[\mathbf{G}_{12}^*(\theta|x)]_{ij} = E[\mathrm{e}^{-\theta\tau(0)}, J(\tau(0)) = j | U(0) = x, J(0) = i], (i \in \mathfrak{I}_1, j \in \mathfrak{I}_2),$$

$$[\mathbf{G}_{22}^*(\theta|x)]_{ij} = E[\mathrm{e}^{-\theta\tau(0)}, J(\tau(0)) = j | U(0) = x, J(0) = i], (i \in \mathfrak{I}_2, j \in \mathfrak{I}_2).$$

Then, from Ahn and Ramaswami [5], it is known that

$$\boldsymbol{\Psi}^*(\theta) = \left[\left(\mathbf{P}_{11} - \frac{\theta}{\lambda}\boldsymbol{\Gamma}_1^{-1}\right)\boldsymbol{\Psi}^*(\theta) + \mathbf{P}_{12}\right]\left[\mathbf{I} - \frac{\mathbf{H}^*(\theta)}{\lambda}\right]^{-1}, \tag{1.4}$$

$$\mathbf{G}_{12}^*(\theta|x) = \boldsymbol{\Psi}^*(\theta)\mathbf{G}_{22}^*\theta|x,\ (x > 0), \tag{1.5}$$

and

$$\mathbf{G}_{22}^*(\theta|x) = \mathrm{e}^{\mathbf{H}^*(\theta)x},\ x > 0, \tag{1.6}$$

in which

$$\mathbf{H}^*(\theta) = \boldsymbol{\Gamma}_2^{-1}[\mathbf{Q}_{22} - \theta\mathbf{I} + \mathbf{Q}_{21}\boldsymbol{\Psi}^*(\theta)]. \tag{1.7}$$

To analyze the idle period of a Type L system, it is necessary to derive the LST of the first passage time from level 0 to x. Conforming to the sets of UMC phases that belong to $\mathfrak{I}_1^{\text{idle}}$ and $\mathfrak{I}_2^{\text{idle}}$, the infinitesimal generator $\mathbf{Q}$ can be partitioned as

$$\mathbf{Q} = \begin{pmatrix} \mathbf{Q}_{11}^{\mathrm{L}} & \mathbf{Q}_{12}^{\mathrm{L}} \\ \mathbf{Q}_{21}^{\mathrm{L}} & \mathbf{Q}_{22}^{\mathrm{L}} \end{pmatrix}.$$

During the idle period, if the UMC phase is $i \in \mathfrak{I}_1^{\text{idle}}$, then the fluid level increases with slope $v_i > 0$ and remains without change if $i \in \mathfrak{I}_2^{\text{idle}}$. Let us define the diagonal matrix $\mathbf{R}_{\mathrm{L}} = \{v_i\}$, $(i \in \mathfrak{I}_1^{\text{idle}})$ and the following LSTs:

$$[\mathbf{T}_{11}^*(x,\theta)]_{ij} = E[\mathrm{e}^{-\theta\tau(x)}, J(\tau(x)) = j | U(0) = 0, J(0) = i], i \in \mathfrak{I}_1^{\text{idle}}, j \in \mathfrak{I}_1^{\text{idle}},$$

$$[\mathbf{T}_{21}^*(x,\theta)]_{ij} = E[\mathrm{e}^{-\theta\tau(x)}, J(\tau(x)) = j | U(0) = 0, J(0) = i], i \in \mathfrak{I}_2^{\text{idle}}, j \in \mathfrak{I}_1^{\text{idle}}.$$

Then $\mathbf{T}_{11}^*(x,\theta)$ and $\mathbf{T}_{21}^*(x,\theta)$ are given by

$$\mathbf{T}_{21}^*(x,\theta) = (\theta\mathbf{I} - \mathbf{Q}_{22}^{\mathrm{L}})^{-1}\mathbf{Q}_{21}^{\mathrm{L}}\mathbf{T}_{11}^*(x,\theta), \tag{1.8}$$

$$\mathbf{T}_{11}^*(x,\theta) = \mathrm{e}^{\mathbf{Q}_{\mathrm{L}}^*(\theta)x}, \tag{1.9}$$

in which

$$\mathbf{Q}_{\mathrm{L}}^*(\theta) = \mathbf{R}_{\mathrm{L}}^{-1}[-\theta\mathbf{I} + \mathbf{Q}_{11}^{\mathrm{L}} + \mathbf{Q}_{12}^{\mathrm{L}}(\theta\mathbf{I} - \mathbf{Q}_{22}^{\mathrm{L}})^{-1}\mathbf{Q}_{21}^{\mathrm{L}}]. \tag{1.10}$$

Main Results: Derivation of Factorizations

In this section, we derive the factorizations for each type. For both types, we commonly use the following notation:

m: Number of UMC phases
S: Amount of fluid brought in by an arrival (jump size) in Type V system
$s(x)$: Probability density function (PDF) of S
$S^*(\theta)$: LST of $S(x)$
$E(S)$: Expected value of S
$\pi_i = \lim_{t\to\infty} \Pr[J(t) = i]$, $(1 \le i \le m)$
$\boldsymbol{\pi} = \{\pi_1, \pi_2, \ldots, \pi_m\}$: Steady-state phase probability vector of UMC process
$\xi(t) = \begin{cases} 0, & \text{if the system is idle at time } t \\ 1, & \text{if the system is busy at time } t \end{cases}$
$\mathbf{e}$: Column vector of 1s

We also define the following probabilities:

$$U_{\text{idle},i}(x,t) = \Pr[U(t) \le x, J(t) = i, \xi(t) = 0],\ x \ge 0,$$
$$U_{\text{busy},i}(x,t) = \Pr[U(t) \le x, J(t) = i, \xi(t) = 1],\ x > 0,$$

and steady-state quantities,

$$U_{\text{idle},i}(x) = \lim_{t\to\infty} U_{\text{idle},i}(x,t), \quad U_{\text{busy},i}(x) = \lim_{t\to\infty} U_{\text{busy},i}(x,t).$$

Defining $\mathbf{u}^*_{\text{idle}}(\theta)$ and $\mathbf{u}^*_{\text{busy}}(\theta)$ as the respective vector LSTs of the fluid level during an idle period and a busy period, the vector LST $\mathbf{u}^*(\theta)$ of the fluid level at an arbitrary time can be obtained from $\mathbf{u}^*(\theta) = \mathbf{u}^*_{\text{idle}}(\theta) + \mathbf{u}^*_{\text{busy}}(\theta)$.

We note that the system becomes stable if and only if the average outflow is greater than the average inflow. For convenience, let $\boldsymbol{\pi}_{\text{idle}} = \mathbf{u}^*_{\text{idle}}(\theta)|_{\theta=0}$ and $\boldsymbol{\pi}_{\text{busy}} = \boldsymbol{\pi} - \boldsymbol{\pi}_{\text{idle}}$. We then have the following stability conditions for each type:

$$\text{(Type V)} \qquad \boldsymbol{\pi}_{\text{idle}} \mathbf{D}_1 \mathbf{e} E(S) + \boldsymbol{\pi}_{\text{busy}} \mathbf{R} \mathbf{e} < 0, \tag{1.11a}$$

$$\text{(Type L)} \qquad \boldsymbol{\pi}_{\text{idle}} \begin{pmatrix} \mathbf{R}_{\text{L}} & \mathbf{0} \\ \mathbf{0} & \mathbf{0} \end{pmatrix} \mathbf{e} + \boldsymbol{\pi}_{\text{busy}} \mathbf{R} \mathbf{e} < 0. \tag{1.11b}$$

Factorization for Type V Systems

In this section, we derive the factorization for the Type V system. We have the following theorem.

Theorem 1.1. *For the Type V MAP-modulated fluid flow models under generalized server vacations, we have the following factorization property:*

$$\mathbf{u}^*(\theta) = \mathbf{u}^*_{\rm idle}(\theta) \cdot \mathbf{W}^*_V(\theta), \tag{1.12}$$

where

$$\mathbf{W}^*_V(\theta) = [\theta\mathbf{R} - \mathbf{D}_1 + \mathbf{D}_1 S^*(\theta)]\,(\theta\mathbf{R} - \mathbf{Q})^{-1}. \tag{1.13}$$

Proof. Let $\phi_{{\rm B},i}(x,t)$ be the rate (number of occurrences per unit time) at which the system becomes busy with fluid level x and UMC phase i at time t. We also define $u_{{\rm busy},i}(0,t)$ as the rate at which the busy period ends with UMC phase i at time t. Defining $u_{{\rm idle},i}(x,t) = \frac{\rm d}{{\rm d}x}U_{{\rm idle},i}(x,t)$ and $u_{{\rm busy},i}(x,t) = \frac{\rm d}{{\rm d}x}U_{{\rm busy},i}(x,t)$, it is not difficult to set up the system equations that represent the level changes during an infinitesimal time Δt as follows:

$$\begin{aligned} U_{{\rm idle},i}(0,t+\Delta t) &= U_{{\rm idle},i}(0,t)[1 + (\mathbf{D}_0)_{ii}\Delta t] \\ &\quad + \sum_{\substack{j=1\\(j\neq i)}}^{m} U_{{\rm idle},j}(0,t)(\mathbf{D}_0)_{ji}\Delta t + u_{{\rm busy},i}(0,t)(-r_i)\Delta t + o(\Delta t), \end{aligned} \tag{1.14}$$

$$\begin{aligned} u_{{\rm idle},i}(x,t+\Delta t) &= u_{{\rm idle},i}(x,t)[1 + (\mathbf{D}_0)_{ii}\Delta t] - \phi_{{\rm B},i}(x,t) \\ &\quad + \sum_{\substack{j=1\\(j\neq i)}}^{m} u_{{\rm idle},j}(x,t)(\mathbf{D}_0)_{ji}\Delta t + \sum_{j=1}^{m} U_{{\rm idle},j}(0,t)(\mathbf{D}_1)_{ji}s(x)\Delta t \\ &\quad + \sum_{j=1}^{m}\int_{u=0}^{x} u_{{\rm idle},j}(x-u,t)(\mathbf{D}_1)_{ji}s(u){\rm d}u\Delta t\Delta t + o(\Delta t),\ x>0, \end{aligned} \tag{1.15}$$

$$\begin{aligned} u_{{\rm busy},i}(x,t+\Delta t) &= u_{{\rm busy},i}(x - r_i\Delta t,t)[1 + (\mathbf{Q})_{ii}\Delta t] + \phi_{{\rm B},i}(x,t) \\ &\quad + \sum_{\substack{j=1\\(j\neq i)}}^{m} u_{{\rm busy},j}(x - r_j\Delta t,t)(\mathbf{Q})_{ji}\Delta t\Delta t + o(\Delta t),\ x>0. \end{aligned} \tag{1.16}$$

We note that $u_{{\rm busy},i}(0,t) = 0$ for $i \in \Im_1$ since the busy period cannot be finished with a UMC phase in $\Im_1$.

Let $\phi_{{\rm B},i}(x) = \lim_{t\to\infty}\phi_{{\rm B},i}(x,t)$ and define a vector $\boldsymbol{\phi}_{\rm B}(x) = \{\phi_{{\rm B},1}(x),\ldots,\phi_{{\rm B},m}(x)\}$. We also define $\mathbf{U}_{\rm idle}(x)$ and $\mathbf{U}_{\rm busy}(x)$ as the vectors of $U_{{\rm idle},i}(x)$ and $U_{{\rm busy},i}(x)$. Using (1.14)–(1.16), we have the following vector steady-state equations:

$$\mathbf{0} = \mathbf{U}_{\rm idle}(0)\mathbf{D}_0 - \mathbf{u}_{\rm busy}(0)\mathbf{R}, \tag{1.17}$$

$$\begin{aligned} \mathbf{0} &= \mathbf{u}_{\rm idle}(x)\mathbf{D}_0 + \mathbf{U}_{\rm idle}(0)\mathbf{D}_1 s(x) - \boldsymbol{\phi}_{\rm B}(x) \\ &\quad + \int_{u=0}^{x} \mathbf{u}_{\rm idle}(x-u)\mathbf{D}_1 s(u){\rm d}u,\ x>0, \end{aligned} \tag{1.18}$$

$$\frac{\mathrm{d}}{\mathrm{d}x}\mathbf{u}_{\mathrm{busy}}(x)\mathbf{R} = \mathbf{u}_{\mathrm{busy}}(x)\mathbf{Q} + \boldsymbol{\phi}_{\mathrm{B}}(x),\ x > 0, \tag{1.19}$$

where $\mathbf{u}_{\mathrm{idle}}(x) = \frac{\mathrm{d}}{\mathrm{d}x}\mathbf{U}_{\mathrm{idle}}(x)$ and $\mathbf{u}_{\mathrm{busy}}(x) = \frac{\mathrm{d}}{\mathrm{d}x}\mathbf{U}_{\mathrm{busy}}(x)$.

Let us define the following Laplace transform (LT):

$$\boldsymbol{\phi}_{\mathrm{B}}^*(\theta) = \int_0^\infty \mathrm{e}^{-\theta x}\boldsymbol{\phi}_{\mathrm{B}}(x)\mathrm{d}x.$$

Taking the LT of (1.18) and using (1.17) we obtain

$$\mathbf{0} = \mathbf{u}_{\mathrm{idle}}^*(\theta)\mathbf{D}_0 + \mathbf{u}_{\mathrm{idle}}^*(\theta)\mathbf{D}_1 S^*(\theta) - \mathbf{u}_{\mathrm{busy}}(0)\mathbf{R} - \boldsymbol{\phi}_{\mathrm{B}}^*(\theta). \tag{1.20}$$

Taking the LT of (1.19) yields

$$\theta\mathbf{u}_{\mathrm{busy}}^*(\theta)\mathbf{R} - \mathbf{u}_{\mathrm{busy}}(0)\mathbf{R} = \mathbf{u}_{\mathrm{busy}}^*(\theta)\mathbf{Q} + \boldsymbol{\phi}_{\mathrm{B}}^*(\theta). \tag{1.21}$$

Then, adding (1.20) and (1.21) completes the proof. □

Factorization for Type L Systems

For Type L systems we have the following theorem.

Theorem 1.2. *For a Type L MAP-modulated fluid flow model under generalaized server vacations, we have the following factorization:*

$$\mathbf{u}^*(\theta) = \mathbf{u}_{\mathrm{idle}}^*(\theta)\cdot\mathbf{W}_{\mathrm{L}}^*(\theta), \tag{1.22}$$

where

$$\mathbf{W}_{\mathrm{L}}^*(\theta) = \theta\left[\mathbf{R} - \begin{pmatrix}\mathbf{R}_{\mathrm{L}} & \mathbf{0}\\ \mathbf{0} & \mathbf{0}\end{pmatrix}\right]\cdot(\theta\mathbf{R} - \mathbf{Q})^{-1}. \tag{1.23}$$

Proof. We have the following system equations:

$$\sum_{j\in\Im_2} U_{\mathrm{idle},j}(0,t)(\mathbf{Q})_{ji} = u_{\mathrm{idle},i}(0,t)v_i, i\in\Im_1, \tag{1.24}$$

$$\begin{aligned} U_{\mathrm{idle},i}(0,t+\Delta t) &= U_{\mathrm{idle},i}(0,t)[1+(\mathbf{Q})_{ii}\Delta t] + u_{\mathrm{busy},i}(0,t)(-r_i)\Delta t \\ &\quad + \sum_{\substack{j\in\Im_2\\(j\neq i)}} U_{\mathrm{idle},j}(0,t)(\mathbf{Q})_{ji}\Delta t + o(\Delta t),\ i\in\Im_2, \end{aligned} \tag{1.25}$$

$$u_{\mathrm{idle},i}(x,t+\Delta t) = u_{\mathrm{idle},i}(x-v_i\Delta t,t)[1+(\mathbf{Q})_{ii}\Delta t] + \sum_{\substack{j\in\Im_1\\(j\neq i)}} u_{\mathrm{idle},j}(x-v_j\Delta t,t)(\mathbf{Q})_{ji}\Delta t + \sum_{j\in\Im_2} u_{\mathrm{idle},j}(x,t)(\mathbf{Q})_{ji}\Delta t - \phi_{\mathrm{B},i}(x,t)\Delta t + o(\Delta t),\ i\in\Im_1, x>0, \tag{1.26}$$

$$u_{\mathrm{idle},i}(x,t+\Delta t) = u_{\mathrm{idle},i}(x,t)[1+(\mathbf{Q})_{ii}\Delta t] + \sum_{j\in\Im_1} u_{\mathrm{idle},j}(x-v_j\Delta t,t)(\mathbf{Q})_{ji}\Delta t + \sum_{\substack{j\in\Im_2\\(j\neq i)}} u_{\mathrm{idle},j}(x,t)(\mathbf{Q})_{ji}\Delta t - \phi_{\mathrm{B},i}(x,t)\Delta t + o(\Delta t),\ i\in\Im_2, x>0, \tag{1.27}$$

$$u_{\mathrm{busy},i}(x,t+\Delta t) = u_{\mathrm{busy},i}(x-r_i\Delta t,t)[1+(\mathbf{Q})_{ii}\Delta t] + \sum_{\substack{j=1\\(j\neq i)}}^{m} u_{\mathrm{busy},j}(x-r_j\Delta t,t)(\mathbf{Q})_{ji}\Delta t + \phi_{\mathrm{B},i}(x,t)\Delta t + o(\Delta t),\ x>0. \tag{1.28}$$

We note $U_{\mathrm{idle},i}(0)=0$ for $i\in\Im_1$ and $u_{\mathrm{busy},i}(0)=0$ for $i\in\Im_1$. From (1.24)–(1.28) we have the following steady-state vector equations:

$$\mathbf{0} = \mathbf{U}_{\mathrm{idle}}(0)\mathbf{Q} - \mathbf{u}_{\mathrm{busy}}(0)\mathbf{R} - \mathbf{u}_{\mathrm{idle}}(0)\begin{pmatrix}\mathbf{R}_{\mathrm{L}} & \mathbf{0}\\ \mathbf{0} & \mathbf{0}\end{pmatrix}, \tag{1.29}$$

$$\frac{\mathrm{d}}{\mathrm{d}x}\mathbf{u}_{\mathrm{idle}}(x)\begin{pmatrix}\mathbf{R}_{\mathrm{L}} & \mathbf{0}\\ \mathbf{0} & \mathbf{0}\end{pmatrix} = \mathbf{u}_{\mathrm{idle}}(x)\mathbf{Q} - \boldsymbol{\phi}_{\mathrm{B}}(x),\ x>0, \tag{1.30}$$

$$\frac{\mathrm{d}}{\mathrm{d}x}\mathbf{u}_{\mathrm{busy}}(x)\mathbf{R} = \mathbf{u}_{\mathrm{busy}}(x)\mathbf{Q} + \boldsymbol{\phi}_{\mathrm{B}}(x),\ x>0. \tag{1.31}$$

Taking the LT of (1.30) and using (1.29) we obtain

$$\theta\mathbf{u}^*_{\mathrm{idle}}(\theta)\begin{pmatrix}\mathbf{R}_{\mathrm{L}} & \mathbf{0}\\ \mathbf{0} & \mathbf{0}\end{pmatrix} = \mathbf{u}^*_{\mathrm{idle}}(\theta)\mathbf{Q} - \boldsymbol{\phi}^*_{\mathrm{B}}(\theta) - \mathbf{u}_{\mathrm{busy}}(0)\mathbf{R}. \tag{1.32}$$

Taking the LT of (1.31) yields

$$\theta\mathbf{u}^*_{\mathrm{busy}}(\theta)\mathbf{R} - \mathbf{u}_{\mathrm{busy}}(0)\mathbf{R} = \mathbf{u}^*_{\mathrm{busy}}(\theta)\mathbf{Q} + \boldsymbol{\phi}^*_{\mathrm{B}}(\theta). \tag{1.33}$$

Adding (1.32) and (1.33) completes the proof. □

Mean Length of a Busy Period

In this section we derive the mean length $E(B)$ of the busy period of both types of systems. The fluid level and the UMC phase at the start of the busy period are all we need for this purpose.

Let us define the following probability:

$[\mathbf{U}_{\mathrm{B}}(x)]_{ij} = \Pr$ (at the end of an idle period, the fluid level is less than or equal to x and the UMC phase is j, under the condition that the UMC phase is i at the start of the idle period).

We note that $\mathbf{U}_{\mathrm{B}}(x)$ differs from system to system depending on the vacation type.

Let $\mathbf{K}$ be the probability matrix that represents the change in the UMC phase during a cycle that is defined as the time interval between two consecutive idle-period starting points. Then we have

$$\mathbf{K} = \int_0^\infty \mathrm{d}\mathbf{U}_{\mathrm{B}}(x)\mathbf{G}(x), \tag{1.34}$$

where

$$\mathbf{G}(x) = \begin{pmatrix} \mathbf{0} & \mathbf{G}^*_{12}(\theta|x)|_{\theta=0} \\ \mathbf{0} & \mathbf{G}^*_{22}(\theta|x)|_{\theta=0} \end{pmatrix}.$$

Equation (1.34) is obvious because $\mathbf{G}(x)$ represents the change in the UMC phases during a busy period starting with level x. If we define $\boldsymbol{\kappa}$ as the stationary probability vector of the UMC phase at the start of an idle period, then we have

$$\boldsymbol{\kappa} = \boldsymbol{\kappa}\mathbf{K}, \quad \boldsymbol{\kappa}\mathrm{e} = 1. \tag{1.35}$$

Let $E(B)$ be the mean length of a busy period. Then we have

$$E(B) = \boldsymbol{\kappa}\int_0^\infty \mathrm{d}\mathbf{U}_{\mathrm{B}}(x)\mathbf{g}(x), \tag{1.36}$$

where

$$\mathbf{g}(x) = \begin{pmatrix} \mathbf{0} & -\dfrac{\mathrm{d}}{\mathrm{d}\theta}\mathbf{G}^*_{12}(\theta|x)|_{\theta=0}\mathbf{e} \\ \mathbf{0} & -\dfrac{\mathrm{d}}{\mathrm{d}\theta}\mathbf{G}^*_{22}(\theta|x)|_{\theta=0}\mathbf{e} \end{pmatrix}.$$

The mean length $E(I)$ of an idle period differs from system to system depending upon the vacation type. Then the probability ρ that the system is busy can be obtained from

$$\rho = \frac{E(B)}{E(I)+E(B)}. \tag{1.37}$$

Moment Formulas

In this section, we derive the recursive moment formula for each type. We will use the notation as follows:

$$\mathbf{M}^{(n)} = \frac{\mathrm{d}^n}{\mathrm{d}\theta^n}\mathbf{M}^*(\theta)\bigg|_{\theta=0}, \quad \mathbf{M}^{(0)} = \mathbf{M} = \mathbf{M}^*(\theta)|_{\theta=0}.$$

Moments Formula for Type V Systems

Using (1.12) and (1.13), we have the following theorem.

Theorem 1.3. *The moment formula for a Type V system becomes*

$$\begin{aligned}\mathbf{u}^{(n)}\mathbf{e} = \frac{1}{\boldsymbol{\pi}\mathbf{Re}}\Bigg\{ & \mathbf{u}_{\text{idle}}^{(n)}[\mathbf{R}-\mathbf{D}_1E(S)]\mathbf{e} + n\mathbf{u}^{(n-1)}\mathbf{R}(\mathbf{e}\boldsymbol{\pi}-\mathbf{Q})^{-1}\mathbf{Re} \\ & - n\mathbf{u}_{\text{idle}}^{(n-1)}[\mathbf{R}-\mathbf{D}_1E(S)](\mathbf{e}\boldsymbol{\pi}-\mathbf{Q})^{-1}\mathbf{Re} \\ & + \frac{1}{1+n}\sum_{k=2}^{n+1} I_{[n+1\geq k]}\binom{n+1}{k}(-1)^k E(S^k)\mathbf{u}_{\text{idle}}^{(n+1-k)}\mathbf{D}_1\mathbf{e} \\ & - \sum_{k=2}^{n} I_{[n\geq k]}\binom{n}{k}(-1)^k E(S^k)\mathbf{u}_{\text{idle}}^{(n-k)}\mathbf{D}_1(\mathbf{e}\boldsymbol{\pi}-\mathbf{Q})^{-1}\mathbf{Re}\Bigg\}, \end{aligned} \tag{1.38}$$

where $I_{[A]}$ is an indicator function that takes 1 if A is true or 0 if A is false.

Proof. From (1.12) we obtain

$$\begin{aligned}&\mathbf{u}^{(n)}(-\mathbf{Q}) + n\mathbf{u}^{(n-1)}\mathbf{R} \\ &\quad = n\mathbf{u}_{\text{idle}}^{(n-1)}[\mathbf{R}-\mathbf{D}_1E(S)] + \sum_{k=2}^{n} I_{[n\geq k]}(-1)^k E(S^k)\binom{n}{k}\mathbf{u}_{\text{idle}}^{(n-k)}\mathbf{D}_1. \end{aligned} \tag{1.39}$$

Using $\boldsymbol{\pi}(\mathbf{e}\boldsymbol{\pi}-\mathbf{Q})^{-1} = \boldsymbol{\pi}$ in (1.39) we obtain

$$\begin{aligned}&\mathbf{u}^{(n)} + n\mathbf{u}^{(n-1)}\mathbf{R}(\mathbf{e}\boldsymbol{\pi}-\mathbf{Q})^{-1} \\ &\quad = \mathbf{u}^{(n)}\mathbf{e}\boldsymbol{\pi} + n\mathbf{u}_{\text{idle}}^{(n-1)}[\mathbf{R}-\mathbf{D}_1E(S)](\mathbf{e}\boldsymbol{\pi}-\mathbf{Q})^{-1} \\ &\qquad + \sum_{k=2}^{n} I_{[n\geq k]}(-1)^k E(S^k)\binom{n}{k}\mathbf{u}_{\text{idle}}^{(n-k)}\mathbf{D}_1(\mathbf{e}\boldsymbol{\pi}-\mathbf{Q})^{-1}. \end{aligned} \tag{1.40}$$

Postmultiplying (1.39) by $\mathbf{e}$ and using $(n+1)$ in place of n we also obtain

$$\mathbf{u}^{(n)}\mathbf{R}\mathbf{e} = \mathbf{u}_{\text{idle}}^{(n)}[\mathbf{R} - \mathbf{D}_1 E(S)]\mathbf{e} + \frac{1}{n+1}\sum_{k=2}^{n+1} I_{[n+1\geq k]}(-1)^k E(S^k)\binom{n+1}{k}\mathbf{u}_{\text{idle}}^{(n+1-k)}\mathbf{D}_1\mathbf{e}. \tag{1.41}$$

Postmultiplying (1.40) by $\mathbf{Re}$ and using (1.41) completes the proof. □

Moment Formula for Type L Systems

For Type L systems we have the following theorem.

Theorem 1.4. *The moment formula for a Type L system is given by*

$$\mathbf{u}^{(n)}\mathbf{e} = \frac{1}{\boldsymbol{\pi}\mathbf{Re}}\left\{\mathbf{u}_{\text{idle}}^{(n)}\left[\mathbf{R} - \begin{pmatrix}\mathbf{R}_{\text{L}} & \mathbf{0}\\ \mathbf{0} & \mathbf{0}\end{pmatrix}\right]\mathbf{e} - n\mathbf{u}_{\text{idle}}^{(n-1)}\left[\mathbf{R} - \begin{pmatrix}\mathbf{R}_{\text{L}} & \mathbf{0}\\ \mathbf{0} & \mathbf{0}\end{pmatrix}\right](\mathbf{e}\boldsymbol{\pi} - \mathbf{Q})^{-1}\mathbf{Re} + n\mathbf{u}^{(n-1)}\mathbf{R}(\mathbf{e}\boldsymbol{\pi} - \mathbf{Q})^{-1}\mathbf{Re}\right\}. \tag{1.42}$$

Proof. From (1.22) we obtain

$$\mathbf{u}^{(n)}(-\mathbf{Q}) + n\mathbf{u}^{(n-1)}\mathbf{R} = n\mathbf{u}_{\text{idle}}^{(n-1)}\left[\mathbf{R} - \begin{pmatrix}\mathbf{R}_{\text{L}} & \mathbf{0}\\ \mathbf{0} & \mathbf{0}\end{pmatrix}\right]. \tag{1.43}$$

Using $\boldsymbol{\pi}(\mathbf{e}\boldsymbol{\pi} - \mathbf{Q})^{-1} = \boldsymbol{\pi}$ in (1.43) we obtain

$$\mathbf{u}^{(n)} + n\mathbf{u}^{(n-1)}\mathbf{R}(\mathbf{e}\boldsymbol{\pi} - \mathbf{Q})^{-1} = n\mathbf{u}_{\text{idle}}^{(n-1)}\left[\mathbf{R} - \begin{pmatrix}\mathbf{R}_{\text{L}} & \mathbf{0}\\ \mathbf{0} & \mathbf{0}\end{pmatrix}\right](\mathbf{e}\boldsymbol{\pi} - \mathbf{Q})^{-1} + \mathbf{u}^{(n)}\mathbf{e}\boldsymbol{\pi}. \tag{1.44}$$

Postmultiplying (1.43) by $\mathbf{e}$ and using $(n+1)$ in place of n yields

$$\mathbf{u}^{(n)}\mathbf{Re} = \mathbf{u}_{\text{idle}}^{(n)}\left[\mathbf{R} - \begin{pmatrix}\mathbf{R}_{\text{L}} & \mathbf{0}\\ \mathbf{0} & \mathbf{0}\end{pmatrix}\right]\mathbf{e}. \tag{1.45}$$

Postmultiplying (1.44) by $\mathbf{Re}$ and using (1.45) completes the proof. □

Application Examples

In this section, we present application examples. We consider the three example systems mentioned in the section "Example systems." All we need to do to obtain $\mathbf{u}^*(\theta)$ is to obtain the LST $\mathbf{u}^*_{\text{idle}}(\theta)$ of the fluid level at an arbitrary time during an idle period as shown in (1.12) and (1.22). To obtain the mean length $E(B)$ of the busy period and the probability ρ that the system is busy, we need only obtain the probability matrix $\mathbf{U}_{\text{B}}(x)$ as shown in (1.36) and (1.37).

Type V Systems

We note that the fluid level process during an idle period of a Type V system is identical to that of the MAP/G/1 queue under server vacations.

Example 1.1 (Multiple vacations). From Lee et al. [17] we have

$$\mathbf{u}^*_{\text{idle}}(\theta) = \frac{\boldsymbol{\kappa}}{E(C)} \left[(\mathbf{I}-\mathbf{V}_0)^{-1}(\mathbf{V}(z)-\mathbf{I})(\mathbf{D}_0+\mathbf{D}_1 z)^{-1}\right]_{z=S^*(\theta)}, \tag{1.46}$$

where $\mathbf{V}(z)$ is the matrix-generating function of the number of jumps (arrivals) during a vacation [with DF $V(x)$], which is given by

$$\mathbf{V}(z) = \int_0^\infty \mathrm{e}^{(\mathbf{D}_0+\mathbf{D}_1 z)x}\mathrm{d}V(x).$$

$\mathbf{U}_{\text{B}}(x)$ is given by

$$\mathbf{U}_{\text{B}}(x) = (\mathbf{I}-\mathbf{V}_0)^{-1}\left[\sum_{n=1}^{\infty}\mathbf{V}_n S^{(n)}(x)\right],\ x>0. \tag{1.47}$$

Example 1.2 (Single vacation). From the result of Lee et al. [17] we have

$$\mathbf{u}^*_{\text{idle}}(\theta) = \frac{\boldsymbol{\kappa}}{E(C)} \left[\mathbf{V}_0(-\mathbf{D}_0)^{-1}+[\mathbf{V}(z)-\mathbf{I}](\mathbf{D}_0+\mathbf{D}_1 z)^{-1}\right]_{z=S^*(\theta)}, \tag{1.48}$$

and

$$\mathbf{U}_{\text{B}}(x) = \mathbf{V}_0(-\mathbf{D}_0)^{-1}\mathbf{D}_1 S(x) + \sum_{n=1}^{\infty}\mathbf{V}_n S^{(n)}(x), (x>0). \tag{1.49}$$

Example 1.3 (D-policy). From the result of Lee et al. [18], we have

$$\mathbf{u}^*_{\text{idle}}(\theta) = \frac{\boldsymbol{\kappa}}{E(C)} \sum_{n=0}^{\infty}\int_0^D \left[(-\mathbf{D}_0)^{-1}\mathbf{D}_1\right]^n(-\mathbf{D}_0)^{-1}\mathrm{e}^{-\theta x}\mathrm{d}S^{(n)}(x) \tag{1.50}$$

and

$$\mathbf{U}_{\mathrm{B}}(x) = (-\mathbf{D}_0)^{-1}\mathbf{D}_1 \mathrm{d}S(x) + \sum_{k=2}^{\infty}[(-\mathbf{D}_0)^{-1}\mathbf{D}_1]^k \int_{0+}^{D} S(x-y)\mathrm{d}S^{(k-1)}(y), \; x > D. \tag{1.51}$$

Type L Systems

To use (1.22), it is necessary to derive the LST $\mathbf{u}^*_{\text{idle}}(\theta)$ of the fluid level at an arbitrary time during an idle period for each vacation type.

Example 1.1 (Multiple vacations). Let us define the following probability:

$[\tilde{\mathbf{V}}(x,t)]_{ij}$ = Pr (the amount of fluid is less than or equal to x, the UMC phase is j at the end of a single vacation, and the length of the vacation is less than or equal to t under the condition that the vacation starts with UMC phase is i), $x \geq 0, t \geq 0$.

Let us also define the following quantities:

$$\mathbf{V}_{\mathrm{F}}(x) = \int_{t=0}^{\infty} \mathrm{d}_t \tilde{\mathbf{V}}(x,t), \; x \geq 0, \tag{1.52}$$

$$\mathbf{V}^*_{\mathrm{F}}(\theta) = \int_{x=0}^{\infty} \mathrm{e}^{-\theta x} \mathrm{d}_x \mathbf{V}_{\mathrm{F}}(x). \tag{1.53}$$

We note that $\mathbf{V}^*_{\mathrm{F}}(\theta)$ is the LST of the total amount of fluid that comes into the system during a vacation. Then we have the following theorem.

Theorem 1.5. *We have*

$$\mathbf{V}^*_{\mathrm{F}}(\theta) = \int_0^{\infty} \mathrm{e}^{\mathbf{Q}^*_{\Lambda}(\theta)t} \mathrm{d}V(t), \tag{1.54}$$

where

$$\mathbf{Q}^*_{\Lambda}(\theta) = \begin{pmatrix} -\theta \mathbf{R}_{\mathrm{L}} & \mathbf{0} \\ \mathbf{0} & \mathbf{0} \end{pmatrix} + \mathbf{Q}. \tag{1.55}$$

Proof. Let $U_I(t)$ be the amount of fluid that comes into the system during a time interval $(0,t]$ contained in the idle period, $U_I(0) = 0$. Let us define the following joint probability:

$$[\mathbf{F}(x,t)]_{ij} = \Pr[U_I(t) \leq x, J(t) = j | U_I(0) = 0, J(0) = i].$$

We then have the following system equations:

$$
\begin{aligned}
&[\mathbf{F}(x,t+\Delta t)]_{ij}\\
&\quad = [\mathbf{F}(x-v_j\Delta t,t)]_{ij}(1+q_{jj}\Delta t) + \sum_{\substack{k\in\mathfrak{I}_1^{\text{idle}}\\(k\neq j)}} [\mathbf{F}(x-v_k\Delta t,t)]_{ik}q_{kj}\Delta t\\
&\qquad + \sum_{k\in\mathfrak{I}_2^{\text{idle}}} [\mathbf{F}(x,t)]_{ik}q_{kj}\Delta t,\ j\in\mathfrak{I}_1^{\text{idle}},
\end{aligned}
\tag{1.56}
$$

$$
\begin{aligned}
&[\mathbf{F}(x,t+\Delta t)]_{ij}\\
&\quad = [\mathbf{F}(x,t)]_{ij}(1+q_{jj}\Delta t) + \sum_{k\in\mathfrak{I}_1^{\text{idle}}} [\mathbf{F}(x-v_k\Delta t,t)]_{ik}q_{kj}\Delta t\\
&\qquad + \sum_{\substack{k\in\mathfrak{I}_2^{\text{idle}}\\(k\neq j)}} [\mathbf{F}(x,t)]_{ik}q_{kj}\Delta t,\ j\in\mathfrak{I}_2^{\text{idle}}.
\end{aligned}
\tag{1.57}
$$

Expressing (1.56) and (1.57) in matrix form we obtain

$$
\frac{\partial}{\partial t}\mathbf{F}(x,t) + \frac{\partial}{\partial x}\mathbf{F}(x,t)\begin{pmatrix}\mathbf{R}_{\text{L}} & \mathbf{0}\\ \mathbf{0} & \mathbf{0}\end{pmatrix} = \mathbf{F}(x,t)\mathbf{Q}. \tag{1.58}
$$

Let us define the following matrix LST:

$$
\mathbf{F}^*(\theta,t) = \int_{x=0}^{\infty} \mathrm{e}^{-\theta x}\mathrm{d}_x\mathbf{F}(x,t). \tag{1.59}
$$

Taking the LST of both sides of (1.58) we obtain

$$
\frac{\mathrm{d}}{\mathrm{d}t}\mathbf{F}^*(\theta,t) = \mathbf{F}^*(\theta,t)\left[\begin{pmatrix}-\theta\mathbf{R}_{\text{L}} & \mathbf{0}\\ \mathbf{0} & \mathbf{0}\end{pmatrix} + \mathbf{Q}\right]. \tag{1.60}
$$

The solution to (1.60) with initial condition $\mathbf{F}(0,0) = \mathbf{I}$ is given by $\mathbf{F}^*(\theta,t) = \mathrm{e}^{\mathbf{Q}_\Lambda^*(\theta)t}$. Then $\mathbf{V}_{\text{F}}^*(\theta) = \int_0^\infty \mathbf{F}^*(\theta,t)\mathrm{d}V(t)$ completes the proof. □

Now the LST $\mathbf{u}_{\text{idle}}^*(\theta)$ is given in the following theorem.

Theorem 1.6. *We have*

$$
\mathbf{u}_{\text{idle}}^*(\theta) = \frac{\boldsymbol{\kappa}}{E(C)}[\mathbf{I}-\mathbf{V}_{\text{F}}(0)]^{-1}[\mathbf{V}_{\text{F}}^*(\theta)-\mathbf{I}][\mathbf{Q}_\Lambda^*(\theta)]^{-1}, \tag{1.61}
$$

where $\mathbf{V}_{\text{F}}(0) = \begin{pmatrix}\mathbf{0} & \mathbf{0}\\ \mathbf{0} & \int_0^\infty \mathrm{e}^{\mathbf{Q}_{22}^{\text{L}}t}\mathrm{d}V(t)\end{pmatrix}$ *denotes the probability that no fluid arrives during a vacation and* $E(C)$ *is the mean length of a cycle.*

Proof. The fluid level at an arbitrary time during the idle period is equal to the amount of fluid that comes into the system during an elapsed time of an arbitrary vacation, and its vector LST becomes

$$\int_{t=0}^{\infty} \mathrm{e}^{\mathbf{Q}_{\Lambda}^{*}(\theta)t} \frac{1-V(t)}{E(V)} \mathrm{d}t = \frac{[\mathbf{V}_{\mathrm{F}}^{*}(\theta)-\mathbf{I}][\mathbf{Q}_{\Lambda}^{*}(\theta)]^{-1}}{E(V)}, \tag{1.62}$$

where $E(V)$ is the mean length of a vacation.

We note that the UMC phase at the start of an arbitrary vacation is given by

$$\frac{\boldsymbol{\kappa}[\mathbf{I}-\mathbf{V}_{\mathrm{F}}(0)]^{-1}}{\boldsymbol{\kappa}[\mathbf{I}-\mathbf{V}_{\mathrm{F}}(0)]^{-1}\mathbf{e}}. \tag{1.63}$$

Let ρ be the probability that the system is busy. Using (1.62) and (1.63) we then have

$$\mathbf{u}_{\mathrm{idle}}^{*}(\theta) = (1-\rho)\frac{\boldsymbol{\kappa}[\mathbf{I}-\mathbf{V}_{\mathrm{F}}(0)]^{-1}[\mathbf{V}_{\mathrm{F}}^{*}(\theta)-\mathbf{I}][\mathbf{Q}_{\Lambda}^{*}(\theta)]^{-1}}{E(V)\boldsymbol{\kappa}[\mathbf{I}-\mathbf{V}_{\mathrm{F}}(0)]^{-1}\mathbf{e}}. \tag{1.64}$$

We note that $E(V)\boldsymbol{\kappa}[\mathbf{I}-\mathbf{V}_{\mathrm{F}}(0)]^{-1}\mathbf{e}$ is the mean length of an idle period. Thus, the mean length of a cycle becomes

$$E(C) = \frac{E(V)\boldsymbol{\kappa}[\mathbf{I}-\mathbf{V}_{\mathrm{F}}(0)]^{-1}\mathbf{e}}{1-\rho}. \tag{1.65}$$

Now, using (1.65) in (1.64) completes the proof. □

The idle period ends only when there is a positive amount of fluid in the system at the end of a vacation. Thus, we have

$$\mathbf{U}_{\mathrm{B}}(x) = [\mathbf{I}-\mathbf{V}_{\mathrm{F}}(0)]^{-1}\mathbf{V}_{\mathrm{F}}(x),\ x>0. \tag{1.66}$$

Example 1.2 (Single vacation). For this vacation type, $\mathbf{u}_{\mathrm{idle}}^{*}(\theta)$ is given in the following theorem.

Theorem 1.7. *We have*

$$\mathbf{u}_{\mathrm{idle}}^{*}(\theta) = \frac{\boldsymbol{\kappa}}{E(C)}\left\{\mathbf{V}_{\mathrm{F}}(0)\mathbf{T}_{\mathrm{D}} + [\mathbf{V}_{\mathrm{F}}^{*}(\theta)-\mathbf{I}][\mathbf{Q}_{\Lambda}^{*}(\theta)]^{-1}\right\}, \tag{1.67a}$$

where

$$\mathbf{T}_{\mathrm{D}} = \begin{pmatrix} \mathbf{0} & \mathbf{0} \\ \mathbf{0} & \int_0^{\infty} \mathrm{e}^{\mathbf{Q}_{22}^{\mathrm{L}}x}\mathrm{d}x \end{pmatrix} = \begin{pmatrix} \mathbf{0} & \mathbf{0} \\ \mathbf{0} & (-\mathbf{Q}_{22}^{\mathrm{L}})^{-1} \end{pmatrix}. \tag{1.67b}$$

Proof. We note that an idle period consists of a vacation and a possible dormant period. Let $(\mathbf{T}_{\mathrm{D}})_{ij}$ be the mean time the dormant process stays in phase j under the condition that it starts with phase i. Then we have (1.67b) because the dormant period terminates as soon as a $\mathfrak{I}_2 \to \mathfrak{I}_1$ transition occurs.

Noting that the mean length of a dormant period is given by $\boldsymbol{\kappa}\mathbf{V}_{\mathrm{F}}(0)\mathbf{T}_{\mathrm{D}}\mathbf{e}$, it is not difficult to see that the mean length $E(I)$ of an idle period becomes

$$E(I) = E(V) + \boldsymbol{\kappa}\mathbf{V}_{\mathrm{F}}(0)\mathbf{T}_{\mathrm{D}}\mathbf{e}. \tag{1.68}$$

An arbitrary time point during an idle period is contained either in a vacation [with probability $\frac{E(V)}{E(I)}$] or in a dormant period [with probability $\kappa\mathbf{V}_{\mathrm{F}}(0)\mathbf{T}_{\mathrm{D}}\mathbf{e}/E(I)$]. The phase probability vector at an arbitrary time during a dormant period is given by

$$\frac{\kappa\mathbf{V}_{\mathrm{F}}(0)\mathbf{T}_{\mathrm{D}}}{\kappa\mathbf{V}_{\mathrm{F}}(0)\mathbf{T}_{\mathrm{D}}\mathbf{e}}. \tag{1.69}$$

Then, using (1.62), (1.68), and (1.69), we have that $\mathbf{u}^*_{\mathrm{idle}}(\theta)$ becomes

$$\mathbf{u}^*_{\mathrm{idle}}(\theta) = \frac{(1-\rho)\boldsymbol{\kappa}}{E(I)}\left\{\mathbf{V}_{\mathrm{F}}(0)\mathbf{T}_{\mathrm{D}} + [\mathbf{V}^*_{\mathrm{F}}(\theta) - \mathbf{I}][\mathbf{Q}^*_{\Lambda}(\theta)]^{-1}\right\}. \tag{1.70}$$

Now, using $E(C) = \frac{E(I)}{(1-\rho)}$ in (1.70) completes the proof. □

In this system, if there exists any fluid at the end of a vacation, then the server becomes busy immediately. If not, then the busy period starts with zero fluid. Thus we have

$$\mathbf{U}_{\mathrm{B}}(x) = \begin{cases} \mathbf{V}_{\mathrm{F}}(0)\mathbf{T}_{\mathrm{D}}\begin{pmatrix} \mathbf{0} & \mathbf{0} \\ \mathbf{Q}^{\mathrm{L}}_{21} & \mathbf{0} \end{pmatrix}, & x = 0, \\ \mathbf{V}_{\mathrm{F}}(x), & x > 0. \end{cases} \tag{1.71}$$

Example 1.3 (D-policy). Let us define the following probability:

$[\mathbf{I}^*(x,t)]_{ij}\mathrm{d}x = \Pr$ [the fluid level process during an idle period visits the level $(x, x+\mathrm{d}x]$ with UMC phase j at time t under the condition that the idle period starts with UMC phase i at time 0), x>0, $[\mathbf{I}^*(0,0) = \mathbf{I}]$.

Let $[\mathbf{I}^*(x)]_{ij} = \int_{t=0}^{\infty}[\mathbf{I}^*(x,t)]_{ij}\mathrm{d}t$ and $\mathbf{I}^*(x)$ be the matrix of $[\mathbf{I}^*(x)]_{ij}$. Let us partition $\mathbf{I}^*(x)$ according to $\mathfrak{I}^{\mathrm{idle}}_1$ and $\mathfrak{I}^{\mathrm{idle}}_2$ as $\mathbf{I}^*(x) = \begin{pmatrix} \mathbf{I}^*_{11}(x) & \mathbf{I}^*_{12}(x) \\ \mathbf{I}^*_{21}(x) & \mathbf{I}^*_{22}(x) \end{pmatrix}$. We then have the following theorem.

Theorem 1.8. *We have*

$$\mathbf{I}^*(x) = \begin{pmatrix} \mathbf{T}_{11}(x)\mathbf{R}_{\mathrm{L}}^{-1} & \mathbf{T}_{11}(x)\mathbf{R}_{\mathrm{L}}^{-1}\mathbf{Q}^{\mathrm{L}}_{12} \\ \mathbf{T}_{21}(x)\mathbf{R}_{\mathrm{L}}^{-1} & \mathbf{T}_{21}(x)\mathbf{R}_{\mathrm{L}}^{-1}\mathbf{Q}^{\mathrm{L}}_{12} \end{pmatrix}, \; x > 0, \tag{1.72}$$

$$\mathbf{I}^*(0) = \mathbf{I}, \tag{1.73}$$

where, using (1.8)–(1.10), $\mathbf{T}_{11}(x)$ *and* $\mathbf{T}_{21}(x)$ *are given by*

$$\mathbf{T}_{21}(x) = \mathbf{T}^*_{21}(x,\theta)|_{\theta=0} = (\mathbf{I} - \mathbf{Q}^{\mathrm{L}}_{22})^{-1}\mathbf{Q}^{\mathrm{L}}_{21}\mathbf{T}_{11}(x), \tag{1.74}$$

$$\mathbf{T}_{11}(x) = \mathbf{T}^*_{11}(x,\theta)|_{\theta=0} = \mathrm{e}^{\mathbf{Q}_{\mathrm{L}}x}, \tag{1.75}$$

in which

$$\mathbf{Q}_{\mathrm{L}} = \mathbf{Q}^*_{\mathrm{L}}(\theta)|_{\theta=0} = \mathbf{R}_{\mathrm{L}}^{-1}\left[\mathbf{Q}^{\mathrm{L}}_{11} + \mathbf{Q}^{\mathrm{L}}_{12}(-\mathbf{Q}^{\mathrm{L}}_{22})^{-1}\mathbf{Q}^{\mathrm{L}}_{21}\right]. \tag{1.76}$$

Proof. Let $[\mathrm{d}_t\mathbf{T}(x,t)]_{ij}$ be the probability that the fluid level process during the idle period visits fluid level x for the first time in the time interval $(t, t+\mathrm{d}t]$, and the UMC phase is j in the visiting epoch under the condition that the idle period starts with UMC phase i at time 0. Since the rate of change in fluid level is v_j when the UMC phase is $j \in \Im^{\mathrm{idle}}_1$, we have

$$[\mathbf{I}^*_{11}(x,t)]_{ij}\mathrm{d}x = [\mathbf{I}^*_{11}(x,t)]_{ij}v_j\mathrm{d}t = [\mathrm{d}_t\mathbf{T}_{11}(x,t)]_{ij}, \tag{1.77}$$

which means

$$[\mathbf{I}^*_{11}(x,t)]_{ij}\mathrm{d}t = \frac{1}{v_j}[\mathrm{d}_t\mathbf{T}_{11}(x,t)]_{ij}. \tag{1.78}$$

Integrating both sides of (1.78) with respect to t we obtain

$$[\mathbf{I}^*_{11}(x)]_{ij} = \frac{1}{v_j}\left[\int_0^\infty \mathrm{d}_t\mathbf{T}_{11}(x,t)\right]_{ij} = \frac{1}{v_j}[\mathbf{T}^*_{11}(x,\theta)|_{\theta=0}]_{ij} \tag{1.79}$$

and

$$[\mathbf{I}^*_{21}(x)]_{ij} = \frac{1}{v_j}\left[\int_0^\infty \mathrm{d}_t\mathbf{T}_{21}(x,t)\right]_{ij} = \frac{1}{v_j}[\mathbf{T}^*_{21}(x,\theta)|_{\theta=0}]_{ij}. \tag{1.80}$$

Equations (1.79) and (1.80) can be written in matrix forms as follows:

$$\mathbf{I}^*_{11}(x) = \mathbf{I}^*_{11}(x)\mathbf{R}_{\mathrm{L}}^{-1}, \tag{1.81}$$

$$\mathbf{I}^*_{21}(x) = \mathbf{I}^*_{21}(x)\mathbf{R}_{\mathrm{L}}^{-1}. \tag{1.82}$$

Noting that $\mathbf{Q}^{\mathrm{L}}_{12}$ represents the rate of transition from $\Im^{\mathrm{idle}}_1$ to $\Im^{\mathrm{idle}}_2$ we have

$$\mathbf{I}^*_{12}(x) = \mathbf{I}^*_{11}(x)\mathbf{Q}^{\mathrm{L}}_{12}, \tag{1.83}$$

$$\mathbf{I}^*_{22}(x) = \mathbf{I}^*_{21}(x)\mathbf{Q}^{\mathrm{L}}_{12}. \tag{1.84}$$

Using (1.81)–(1.84) completes the proof. □

With (1.72), $\mathbf{u}^*_{\mathrm{idle}}(\theta)$ is given by the following theorem.

Theorem 1.9. *We have*

$$\mathbf{u}^*_{\mathrm{idle}}(\theta) = \frac{\boldsymbol{\kappa}}{E(C)}\left[\mathbf{I} + \int_{0+}^{D} \mathrm{e}^{-\theta x}\mathbf{I}^*(x)\mathrm{d}x\right]\begin{pmatrix}\mathbf{I} & \mathbf{0}\\ \mathbf{0} & (-\mathbf{Q}^{\mathrm{L}}_{22})^{-1}\end{pmatrix}. \tag{1.85}$$

Proof. Let $E(I)$ be the mean length of an idle period. We note that $U_{\mathrm{idle},j}(x) = \frac{(1-\rho)E(T_{x,j})}{E(I)}$, where $E(T_{x,j})$ is the total sojourn time in phase j with a fluid level less than or equal to x during an idle period. If the fluid level process during the idle period visits level x with a UMC phase in $\mathfrak{I}_2$ (with probability $[\mathbf{I}^*(x)]_{ij}$), it stays there for $(-\mathbf{Q}^{\mathrm{L}}_{22})^{-1}$ on average. Thus, we have

$$\mathbf{U}_{\mathrm{idle}}(0) = (1-\rho)\frac{\boldsymbol{\kappa}\begin{pmatrix}\mathbf{I} & \mathbf{0}\\ \mathbf{0} & (-\mathbf{Q}^{\mathrm{L}}_{22})^{-1}\end{pmatrix}}{E(I)}, \tag{1.86}$$

$$\begin{aligned}\mathbf{u}_{\mathrm{idle}}(x) &= \frac{\mathrm{d}}{\mathrm{d}x}\mathbf{U}_{\mathrm{idle}}(x)\\ &= (1-\rho)\frac{\boldsymbol{\kappa}\mathbf{I}^*(x)\begin{pmatrix}\mathbf{I} & \mathbf{0}\\ \mathbf{0} & (-\mathbf{Q}^{\mathrm{L}}_{22})^{-1}\end{pmatrix}}{E(I)}, \quad 0 < x \le D.\end{aligned} \tag{1.87}$$

Then taking the LT of (1.87) and using $E(C) = \frac{E(I)}{(1-\rho)}$ completes the proof. □

This system becomes busy only when the fluid level reaches D. Thus we obtain

$$\begin{aligned}\mathbf{U}_{\mathrm{B}}(x) &= \begin{pmatrix}\mathbf{T}_{11}(x) & \mathbf{0}\\ \mathbf{T}_{21}(x) & \mathbf{0}\end{pmatrix}\\ &= \begin{pmatrix}\mathrm{e}^{\mathbf{Q}_{\mathrm{L}}x} & \mathbf{0}\\ (-\mathbf{Q}^{\mathrm{L}}_{22})^{-1}\mathbf{Q}^{\mathrm{L}}_{21}\mathrm{e}^{\mathbf{Q}_{\mathrm{L}}x} & \mathbf{0}\end{pmatrix}, \quad x = D.\end{aligned} \tag{1.88}$$

Control of System Factors and Cost Optimization

In this section, we control some factors and see the effects on system performance. We also present a cost optimization.

Control of Outflow Rate vs. Control of Jump Size (Type V)

In a Type V system, the probability ρ that the system is busy is affected by two factors: the outflow rate during the busy period and the jump size (offered load)

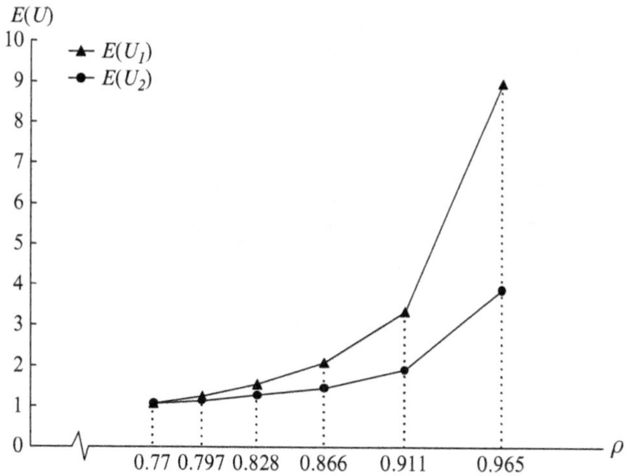

Fig. 1.3 Mean fluid levels with changes in outflow rate and jump size of Type V system with multiple vacations

during the idle period. In this section, we explore the effects of these factors on the performance of a Type V system. For this purpose we consider the multiple-vacation system.

Let us consider the parameter matrices as follows:

$$\mathbf{D}_0 = \begin{pmatrix} -8 & 1 & 2 & 2 \\ 1 & -9.5 & 1 & 3 \\ 1 & 2 & -10 & 2 \\ 1 & 1 & 1 & -10 \end{pmatrix}, \quad \mathbf{D}_1 = \begin{pmatrix} 1 & 0.5 & 0.5 & 1 \\ 2 & 0.5 & 1 & 1 \\ 1 & 2 & 1 & 1 \\ 3 & 2 & 1 & 1 \end{pmatrix}, \mathbf{R} = \begin{pmatrix} 3 & 0 & 0 & 0 \\ 0 & 2 & 0 & 0 \\ 0 & 0 & -4 & 0 \\ 0 & 0 & 0 & -5 \end{pmatrix}.$$

We first note that each r_4 and $E(S)$ affects the probability ρ that the system is busy. With this in mind, we change r_4 (system 1) and $E(S)$ of the jump size (system 2), with all other parameter values remaining the same. We then compute the mean fluid levels $E(U_1)$ and $E(U_2)$ of systems 1 and 2 using (1.38) and (1.46) as ρ varies. Then we can see the relative effects of the two factors on system performance.

As can be seen in Fig. 1.3, the change in the service rate results in a higher mean fluid level than the change in the mean jump size under the same ρ. This implies that controlling the service rate may benefit the system more than controlling the offered load does.

Cost Optimization

In this section, we consider a cost optimization model and demonstrate how we can determine the optimal threshold value that minimizes the long-run average

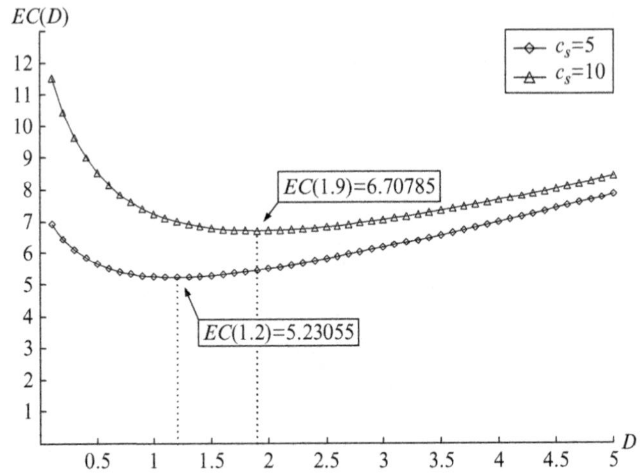

Fig. 1.4 Optimal thresholds for Type L system under the *D*-policy

operating cost. For this purpose we take a Type L system under the *D*-policy as an example system. Readers are referred to Baek et al. [9] for more details on the system.

In addition to the parameter matrices in the section "Control of Outflow Rate vs. Control of Jump Size (Type V)," we additionally assume $\mathbf{R}_\mathrm{L} = \begin{pmatrix} 8 & 0 \\ 0 & 7 \end{pmatrix}$ with $\mathfrak{I}_1^{\mathrm{idle}} = \{1,2\}, \mathfrak{I}_2^{\mathrm{idle}} = \{3,4\}$.

The *D*-policy is beneficial when startup (reactivation) costs of the server are very high. When the *D*-policy is employed, the mean cycle length becomes larger, which means fewer startups per unit time. In this way the *D*-policy reduces the setup costs of the system per unit time. Instead, the system maintains a higher level of fluid, which increases the fluid holding costs. All these considerations require the determination of the optimal threshold value of *D*.

Let us consider a linear cost function as follows:

$$EC(D) = \frac{c_\mathrm{s}}{E(C)} + c_\mathrm{h} \cdot E(U), \tag{1.89}$$

where c_s is a one-time startup cost to turn the idle server on and c_h is a holding cost for maintaining a unit amount of fluid per unit time. Then $EC(D)$ becomes the average operating cost per unit time. We note that cost function (1.89) is frequently used in stochastic optimization models related to queueing systems.

We consider two startup costs, ($c_\mathrm{s} = 5, 10$) and a holding cost ($c_\mathrm{h} = 2$). Using (1.85) in (1.42) we can compute $E(U)$. Using (1.88) and (1.37) we can compute $E(C) = E(I) + E(B)$. Figure 1.4 shows the values of $EC(D)$ as a function of D and the optimal threshold values $D^* = 1.2$ and $D^* = 1.9$ for both cases.

Conclusions

In this chapter, we derived factorizations of the fluid level distribution for a MAP-modulated fluid flow model. We also presented recursive moment formulas. We demonstrated how our factorizations could be used to derive the fluid level distribution for some example systems. We explored the effects of the service rate and fluid size on the system performance and presented a cost optimization example.

Acknowledgements Ho Woo Lee was supported by the Basic Science Research Program through the National Research Foundation (NRF) of Korea funded by the Ministry of Education, Science, and Technology (Grant No. 2010-0010023).

Soohan Ahn was partially supported by the Basic Science Research Program through the NRF of Korea funded by the Ministry of Education, Science, and Technology (Grant No. 2010-0021831).

References

1. Aggarwal, V., Gautam, N., Kumara, S.R.T., Greaves, M.: Stochastic fluid flow models for determining optimal switching thresholds. Perform. Eval. **59**(1), 19–46 (2005)
2. Ahn, S.: A transient analysis of Markov fluid models with jumps. J. Kor. Stat. Soc. **38**(4), 351–366 (2009)
3. Ahn, S., Ramaswami, V.: Fluid flow models and queues: a connection by stochastic coupling. Stoch. Models **19**(3), 325–348 (2003)
4. Ahn, S., Ramaswami, V.: Transient analysis of fluid flow models via stochastic coupling to a queue. Stoch. Models **20**(1), 71–101 (2004)
5. Ahn, S., Ramaswami, V.: Efficient algorithms for transient analysis of stochastic fluid flow models. J. Appl. Probab. **42**, 531–549 (2005)
6. Anick, D., Mitra, D., Sondhi, M.: Stochastic theory of a data handling system with multiple sources. Bell Syst. Tech. J. **61**, 1871–1894 (1982)
7. Asmussen, S.: Stationary distributions for fluid flow models with or without Brownian noise. Stoch. Models **11**(1), 21–49 (1995)
8. Baek, J.W., Lee, H.W., Lee, S.W., Ahn, S.: A factorization property for BMAP/G/1 vacation queues under variable service speed. Ann. Oper. Res. **160**, 19–29 (2008)
9. Baek, J.W., Lee, H.W., Lee, S.W., Ahn, S.: A Markov-modulated fluid flow queueing model under D-policy. Numer. Lin. Algebra **18**(6), 993–1010 (2011)
10. Chang, S.H., Takine, T., Chae, K.C., Lee, H.W.: A unified queue length formula for BMAP/G/1 queue with generalized vacations. Stoch. Models **18**(3), 369–386 (2002)
11. Da Silva Soares, A., Latouche, G.: Fluid queues with level dependent evolution. Eur. J. Oper. Res. **196**(3), 1041–1048 (2009)
12. Doshi, B.T.: Queueing systems with vacations: survey. Queue. Syst. **1**(1), 29–66 (1986)
13. Fuhrmann, S.W., Cooper, R.B.: Stochastic decompositions in the M/G/1 queue with generalized vacations. Oper. Res. **33**, 1117–1129 (1985)
14. Heyman, D.P.: T-policy for the M/G/1 queue. Manage. Sci. **23**(7), 775–778 (1977)
15. Kulkarni, V.G., Yan K.: A fluid model with upward jumps at the boundary. Queue. Syst. **56**(2), 103–117 (2007)
16. Lee, H.W., Baek, J.W.: BMAP/G/1 queue under D-policy: queue length analysis. Stoch. Models **21**(2–3), 1–21 (2005)
17. Lee, H.W., Ahn, B.Y., Park, N.I.: Decompositions of the queue length distributions in the MAP/G/1 queue under multiple and single vacations with N-policy. Stoch. Models **17**(2), 157–190 (2001)

18. Lee, H.W., Cheon, S.H., Lee, E.Y., Chae, K.C.: Workload and waiting time analysis of MAP/G/1 queue under *D*-policy. Queue. Syst. **48**, 421–443 (2004)
19. Lucantoni, D.M., Meier-Hellstern, K.S., Neuts, M.F.: A single server queue with server vacations and a class of non-renewal arrival processes. Adv. Appl. Probab. **22**(3), 676–709 (1990)
20. Malhotra, R., Mandjes, M., Scheinhardt W., van den Berg, J.L.: A feedback fluid queue with two congestion control thresholds. Math. Meth. Oper. Res. **70**, 149–169 (2009)
21. Mandjes, M., Mitra, D., Scheinhardt, W.R.W.: Models of network access using feedback fluid queues. Queue. Syst. **44**(4), 365–398 (2003)
22. Mitra, D.: Stochastic theory of a fluid model of producers and consumers coupled by a buffer. Adv. Appl. Probab. **20**(3), 646–676 (1988)
23. Van Foreest, N., Mandjes, M., Scheinhardt, W.: Analysis of a feedback fluid model for heterogeneous TCP sources. Comm. Stat. Stoch. Model **19**, 299–324 (2003)

Chapter 2
A Compressed Cyclic Reduction for QBD processes with Low-Rank Upper and Lower Transitions

Dario A. Bini, Paola Favati, and Beatrice Meini

Introduction

A quasi-birth-and-death (QBD) process [10] is a Markov chain associated with a probability transition matrix

$$P = \begin{bmatrix} B_0 & B_1 & & & 0 \\ B_{-1} & A_0 & A_1 & & \\ & A_{-1} & A_0 & A_1 & \\ & & A_{-1} & A_0 & \ddots \\ 0 & & & \ddots & \ddots \end{bmatrix}, \tag{2.1}$$

where B_0, B_1, and A_i, $i = -1, 0, 1$, are $m \times m$ matrices, m being the phase space dimension. In the numerical solution of QBD processes, a crucial step is the computation of the minimal nonnegative solution G of the quadratic matrix equation

$$X = A_{-1} + A_0 X + A_1 X^2. \tag{2.2}$$

To this end, many numerical methods, with different properties, have been proposed in recent years (see, for instance, [2, 4–6]). Most of these algorithms are designed to deal with the general case where the block coefficients A_{-1}, A_0, and A_1 have no special structure.

D.A. Bini (✉) • B. Meini
Dipartimento di Matematica, Università di Pisa, Italy
e-mail: bini@dm.unipi.it; meini@dm.unipi.it

P. Favati
Istituto di Informatica e Telematica, CNR, Pisa, Italy
e-mail: paola.favati@iit.cnr.it

G. Latouche et al. (eds.), *Matrix-Analytic Methods in Stochastic Models*, Springer Proceedings in Mathematics & Statistics 27, DOI 10.1007/978-1-4614-4909-6_2,

However, there are important applications where the block coefficients exhibit a structure that can be exploited to efficiently compute G. For instance, if A_{-1} has only one nonzero column, then G also has only one nonzero column, which can be computed using an explicit formula. We refer the reader to Neuts [12], to Sect. 10.4 of Latouche and Ramaswami [10], and to a few articles [7, 11, 15, 16] for some examples of queues where A_{-1} or A_1 has rank one, and this property is used to provide a simple expression for G and for the steady state vector.

More recently, some interest has been demonstrated in specific cases where the blocks A_{-1} and/or A_1 have many zero columns and rows, respectively. The interest in these cases is motivated by QBD processes with restricted transitions to higher (or lower) levels encountered in certain applications [8, 13, 14]. In particular, in [14] Pérez and Van Houdt exploit these properties of the matrix A_{-1} or A_1 to formulate the QBD process in terms of an M/G/1- or GI/M/1-type Markov chain, where the block matrices have a size equal to the number of nonzero columns of A_{-1} or nonzero rows of A_1. In [8] Grassman and Tavakoli show how the structure of A_{-1} is used to reduce the computational cost of certain fixed-point iterations for computing G.

In this chapter, we consider the more general case where the matrices A_{-1} and A_1 have small rank with respect to their size. This assumption is in particular satisfied in the case of restricted transitions to higher and lower levels. We exploit these rank properties to improve the efficiency of known algorithms for the computation of G. More specifically, we consider the cyclic reduction (CR) algorithm [2, 3] and show that if the sum of the ranks of A_{-1} and A_1 is equal to $r < m$, then the CR step can be implemented with $O(r^3)$ arithmetic operations (ops), instead of the $O(m^3)$ ops required in the general case. This fact leads to a dramatic acceleration in the CPU time when r is much smaller than m. The same acceleration can be obtained for the logarithmic reduction (LR) algorithm of Latouche and Ramaswami [10]. The new algorithms keep the nice properties of numerical stability of the original algorithms because they avoid numerical cancellation. In fact, assuming that the low-rank decomposition of A_{-1} and A_1 is formed by nonnegative matrices, we prove that the nonnegativity of the matrices involved in the algorithm is preserved and that all the operations consist of multiplications of nonnegative matrices and inversions of M-matrices. In the case where the low-rank decomposition is not given by nonnegative matrices, the algorithm can still be applied and keeps the same computational cost and convergence properties, but we cannot guarantee its numerical stability.

We consider also the case where only one matrix, among A_{-1} and A_1, has a small rank. We provide a version of CR where the number of ops required by the iteration is substantially smaller, even though it is still of the same order $O(m^3)$.

Our algorithms have been compared to the existing algorithms available in the literature. The many numerical experiments that we have performed show that they behave much better than the available methods. The larger the size of the matrix with respect to the rank, the larger the gain with respect to the customary algorithms.

The chapter is organized as follows. In the section "Cyclic Reduction for QBD processes" the customary CR algorithm is recalled together with its convergence and applicability properties. The new algorithm for the case of low-rank downward and upward transitions is presented in the section "Case of Low-Rank Downward and Upward Transitions"; its numerical stability and computational complexity are also discussed. In the section "Case of Low-Rank Downward or Upward Transitions" we present an algorithm for the case where only one transition between the downward and upward transitions has low rank. Finally, in the section "Numerical Experiments" some numerical experiments are reported, showing the effectiveness of the proposed algorithms in terms of computational cost.

For the count of the arithmetic operations we use the following classical results: the LU factorization of a $p \times p$ matrix A costs $\frac{2}{3}p^3$ ops, the solution of q linear systems $AX = B$, given the LU factorization of the nonsingular matrix A, costs $2p^2q$ ops, the inversion of A costs $2p^3$ ops, and the multiplication of a $p \times q$ matrix by a $q \times s$ matrix costs $2pqs$ ops.

Throughout the chapter we assume that the matrix P is irreducible, the matrix $A_{-1} + A_0 + A_1$ is irreducible, and the QBD process is not null recurrent.

Cyclic Reduction for QBD processes

The CR algorithm provides an effective method for computing the matrix G. It consists in generating a sequence of matrices $A_i^{(k)}$, $i = -1, 0, 1$, and $\widehat{A}_0^{(k)}$ according to the following equations [2, 3]:

$$
\begin{aligned}
A_1^{(k+1)} &= A_1^{(k)}(I - A_0^{(k)})^{-1}A_1^{(k)},\\
A_0^{(k+1)} &= A_0^{(k)} + A_1^{(k)}(I - A_0^{(k)})^{-1}A_{-1}^{(k)} + A_{-1}^{(k)}(I - A_0^{(k)})^{-1}A_1^{(k)},\\
A_{-1}^{(k+1)} &= A_{-1}^{(k)}(I - A_0^{(k)})^{-1}A_{-1}^{(k)},\\
\widehat{A}_0^{(k+1)} &= \widehat{A}_0^{(k)} + A_1^{(k)}(I - A_0^{(k)})^{-1}A_{-1}^{(k)},
\end{aligned}
\tag{2.3}
$$

with $A_1^{(0)} = A_1$, $A_0^{(0)} = A_0$, $A_{-1}^{(0)} = A_{-1}$, $\widehat{A}_0^{(0)} = A_0$, for $k \geq 0$, where we assume that $I - A_0^{(k)}$ is nonsingular.

An approximation of G is provided by $(I - \widehat{A}_0^{(k)})^{-1}A_{-1}$, for a sufficiently large value of k, since $G = \lim_{k\to\infty}(I - \widehat{A}_0^{(k)})^{-1}A_{-1}$, according to the following convergence and applicability properties [2, Theorems 7.5, 7.6].

Theorem 2.1. *If the QBD processes is positive recurrent, then* $\det(I - A_0^{(k)}) \neq 0$ *and* $\det(I - \widehat{A}_0^{(k)}) \neq 0$, *so that the CR can be carried out with no breakdown; the matrices* $I - A_0^{(k)}$ *and* $I - \widehat{A}_0^{(k)}$ *are (nonsingular) M-matrices,* $A_i^{(k)}$, $i = -1, 0, 1$, *are nonnegative, and* $A_{-1}^{(k)} + A_0^{(k)} + A_1^{(k)}$ *is stochastic for* $k \geq 0$. *Moreover, the following limits exist:*

$$\lim_k A_0^{(k)} = A_0^{(\infty)}, \quad \lim_k \widehat{A}_0^{(k)} = \widehat{A}_0^{(\infty)},$$

$$\lim_k A_{-1}^{(k)} = (I - A_0^{(\infty)})eg^T, \quad \lim_k A_1^{(k)} = 0,$$

$$\lim_k (I - \widehat{A}_0^{(k)})^{-1} A_{-1} = G, \tag{2.4}$$

where $\widehat{A}_0^{(\infty)}$ *is the minimal nonnegative solution of*

$$X = A_0 + A_1(I - X)^{-1}A_{-1}, \tag{2.5}$$

$e = (1, \ldots, 1)^T$, $g \geq 0$ *is such that* $g^T G = g^T$, $g^T e = 1$, *and all the sequences in equations* (2.4) *quadratically converge to their limits.*

In the general case each step of the CR algorithm requires the solution of $2m$ linear systems of size m and four matrix multiplications of order m, with an overall cost of $\frac{38}{3}m^3$ ops. Due to the quadratic convergence, few steps are generally sufficient to reach the desired accuracy for the computation of G.

If the QBD process is transient, a similar applicability and convergence result can be given. We refer the reader to [2] for more details.

Case of Low-Rank Downward and Upward Transitions

Now we consider the case where both downward and upward transitions have low rank, i.e., both the matrices A_{-1} and A_1 have low rank. Denote by U_i and V_i, $i = -1, 1$, matrices of size $m \times r_i$ and $r_i \times m$, respectively, such that $A_i = U_i V_i$, $i = -1, 1$, and set $r = r_{-1} + r_1$, where we assume that r is much smaller than m.

We show that the matrices $A_i^{(k)}$, $i = -1, 0, 1$, generated at the kth step of the CR can be expressed in terms of the matrices U_i and V_i and in terms of small size matrices that depend on the step k.

To this end, define the following two matrices of size $m \times r$ and $r \times m$, respectively:

$$U = \left[U_{-1} \,|\, U_1 \right], \qquad V = \left[\frac{V_{-1}}{V_1} \right]. \tag{2.6}$$

The following result holds:

Theorem 2.2. *Assume that* $A_i = U_i V_i$, $i = -1, 1$, *where* U_i *and* V_i, $i = -1, 1$, *are matrices of size* $m \times r_i$ *and* $r_i \times m$, *respectively. Let* U *and* V *be the matrices defined in* (2.6)*. Then the sequences of matrices generated by cyclic reduction verify the following relations:*

$$A_1^{(k)} = U H_1^{(k)} V,$$

$$A_0^{(k)} = A_0 + U H_0^{(k)} V,$$

$$\begin{aligned} A_{-1}^{(k)} &= UH_{-1}^{(k)}V, \\ \widehat{A}_0^{(k)} &= A_0 + U\widehat{H}_0^{(k)}V, \end{aligned} \tag{2.7}$$

where the $r \times r$ matrices $H_i^{(k)}$, $i = -1, 0, 1$, and $\widehat{H}_0^{(k)}$ are recursively defined, for $k \geq 0$, by

$$\begin{aligned} H_1^{(k+1)} &= H_1^{(k)} Q^{(k)} H_1^{(k)}, \\ H_0^{(k+1)} &= H_0^{(k)} + H_1^{(k)} Q^{(k)} H_{-1}^{(k)} + H_{-1}^{(k)} Q^{(k)} H_1^{(k)}, \\ H_{-1}^{(k+1)} &= H_{-1}^{(k)} Q^{(k)} H_{-1}^{(k)}, \\ \widehat{H}_0^{(k+1)} &= \widehat{H}_0^{(k)} + H_1^{(k)} Q^{(k)} H_{-1}^{(k)}, \end{aligned} \tag{2.8}$$

where

$$Q^{(k)} = (I - Q^{(0)} H_0^{(k)})^{-1} Q^{(0)}, \tag{2.9}$$

with $Q^{(0)} = V\,(I - A_0)^{-1} U$, $H_0^{(0)} = \widehat{H}_0^{(0)} = 0$, and

$$H_{-1}^{(0)} = \begin{bmatrix} I_{r_{-1}} & 0 \\ 0 & 0 \end{bmatrix}, \quad H_1^{(0)} = \begin{bmatrix} 0 & 0 \\ 0 & I_{r_1} \end{bmatrix}.$$

Proof. We prove the result by induction on k. For $k = 0$, Eqs. (2.7) hold by construction. Assume that (2.7)–(2.9) hold for a fixed $k \geq 1$, and prove them for $k+1$. Consider $A_1^{(k+1)}$. From (2.3) it follows that $A_1^{(k+1)} = A_1^{(k)}(I - A_0^{(k)})^{-1} A_1^{(k)}$. By inductive hypothesis, one has $A_1^{(k)} = UH_1^{(k)}V$; therefore $A_1^{(k+1)} = UH_1^{(k+1)}V$, where

$$H_1^{(k+1)} = H_1^{(k)} V (I - A_0^{(k)})^{-1} U H_1^{(k)}. \tag{2.10}$$

Since $I - A_0^{(k)} = I - A_0 - UH_0^{(k)}V$, by applying the Sherman–Woodbury–Morrison formula [9], we have

$$(I - A_0^{(k)})^{-1} = (I - A_0)^{-1} + (I - A_0)^{-1} U (T^{(k)})^{-1} H_0^{(k)} V (I - A_0)^{-1},$$

with $T^{(k)} = I - H_0^{(k)} V (I - A_0)^{-1} U$. The latter matrix is invertible since both the matrices $I - A_0^{(k)}$ and $I - A_0$ are invertible by Theorem 2.1. Observe that

$$V (I - A_0^{(k)})^{-1} U = Q^{(0)} + Q^{(0)} (I - H_0^{(k)} Q^{(0)})^{-1} H_0^{(k)} Q^{(0)} = (I - Q^{(0)} H_0^{(k)})^{-1} Q^{(0)}.$$

Hence $V(I - A_0^{(k)})^{-1} U = Q^{(k)}$, so that from (2.10) one finds that $H_1^{(k+1)} = H_1^{(k)} Q^{(k)} H_1^{(k)}$. We proceed similarly, for the remaining matrix sequences. □

According to the preceding theorem, the matrices $A_{-1}^{(k)}$ and $A_1^{(k)}$ have rank at most r and can be expressed by means of the $r \times r$ matrices $H_i^{(k)}$, $i = -1, 1$. Moreover, the matrices $A_0^{(k)}$ and $\widehat{A}_0^{(k)}$ are at most a rank r correction of the original matrix A_0. These properties allow one to carry out CR relying on (2.8) and (2.9) with a reduced computational cost.

Observe that $G = \lim_k X^{(k)}$, where $X^{(k)}$ is the solution of the linear system $(I - \widehat{A}_0^{(k)})X = A_{-1}$. For the structure of the matrix $\widehat{A}_0^{(k)}$, by applying the Sherman–Woodbury–Morrison formula, we find that

$$(I - \widehat{A}_0^{(k)})^{-1}A_{-1} = (I - A_0)^{-1}A_{-1} + (I - A_0)^{-1}U(\widehat{T}^{(k)})^{-1}\widehat{H}_0^{(k)}V(I - A_0)^{-1}A_{-1}, \tag{2.11}$$

with $\widehat{T}^{(k)} = I - \widehat{H}_0^{(k)}V(I - A_0)^{-1}U$. From (2.11), since $A_{-1} = U_{-1}V_{-1}$, it follows that G has at most rank r_{-1}, and $G = U_G V_{-1}$, where $U_G = \lim_{k\to\infty} U_G^{(k)}$ and

$$\begin{aligned} U_G^{(k)} &= (I - A_0)^{-1}U_{-1} + (I - A_0)^{-1}U(\widehat{T}^{(k)})^{-1}\widehat{H}_0^{(k)}V(I - A_0)^{-1}U_{-1} \\ &= (I + (I - A_0)^{-1}U(\widehat{T}^{(k)})^{-1}\widehat{H}_0^{(k)}V)(I - A_0)^{-1}U_{-1}. \end{aligned}$$

If the matrices U and V are nonnegative, then the sequences $H_i^{(k)}$ are nonnegative as well, and their computation is numerically stable since it involves additions of nonnegative matrices and inversions of M-matrices, as stated by the following theorem.

Theorem 2.3. *Assume that the assumptions of Theorem 2.2 hold. If $U \geq 0$ and $V \geq 0$, then $Q^{(0)} \geq 0$, and the sequences $\{H_i^{(k)}\}_k$, $i = -1, 0, 1$, and $\{\widehat{H}_0^{(k)}\}_k$ are such that $\widehat{H}_0^{(k)} \geq 0$, $H_i^{(k)} \geq 0$, $i = -1, 0, 1$; moreover, the matrix $I - Q^{(0)}H_0^{(k)}$ is a nonsingular M-matrix for any $k \geq 1$.*

Proof. The matrix $Q^{(0)}$ is nonnegative since $U, V \geq 0$ and $I - A_0$ is a nonsingular M-matrix. To prove the remaining part of the theorem, we proceed by induction on k. For $k = 1$, one has $\widehat{H}_0^{(1)} \geq 0$, $H_i^{(1)} \geq 0$, $i = -1, 0, 1$, by construction since $Q^{(0)} \geq 0$. To show that $S = I - Q^{(0)}H_0^{(1)}$ is a nonsingular M-matrix; consider the 2×2 block matrix

$$B = \begin{bmatrix} I - A_0 & -UH_0^{(1)} \\ -V & I \end{bmatrix},$$

and observe that, since $Q^{(0)} = V(I - A_0)^{-1}U$, then S is the Schur complement of $I - A_0$ in B. We show that B is a nonsingular M-matrix, therefore, since the Schur complement in a nonsingular M-matrix is a nonsingular M-matrix [1], also S is a nonsingular M-matrix. The matrix B is a Z-matrix since $I - A_0$ is an M-matrix and $UH_0^{(1)} \geq 0$, $V \geq 0$. Therefore, to show that B is an M-matrix, it is sufficient to find a positive vector w such that $Bw > 0$ [1]. For Theorem 2.1 the matrix $I - A_0^{(1)}$ is a nonsingular M-matrix; therefore, there exists a positive vector r such that $(I - A_0^{(1)})r = s > 0$ [1]. Since $I - A_0^{(1)} = I - A_0 - UH_0^{(1)}V$ in view of (2.7), one finds that

$$B\begin{bmatrix} r \\ Vr + \varepsilon e \end{bmatrix} = \begin{bmatrix} s - \varepsilon UH_0^{(1)}e \\ \varepsilon e \end{bmatrix}, \tag{2.12}$$

where e is the vector of all ones. Since $s > 0$, we may find $\varepsilon > 0$ such that $s - \varepsilon UH_0^{(1)}e > 0$. The vector $w = \left[\begin{smallmatrix} r \\ Vr+\varepsilon e \end{smallmatrix}\right]$ is positive and, with this choice of ε, the right-hand side in (2.12) is positive; therefore, B is a nonsingular M-matrix. Assume that the properties hold for a fixed $k \geq 1$. Since $I - Q^{(0)}H_0^{(k)}$ is a nonsingular M-matrix, its inverse is nonnegative; therefore, $Q^{(k)} \geq 0$ and, from (2.8), $\widehat{H}_0^{(k+1)} \geq 0$, $H_i^{(k+1)} \geq 0$, $i = -1, 0, 1$. To show that $S = I - Q^{(0)}H_0^{(k+1)}$ is a nonsingular M-matrix, we proceed as in the case $k = 1$ by observing that S is the Schur complement of $I - A_0$ in the matrix

$$B^{(k+1)} = \begin{bmatrix} I - A_0 & -UH_0^{(k+1)} \\ -V & I \end{bmatrix}.$$

The latter matrix is a nonsingular M-matrix since it is a Z-matrix and, for a suitable $\varepsilon > 0$, one has $B^{(k+1)}\left[\begin{smallmatrix} r \\ Vr+\varepsilon e \end{smallmatrix}\right] > 0$, where $r > 0$ is such that $(I - A_0^{(k+1)})r > 0$. Such a positive vector r exists since $I - A_0^{(k+1)}$ is a nonsingular M-matrix for Theorem 2.1.

□

Algorithm 1 reports a pseudocode that implements CR by exploiting the low-rank properties of the matrices. In the code, we use the Matlab notation where $A(:, 1:s)$ is the matrix formed by the first s columns of the matrix A. The algorithm should stop the iterative process if the norm of $A_{-1}^{(k)}$ or $A_1^{(k)}$ is sufficiently small, but in practice, since the computation of the norm of $H_i^{(k)}$, $i = -1, 1$, is less expensive, we stop the algorithm if $\|H_{-1}^{(k)}\|_1 < \varepsilon$ or $\|H_1^{(k)}\|_1 < \varepsilon$ for a fixed tolerance ε. On output, the algorithm provides an approximation to the $m \times r_{-1}$ matrix U_G such that $G = U_G V_{-1}$.

To apply formulas (2.8) and (2.9), we must first compute the matrix $Q^{(0)}$ by solving r linear systems with an $m \times m$ matrix and by computing a multiplication between an $r \times m$ matrix and an $m \times r$ matrix, with an overall cost of $\frac{2}{3}m^3 + 2m^2r + 2mr^2$ arithmetic operations. At each step the computation of $Q^{(k)}$ requires one multiplication between two $r \times r$ matrices and the solution of r linear systems of size r, with a cost of $\frac{14}{3}r^3$ ops; the computation of $H_i^{(k)}$ and $\widehat{H}_0^{(k)}$ requires six matrix multiplications of size r, with a cost of $12r^3$. Thus the overall arithmetic cost of the kth step is $\frac{50}{3}r^3$ ops. The computational cost of recovering U_G amounts to $\frac{14}{3}r^3 + 2r^2r_{-1} + 2mrr_{-1}$.

Algorithm 1 Low-Rank CR

Set $k = 0$

Set $H_0^{(0)} = \widehat{H}_0^{(0)} = 0_{r\times r}$, $H_{-1}^{(0)} = \begin{bmatrix} I_{r_{-1}} & 0 \\ 0 & 0 \end{bmatrix}$, $H_1^{(0)} = \begin{bmatrix} 0 & 0 \\ 0 & I_{r_1} \end{bmatrix}$

Compute $W = (I - A_0)^{-1}U$ by solving the linear system $(I - A_0)X = U$

Compute $Q^{(0)} = VW$

while $\min\{\|H_{-1}^{(k)}\|_1, \|H_1^{(k)}\|_1\} \geq \varepsilon$ **do**

  Compute $Y_1 = H_1^{(k)}Q^{(k)}$ and $Y_{-1} = H_{-1}^{(k)}Q^{(k)}$

  Compute $H_1^{(k+1)} = Y_1H_1^{(k)}$ and $H_{-1}^{(k+1)} = Y_{-1}H_{-1}^{(k)}$

  Compute $Z = Y_1H_{-1}^{(k)}$

  Compute $H_0^{(k+1)} = H_0^{(k)} + Z + Y_{-1}H_1^{(k)}$ and $\widehat{H}_0^{(k+1)} = \widehat{H}_0^{(k)} + Z$

  Compute $F = Q^{(0)}H_0^{(k)}$

  Compute $Q_0^{(k+1)}$ by solving the linear system $(I - F)X = Q^{(0)}$.

  Set $k = k + 1$

end while

Compute $F = \widehat{H}_0^{(k)}Q^{(0)}$

Solve the linear system $(I - F)X = \widehat{H}_0^{(k)}$.

return $U_G = W(:, 1 : r_{-1}) + WXQ^{(0)}(:, 1 : r_{-1})$

To sum up, the algorithm consists of a preprocessing stage that costs $\frac{2}{3}m^3$ ops, an iterative stage where each iteration costs $\frac{50}{3}r^3$ ops, and a postprocessing stage that costs $\frac{14}{3}r^3 + 2r^2r_{-1} + 2mrr_{-1}$. It is important to point out that the cost of the iterative part is independent of the size m of the blocks.

Due to the interplay between CR and LR [2], a similar analysis can be carried out for the LR algorithm.

Case of Low-Rank Downward or Upward Transitions

We consider now the case where only one matrix between A_{-1} and A_1 has low rank. More specifically, assume that A_{-1} has low rank, that is, $A_{-1} = U_{-1}V_{-1}$, where U_{-1} and V_{-1} are $m \times r_{-1}$ and $r_{-1} \times m$ matrices with r_{-1} much smaller than m.

Also in this case the CR algorithm can be carried out with a computational cost lower than the cost of the general case. This improvement relies on the following properties:

– The matrix $A_{-1}^{(k)}$ generated at the kth step of CR can be expressed in terms of the matrices U_{-1} and V_{-1} and in terms of small size matrices that depend on step k.
– The matrices $A_0^{(k)}$ and $\widehat{A}_0^{(k)}$ are corrections of rank at most $2r_{-1}$ and r_{-1}, respectively, of the original matrix A_0.

More precisely, the following result provides the recursive equations for $A_0^{(k)}$, $\widehat{A}_0^{(k)}$, $A_{-1}^{(k)}$, while the equation for $A_1^{(k)}$ is left unchanged.

Theorem 2.4. *Let $A_{-1} = U_{-1}V_{-1}$. Then the sequences of matrices $A_{-1}^{(k)}$, $A_0^{(k)}$, and $\widehat{A}_0^{(k)}$ generated by the CR verify the following relations:*

$$\begin{aligned} A_{-1}^{(k)} &= U_{-1}K_{-1}^{(k)}V_{-1}, \\ A_0^{(k)} &= A_0 + U_{-1}W^{(k)} + Z^{(k)}V_{-1}, \\ \widehat{A}_0^{(k)} &= A_0 + Z^{(k)}V_{-1}, \end{aligned} \tag{2.13}$$

where the matrices $K_{-1}^{(k)}$ of size $r_{-1} \times r_{-1}$, $Z^{(k)}$ of size $m \times r_{-1}$, and $W^{(k)}$ of size $r_{-1} \times m$ are recursively defined, for $k \geq 0$, by

$$\begin{aligned} K_{-1}^{(k+1)} &= K_{-1}^{(k)}V_{-1}(I - A_0^{(k)})^{-1}U_{-1}K_{-1}^{(k)} \\ Z^{(k+1)} &= Z^{(k)} + A_1^{(k)}(I - A_0^{(k)})^{-1}U_{-1}K_{-1}^{(k)} \\ W^{(k+1)} &= W^{(k)} + K_{-1}^{(k)}V_{-1}(I - A_0^{(k)})^{-1}A_1^{(k)} \end{aligned} \tag{2.14}$$

with $K_{-1}^{(0)} = I_{r_{-1}}$ and $Z^{(0)} = W^{(0)} = 0$.

Proof. We prove the result by induction on k. For $k = 0$, Eqs. (2.13) hold by construction. Assume that (2.13) and (2.14) hold for a fixed $k \geq 1$, and we prove the result for $k+1$. Consider first $A_{-1}^{(k+1)}$. From (2.3) one has $A_{-1}^{(k+1)} = A_{-1}^{(k)}(I - A_0^{(k)})^{-1}A_{-1}^{(k)}$. By inductive hypothesis, $A_{-1}^{(k)} = U_{-1}K_{-1}^{(k)}V_{-1}$; therefore, $A_{-1}^{(k+1)} = U_{-1}K_{-1}^{(k+1)}V_{-1}$, where $K_{-1}^{(k+1)} = K_{-1}^{(k)}V_{-1}(I - A_0^{(k)})^{-1}U_{-1}K_{-1}^{(k)}$.

Consider now $A_0^{(k+1)}$. From (2.3) one has $A_0^{(k+1)} = A_0^{(k)} + A_1^{(k)}(I - A_0^{(k)})^{-1}A_{-1}^{(k)} + A_{-1}^{(k)}(I - A_0^{(k)})^{-1}A_1^{(k)}$. By inductive hypothesis, $A_0^{(k)} = A_0 + U_{-1}W^{(k)} + Z^{(k)}V_{-1}$; therefore,

$$\begin{aligned} A_0^{(k+1)} &= A_0 + U_{-1}(W^{(k)} + K_{-1}^{(k)}V_{-1}(I - A_0^{(k)})^{-1}A_1^{(k)}) \\ &\quad + (Z^{(k)} + A_1^{(k)}(I - A_0^{(k)})^{-1}U_{-1}K_{-1}^{(k)})V_{-1}, \end{aligned}$$

that is, $A_0^{(k+1)} = A_0 + U_{-1}W^{(k+1)} + Z^{(k+1)}V_{-1}$. We proceed similarly for the remaining matrix sequences. □

Consider the matrices

$$U^{(k)} = \left[U_{-1} \,\middle|\, Z^{(k)}\right], \qquad V^{(k)} = \left[\frac{W^{(k)}}{V_{-1}}\right]$$

of size $m \times 2r_{-1}$ and $2r_{-1} \times m$, respectively. Then we can write

$$A_0^{(k)} = A_0 + U^{(k)}V^{(k)},$$

Algorithm 2 Downward Low-Rank CR

Set $k=0$
Set $K_{-1}^{(0)}=I_{r_{-1}}$, $Z^{(0)}=0_{m\times r_{-1}}$, $W^{(0)}=0_{r_{-1}\times m}$
Compute $B=(I-A_0)^{-1}$, $N_{-1}=BU_{-1}$ and $Q_{-1}=V_{-1}N_{-1}$
while $\min\{\|K_{-1}^{(k)}\|_1,\|A_1^{(k)}\|_1\}\geq\varepsilon$ **do**
  Compute $F=U_{-1}K_{-1}^{(k)}$, $G=(I-A_0^{(k)})^{-1}F$, $M=K_{-1}^{(k)}V_{-1}$
  Compute $K_{-1}^{(k+1)}=MG$
  Compute $L=(I-A_0^{(k)})^{-1}A_1^{(k)}$ and $A_1^{(k+1)}=A_1^{(k)}L$
  Compute $Z^{(k+1)}=Z^{(k)}+A_1^{(k)}G$ and $W^{(k+1)}=W^{(k)}+ML$
  Compute $N=BZ^{(k+1)}$,
  Compute $R_{11}=W^{(k+1)}N_{-1}$, $R_{12}=W^{(k+1)}N$ and $R_{22}=V_{-1}N$
  Set $R=\begin{bmatrix}R_{11} & R_{12}\\ Q_{-1} & R_{22}\end{bmatrix}$, $V^{(k+1)}=\begin{bmatrix}W^{(k+1)}\\ \hline V_{-1}\end{bmatrix}$
  Compute $S=(I-R)^{-1}V^{(k+1)}$ by solving the linear system $(I-R)X=V^{(k+1)}$
  Compute $(I-A_0^{(k+1)})^{-1}=B+[N_{-1}\,|\,N]\,SB$
  Set $k=k+1$
end while
Compute $F=(I-R_{22})^{-1}Q_{-1}$ by solving the linear system $(I-R_{22})X=Q_{-1}$
return $U_G=N_{-1}+NF$

and from the Sherman–Woodbury–Morrison formula we have

$$(I-A_0^{(k)})^{-1}=B+BU^{(k)}(I-V^{(k)}BU^{(k)})^{-1}V^{(k)}B, \tag{2.15}$$

where $B=(I-A_0)^{-1}$. Moreover, also in this case, the desired solution G is given by $G=\lim_k X^{(k)}$, where $X^{(k)}$ is the solution of the linear system $(I-\widehat{A}_0^{(k)})X=A_{-1}$. It can be expressed as $G=U_GV_{-1}$, where $U_G=\lim_{k\to\infty}U_G^{(k)}$ and

$$\begin{aligned}U_G^{(k)}&=(I-A_0)^{-1}U_{-1}+(I-A_0)^{-1}Z^{(k)}(\widehat{T}^{(k)})^{-1}V_{-1}(I-A_0)^{-1}U_{-1}\\&=(I+(I-A_0)^{-1}Z^{(k)}(\widehat{T}^{(k)})^{-1}V_{-1})(I-A_0)^{-1}U_{-1},\end{aligned} \tag{2.16}$$

with $\widehat{T}^{(k)}=I-V_{-1}(I-A_0)^{-1}Z^{(k)}$.

Algorithm 2 shows a pseudocode that implements CR by exploiting the low-rank properties of the matrices in the case of low-rank downward transitions. Also in this case, we have chosen as a stopping criterion the condition $\|K_{-1}^{(k)}\|_1<\varepsilon$ or $\|A_1^{(k)}\|_1<\varepsilon$ for a fixed tolerance ε. On output, the algorithm provides an approximation to the $m\times r_{-1}$ matrix U_G such that $G=U_GV_{-1}$.

Computing the matrices B, N_{-1}, and Q_{-1} requires an $m\times m$ matrix inversion, a multiplication between an $m\times m$ matrix, and an $m\times r_{-1}$ matrix and a multiplication between an $r_{-1}\times m$ matrix and an $m\times r_{-1}$, at an overall cost of $2(m^3+m^2r_{-1}+mr_{-1}^2)$ arithmetic operations. At each step the updating of $K_{-1}^{(k)}$ requires four matrix multiplications at a cost of $2(m^2r_{-1}+3mr_{-1}^2)$ ops, the updating of $A_1^{(k)}$ requires two

$m \times m$ matrices at a cost of $4m^3$ ops, the updating of matrices $Z^{(k)}$ and $W^{(k)}$ requires two matrix multiplications at a cost of $4m^2 r_{-1}$ ops, and, finally, the updating of $(I - A_0^{(k)})^{-1}$ according to formulas (2.15) requires four matrix multiplications at a cost of $2(m^2 r_{-1} + 3mr_{-1}^2)$ ops, the solution of m linear systems of size $2r_{-1}$, at a cost of $\frac{16}{3} r_{-1}^3 + 8mr_{-1}^2$ ops, and two further matrix multiplications at a cost of $8m^2 r_{-1}$ ops. Thus the overall arithmetic cost of the kth step is $4m^3 + 16m^2 r_{-1} + 20mr_{-1}^2 + \frac{16}{3} r_{-1}^3$ ops. The computation of U_G according to formulas (2.16) requires the solution of r_{-1} linear systems of size r_{-1} at a cost of $\frac{8}{3} r_{-1}^3$ ops and a matrix multiplication at a cost of $2mr_{-1}^2$ ops.

Observe that, even in this case, the algorithm consists of a preprocessing stage, an iterative stage, and a postprocessing stage. However, unlike the case where both A_{-1} and A_1 have low rank, each step of the iterative stage has a cost dependent on the size m of the blocks. On the other hand, the number of ops needed at each step of the iterative stage is smaller, by a fixed constant, w.r.t. the cost of the general CR.

If A_1, instead of A_{-1}, were of low rank, then we might apply a similar technique by switching the role of A_{-1} with that of A_1.

Numerical Experiments

We report some numerical experiments that show the gain, in terms of computational time, of the proposed algorithms. We performed the experiments using Matlab on an Intel Xeon 2.80-GHz processor.

The first example is a QBD process where the matrices A_{-1} and A_1 have rank 2 and 3, respectively. The entries are randomly generated in such a way that the matrices U_i and V_i, $i = -1, 1$, are nonnegative. Figure 2.1 reports the CPU time (in seconds) needed by CR, exploiting and without exploiting the low-rank properties, for different values of m leaving unchanged the rank of A_{-1} and A_1. It is clear from the figure that our algorithm outperforms the general algorithm already for small values of block size m.

The second example is a PH/PH/1 queue, where $A_{-1} = (t\alpha) \otimes S$, $A_0 = T \otimes S + (t\alpha) \otimes (s\beta)$, $A_1 = T \otimes (s\beta)$, and

$$T = \begin{bmatrix} 0.5 & 0.4 & & \\ & 0.5 & \ddots & \\ & & \ddots & 0.4 \\ & & & 0.5 \end{bmatrix}, \quad S = \begin{bmatrix} 0.4 & 0.3 & & \\ & 0.4 & \ddots & \\ & & \ddots & 0.3 \\ & & & 0.4 \end{bmatrix}$$

are $n \times n$ matrices, $t = e - Te$, $s = e - Se$, $\alpha = \beta = (1, 0, \ldots, 0)$. In this case $U_{-1} = t \otimes I_n$, $V_{-1} = \alpha \otimes I_n$, $U_1 = I_n \otimes s$, and $V_1 = I_n \otimes \beta$. Therefore, the size of the blocks is $m = n^2$, while the blocks A_{-1} and A_1 have rank n. With this choice of the vectors α and β, the matrices A_{-1} and A_1 have n nonzero columns.

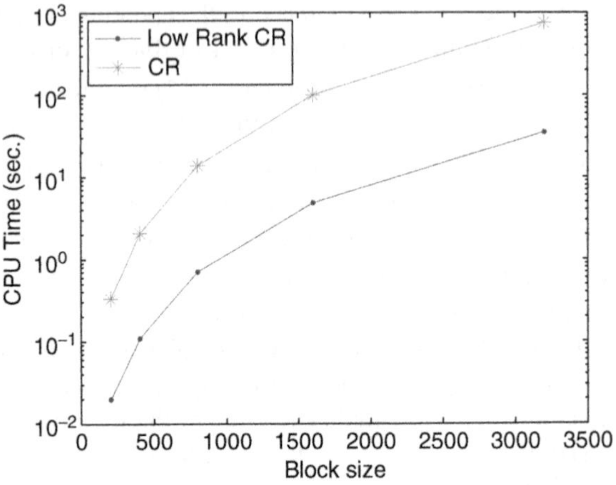

Fig. 2.1 CPU time for CR and low rank CR, for matrices with fixed rank

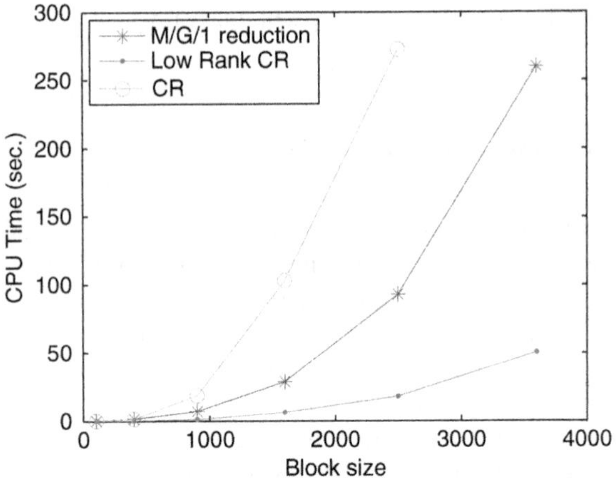

Fig. 2.2 CPU time of CR, low-rank CR, and M/G/1 reduction for a PH/PH/1 queue

Since the matrix A_{-1} has n nonzero columns, we may apply the algorithm proposed in [14] for solving QBD processes with restricted transitions to lower levels. This algorithm, which we call an *M/G/1 reduction algorithm*, consists in solving the QBD process by solving an M/G/1-type Markov chain, with block matrices of size equal to the number s of nonzero columns of A_{-1}, followed by the solution of a Stein matrix equation at a cost of $O((m-s)^3)$ ops. The algorithm of [14] can be applied also to the case where A_1 has a few nonzero rows; in this case the QBD process is reduced to a GI/M/1-type Markov chain.

Figure 2.2 reports the CPU time needed by customary CR, by low-rank CR, and by the M/G/1 reduction algorithm of [14]. Observe that the algorithm of [14] provides an improvement with respect to the general CR and that it is our algorithm that has the minimum computational cost.

Table 2.1 CPU time for low-rank CR and customary CR

	Low-rank CR				
m	Preproc.	CR	G	Total	CR
100	0	1.0e−02	0	1.0e−02	5.0e−02
400	6.0e−02	1.0e−02	6.0e−02	1.3e−01	2.3e+00
900	5.4e−01	1.0e−02	4.3e−01	9.8e+00	2.4e+01
1,600	2.6e+00	1.0e−02	2.0e+00	4.6e+00	1.1e+02
2,500	9.3e+00	1.0e−02	7.5e+00	1.7e+01	4.8e+02
3,600	2.7e+01	1.0e−02	2.2e+01	4.9e+01	1.3e+03

The third example consists of a QBD process where the coefficients are defined by

$$
\begin{aligned}
A_{-1} &= C_0 \otimes D_1,\\
A_0 &= C_0 \otimes D_0 + C_1 \otimes D_1,\\
A_1 &= C_1 \otimes D_0,
\end{aligned}
$$

where (C_0, C_1) and (D_0, D_1) define two Markov arrival processes (MAP's), i.e., C_i, D_i, $i = 0, 1$, are nonnegative matrices such that $C_0 + C_1$ and $D_0 + D_1$ are stochastic. If C_i and D_i, $i = 0, 1$, are low-rank matrices, then also A_{-1} and A_1 are low rank. More specifically, if

$$
\begin{aligned}
C_0 &= U_{C_0} V_{C_0}, \quad C_1 = U_{C_1} V_{C_1},\\
D_0 &= U_{D_0} V_{D_0}, \quad D_1 = U_{D_1} V_{D_1},
\end{aligned}
$$

where U_{C_0}, U_{C_1}, U_{D_0}, U_{D_1} are $n \times h_1$, $n \times h_2$, $n \times h_3$, $n \times h_4$ matrices, respectively, then $A_i = U_i V_i$, $i = -1, 1$, where

$$
\begin{aligned}
U_{-1} &= U_{C_0} \otimes U_{D_1}, \quad V_{-1} = V_{C_0} \otimes V_{D_1}\\
U_1 &= U_{C_1} \otimes U_{D_0}, \quad V_1 = V_{C_1} \otimes V_{D_0}.
\end{aligned}
$$

Therefore, the blocks A_i have size $m = n^2$, while A_{-1} and A_1 have rank $r_{-1} = h_1 h_4$ and $r_1 = h_2 h_3$, respectively. We set $h_1 = 5$, $h_2 = 3$, $h_3 = 7$, and $h_4 = 4$ and tried several values of n. Table 2.1 reports, for different values of $m = n^2$, the CPU time of low-rank CR, where we have distinguished the time needed in the preprocessing stage, the time needed by CR, the time to recover U_G, and the total time; we report also the overall time needed by customary CR. In all the tests both low-rank CR and customary CR performed the same number of iterations and provided an approximation of G to the same accuracy, i.e., having a residual error around 10^{-15}.

The low-rank CR is faster than general CR, and the major cost of low-rank CR is due to the pre- and postprocessing stages, that is, the computation of W and $Q^{(0)}$ (preprocessing) and the computation of U_G (postprocessing) in Algorithm 1. The remaining computation has a negligible cost independent of the size m.

The fourth example is the overflow queueing system described in Example 5.3 of [14]. The queueing system consists of two queues, the first having a finite buffer of size C and the second having an infinite buffer. Customers arriving at the first queue are served on a first-come, first-served (FCFS) basis by a single server, and the customers that find the buffer full are sent to the second queue. The second queue receives only overflow arrivals from the first queue and serves them in FCFS order with a single server. The arrival process at the first queue is a MAP characterized by (m_a, D_0, D_1), and the service time follows a PH distribution characterized by the parameters (m_1, α, T). The service time of the second queue follows a PH distribution with parameters (m_2, β, S). The arrival process at the second queue can be represented by a MAP with parameters (m_0, C_0, C_1), where $m_0 = (C+1)m_a m_1$ and

$$C_0 = \begin{bmatrix} D_0 \otimes I & D_1 \otimes I & 0 & \dots & \dots & 0 \\ I \otimes t\alpha & D_0 \oplus T & D_1 \otimes I & \ddots & \ddots & \vdots \\ 0 & I \otimes t\alpha & D_0 \oplus T & \ddots & \ddots & \vdots \\ \vdots & \ddots & \ddots & \ddots & \ddots & 0 \\ \vdots & \ddots & \ddots & \ddots & D_0 \oplus T & D_1 \otimes I \\ 0 & \dots & \dots & 0 & I \otimes t\alpha & D_0 \oplus T \end{bmatrix}, \quad C_1 = \begin{bmatrix} 0 & \dots & 0 & 0 \\ \vdots & \ddots & \vdots & \vdots \\ 0 & \dots & 0 & D_1 \otimes I \end{bmatrix},$$

with $t = -Te$.

This queueing system can be described by a continuous-time QBD process, where the level represents the number of customers in the second queue. The blocks are $A_{-1} = I_{m_0} \otimes s\beta$, $A_0 = C_0 \otimes I_{m_2} + I_{m_0} \otimes S$, $A_1 = C_1 \otimes I_{m_2}$, where $s = -Se$, with size $m = m_0 m_2$. The number of nonzero rows in A_1 is $r = m_a m_1 m_2$.

The matrices A_{-1} and A_1 can be decomposed as the product of matrices of rank m_0 and r, respectively, as

$$A_{-1} = (I_{m_0} \otimes s)(I_{m_0} \otimes \beta), \quad A_1 = \left(\begin{bmatrix} 0 \\ \vdots \\ 0 \\ I \end{bmatrix} \otimes I_{m_2} \right) \left(\begin{bmatrix} 0 & \cdots & 0 & D_1 \otimes I \end{bmatrix} \otimes I_{m_2} \right).$$

The continuous-time QBD process is transformed into a discrete-time QBD process using standard uniformization.

We have chosen the same parameters as in [14], i.e., the arrival process at the first queue has an arrival rate and squared coefficient of variation (SCV) equal to five, while the service time has mean 1 and SCV equal to 2. Also for the second queue service times have SCV equal to two. Therefore, the first queue is heavily loaded and many customers are overflowed to the second queue. The load of the second queue is a parameter ρ_2 that can vary and the buffer capacity C.

Table 2.2 CPU time for low-rank CR and GI/M/1 reduction for overflow queueing system, with $C = 20$

	Low-rank CR				GI/M/1 reduction				
ρ_2	Preproc.	CR	G	Total	Bandwidth	Construction	CR	Stein eq.	Total
0.2	0	3.0e−02	2.0e−02	5.0e−02	1,296	1.9e+00	1.9e+01	1.9e−01	2.1e+01
0.3	0	4.0e−02	2.0e−02	6.0e−02	881	9.9e−01	5.2e+01	2.0e−01	6.4e+00
0.4	1.0e−02	4.0e−02	1.0e−02	6.0e−02	671	6.1e−01	5.3e+00	1.9e−01	6.1e+00
0.5	1.0e−02	5.0e−02	1.0e−02	7.0e−02	544	4.3e−01	2.5e+00	1.9e−01	3.1e+00
0.6	1.0e−02	5.0e−02	1.0e−02	7.0e−02	459	3.2e−01	1.5e+00	2.0e−01	2.1e+00
0.7	1.0e−02	5.0e−02	1.0e−02	7.0e−02	399	2.7e−01	1.5e+00	2.0e−01	2.0e+00
0.8	1.0e−02	5.0e−02	1.0e−02	7.0e−02	352	2.2e−01	1.5e+00	1.9e−01	1.9e+00
0.9	1.0e−02	6.0e−02	1.0e−02	8.0e−02	316	1.9e−01	1.5e+00	1.9e−01	1.9e+00

We have set $m_a = m_1 = m_2 = 2$ and $C = 20$. In Table 2.2 we report the CPU time needed by low-rank CR, and by the algorithm of [14] based on the reduction to a GI/M/1-type Markov chain for different values of ρ_2. For the latter algorithm we have reported the bandwidth of the GI/M/1-type Markov chain, the time needed to construct its blocks, the time needed by CR, the time to solve the Stein equation, and the total time. The higher computational time of the algorithm of [14] is mainly due to the large bandwidth of the GI/M/1-type Markov chain; in fact, as observed in [14], the bandwidth is larger when the load ρ_2 is closer to zero.

Acknowledgements The authors wish to thank the anonymous referees, and Juan Pérez and Benny Van Houdt for providing the code to construct the matrices A_i, $i = -1, 0, 1$, for the overflow queueing model example.

References

1. Berman, A., Plemmons, R.J.: Nonnegative matrices in the mathematical sciences. In: Classics in Applied Mathematics, vol. 9. Society for Industrial and Applied Mathematics (SIAM), Philadelphia (1994). Revised reprint of the 1979 original
2. Bini, D.A., Latouche, G., Meini, B.: In: Numerical Methods for Structured Markov Chains. Numerical Mathematics and Scientific Computation. Oxford University Press, New York (2005). Oxford Science Publications
3. Bini, D.A., Meini, B.: The cyclic reduction algorithm: From Poisson equation to stochastic processes and beyond. In memoriam of Gene H. Golub. Numer. Algorithms **51**(1), 23–60 (2009)
4. Bini, D.A., Meini, B., Steffé, S., Van Houdt, B.: Structured Markov chains solver: Algorithms. In: SMCtools '06: Proceeding from the 2006 workshop on Tools for solving structured Markov chains Article No. 13 ACM New York, NY, USA (2006)
5. Bini, D.A., Meini, B., Steffé, S., Van Houdt, B.: Structured Markov chains solver: Software tools. In: SMCtools '06: Proceeding from the 2006 workshop on Tools for solving structured Markov chains Article No. 14 Publisher: ACM New York, NY, USA (2006)

6. Bini, D.A., Meini, B., Steffé, S., Van Houdt, B.: Structured Markov chains solver: Tool extension. In: VALUETOOLS '09: Proceedings of the Fourth International ICST Conference on Performance Evaluation Methodologies and Tools Article No. 20 Publisher: ICST (Institute for Computer Sciences, Social-Informatics and Telecommunications Engineering), Brussels, Belgium, Belgium (2009)
7. Carroll, J.L., Van de Liefvoort, A., Lipsky, L.: Solutions of M/G/1/N-type loops with extensions to M/G/1 and GI/M/1 queues. Oper. Res. **30**(3), 490–514 (1982)
8. Grassmann, W.K., Tavakoli, J.: Comparing some algorithms for solving QBD processes exhibiting special structures. INFOR: Inf. Syst. Oper. Res. **48**(3), 133–141 (2010)
9. Hogben, L. (ed.): In: Handbook of Linear Algebra. Discrete Mathematics and Its Applications (Boca Raton). Chapman & Hall/CRC, Boca Raton (2007). Associate editors: Richard Brualdi, Anne Greenbaum, and Roy Mathias
10. Latouche, G., Ramaswami, V.: In: Introduction to Matrix Analytic Methods in Stochastic Modeling. ASA-SIAM Series on Statistics and Applied Probability. Society for Industrial and Applied Mathematics (SIAM), Philadelphia (1999)
11. Neuts, M.F.: Explicit steady-state solutions to some elementary queueing models. Oper. Res. **30**(3), 480–489 (1982)
12. Neuts, M.F.: Matrix-Geometric Solutions in Stochastic Models. Dover, New York (1994). An algorithmic approach, Corrected reprint of 1981 original
13. Pérez, J.F., Van Houdt, B.: The M/G/1-type Markov chain with restricted transitions and its application to queues with batch arrivals. Probability in the Engineering and Informational Sciences **25**(04), 487–517 (2011)
14. Pérez, J.F., Van Houdt, B.: Quasi-birth-and-death processes with restricted transitions and its applications. Perform. Eval. **68**(2), 126–141 (2011)
15. Ramaswami, V., Latouche, G.: A general class of Markov processes with explicit matrix-geometric solution. OR Spektrum **8**, 209–218 (1986)
16. Ramaswami, V., Lucantoni, D.: Algorithmic analysis of a dynamic priority queue. In: Applied Probability–Computer Science: The Interface, vol. II, pp. 157–204. Birkäuser, Basel (1981)

Chapter 3
Bilateral Matrix-Exponential Distributions

Mogens Bladt, Luz Judith R. Esparza, and Bo Friis Nielsen

Introduction

Phase-type (PH) distributions [13, 14] have become a standard assumption in many areas of applied probability since they allow for either explicit or numerical exact solutions in complex stochastic models. A PH distributed random variable can be interpreted as the time to absorption in a Markov jump process with one absorbing state and the rest being transient. This class of distributions is dense in the class of distributions on the positive reals, meaning that they can approximate any nonnegative distribution arbitrarily closely [3].

Multivariate classes of PH distributions have been defined by Assaf et al. [6] and later by Kulkarni [12]. PH distributions have also been extended into the real line by Shanthikumar [17] and by Ahn and Ramaswami [1], defining a class of bilateral PH distributions.

Another generalization of PH distributions is the class of matrix-exponential (ME) distributions (distributions with rational Laplace transforms) that have been studied, for instance, by Asmussen and Bladt [5], Bladt and Neuts [7], and, in the multivariate case, by Bladt and Nielsen [9].

Asmussen and Bladt [5] have studied the class of ME distributions in general, identifying some necessary and sufficient conditions for an ME representation to be minimal. Liefvoort [18] proposed a method that provides insight into the minimal representation problem for PH distributions and characterizes the class of ME distributions of finite order. He and Zhang [11] established some relationships

M. Bladt (✉)
Instituto de Investigaciones en Matematicas Aplicadas y en Sistemas, UNAM, A.P. 20–726, 01000 Mexico, DF, Mexico
e-mail: bladt@sigma.iimas.unam.mx

L.J.R. Esparza • B.F. Nielsen
Department of Informatics and Mathematical Modelling, DTU, 2800 Kgs., Lyngby, Denmark
e-mail: ljre@imm.dtu.dk; bfn@imm.dtu.dk

G. Latouche et al. (eds.), *Matrix-Analytic Methods in Stochastic Models*, Springer Proceedings in Mathematics & Statistics 27, DOI 10.1007/978-1-4614-4909-6_3,

between the Laplace transforms, the distribution functions, and the minimal ME representations of ME distributions. Bodrog et al. [10] and Bladt and Nielsen [8] have characterized of ME distributions, presenting an algorithm to compute their finite-dimensional moments based on a set of required (low-order) moments.

The main purpose of this chapter is to generalize the class of matrix-exponential (univariate and multivariate) distributions into the real space. This shall provide an alternative class to the Gaussian distributions that is tractable in stochastic modeling. We introduce the class of bilateral ME distributions (distributions with rational moment-generating function) for both univariate and multivariate cases as a natural extension of the ME and multivariate ME distributions, respectively.

The remainder of this chapter is organized as follows. In the section "Background" we provide necessary background on PH and ME distributions. Bilateral ME distributions are defined in the section "Univariate Bilateral Matrix-Exponential Distributions." In the section "A Generalization of Phase-Type Distributions," we give a generalization of bilateral PH distributions considering the multivariate case. The minimal order of bilateral ME distributions is analyzed in the section "Order of Bilateral Matrix-Exponential Distributions." The multivariate case of bilateral ME is considered in the section "Multivariate Bilateral Matrix-Exponential Distributions." In the section "Markov Additive Processes with Absorption," as an application, we study terminal distributions of Markov additive processes with absorption. The last section is "Conclusion."

Background

Let $J = \{J(t)\}_{t\geq 0}$ be a continuous-time Markov jump process with state space composed of m transient states $1, 2, \ldots, m$ and one absorbing state $m+1$. Suppose that J has an initial probability vector $(\boldsymbol{\alpha}, \alpha_{m+1})$, where $\boldsymbol{\alpha}$ is a vector of dimension m. The generator matrix is given by

$$\begin{pmatrix} \mathbf{T} & \mathbf{t} \\ \mathbf{0} & 0 \end{pmatrix}, \tag{3.1}$$

where $\mathbf{T}$ is an invertible $m \times m$ matrix satisfying $t_{ii} < 0$ and $t_{ij} \geq 0$ for $i \neq j$ and $\mathbf{t}$ is an m-dimensional column vector such that $\mathbf{t} = -\mathbf{T}\mathbf{e}$, where $\mathbf{e}$ denotes a column vector with 1 at all entries. Then the time to absorption of J, $\tau = \inf\{t \geq 0 : J(t) = m+1\}$, is said to be PH distributed with initial probability vector $\boldsymbol{\alpha}$ and subgenerator matrix $\mathbf{T}$, and we shall write $\tau \sim \mathrm{PH}(\boldsymbol{\alpha}, \mathbf{T})$. The probability density function of τ for $x > 0$ is given by $f(x) = \boldsymbol{\alpha} e^{\mathbf{T}x}\mathbf{t}$ [13, 14]. If $\alpha_{m+1} > 0$, then the distribution of τ has an atom at zero with this probability.

More generally, if X is a nonnegative random variable with a possible atom at zero and an absolute continuous part with density function in the form $b(x) = \boldsymbol{\alpha} e^{\mathbf{T}x}\mathbf{t}$, where $\boldsymbol{\alpha}$ is a row vector, $\mathbf{t}$ is a column vector, and $\mathbf{T}$ is a matrix, then we say that X is matrix-exponentially distributed. The triple $(\boldsymbol{\alpha}, \mathbf{T}, \mathbf{t})$ is called a representation for the distribution of X, and we write $X \sim \mathrm{ME}(\boldsymbol{\alpha}, \mathbf{T}, \mathbf{t})$. Hence, the moment-generating function of X, its moments, and reduced moments can be computed as follows:

$$M_X(s) = \mathbb{E}(e^{sX}) = \alpha_{m+1} + \boldsymbol{\alpha}(-s\mathbf{I} - \mathbf{T})^{-1}\mathbf{t},$$

$$M_i = \mathbb{E}(X^i) = i!\boldsymbol{\alpha}(-\mathbf{T})^{-(i+1)}\mathbf{t},$$

$$\mu_i = \frac{\mathbb{E}(X^i)}{i!} = \boldsymbol{\alpha}(-\mathbf{T})^{-(i+1)}\mathbf{t},$$

where $\mathbf{I}$ is an identity matrix of appropriate dimension.

The moment-generating function of the ME-distributed random variable X is, hence, rational. Also, any random variable with rational moment-generating function has an ME distribution; see Asmussen and Bladt [5] for details.

It is immediate that a PH distribution is ME with the representation $(\boldsymbol{\alpha}, \mathbf{T}, -\mathbf{Te})$. In general, we also may take $0 \leq \boldsymbol{\alpha}\mathbf{e} \leq 1$ and $\mathbf{Te} + \mathbf{t} = 0$ also in the ME case.

Univariate Bilateral Matrix-Exponential Distributions

In this section we generalize the class of ME distributions to a class on the entire real line $(-\infty, \infty)$, which we shall call bilateral ME distributions.

Let X be a random variable with a rational moment-generating function expressed as

$$M_X(s) = \frac{B(s)}{A(s)}, \tag{3.2}$$

where $A(s)$ and $B(s)$ are polynomials in s, $s \in \mathbb{R}$.

Theorem 3.1. *X has a rational moment-generating function if and only if the density function of its absolutely continuous part can be written as*

$$f_X(x) = \boldsymbol{\alpha}_+ e^{\mathbf{T}_+ x}\mathbf{t}_+ \mathbf{1}_{\{x>0\}} + \boldsymbol{\alpha}_- e^{\mathbf{T}_- |x|}\mathbf{t}_- \mathbf{1}_{\{x<0\}}, \tag{3.3}$$

where $\boldsymbol{\alpha}_+$ is a row vector of some dimension m_+, $\mathbf{T}_+$ is a matrix of dimension $m_+ \times m_+$, and $\mathbf{t}_+$ is an m_+-dimensional column vector. Similarly, both the vectors $\boldsymbol{\alpha}_-$ and $\mathbf{t}_-$ and the matrix $\mathbf{T}_-$ are defined by some dimension m_-.

Without loss of generality, we can take $\boldsymbol{\alpha}_+$, $\boldsymbol{\alpha}_-$, $\mathbf{T}_+$, and $\mathbf{T}_-$ real valued such that $0 \leq \boldsymbol{\alpha}_+\mathbf{e} + \boldsymbol{\alpha}_-\mathbf{e} \leq 1$ and $\mathbf{T}_+\mathbf{e} + \mathbf{t}_+ = \mathbf{T}_-\mathbf{e} + \mathbf{t}_- = 0$.

Proof. Let X be a random variable with density given by (3.3); then, its moment-generating function is given by

$$\begin{aligned}
\int_{-\infty}^{\infty} e^{sx} dF(x) &= (1 - \boldsymbol{\alpha}_+\mathbf{e} - \boldsymbol{\alpha}_-\mathbf{e}) + \int_{-\infty}^{\infty} e^{sx}\left(\boldsymbol{\alpha}_+ e^{\mathbf{T}_+ x}\mathbf{t}_+ \mathbf{1}_{\{x>0\}} + \boldsymbol{\alpha}_- e^{\mathbf{T}_- |x|}\mathbf{t}_- \mathbf{1}_{\{x<0\}}\right) dx \\
&= (1 - \boldsymbol{\alpha}_+\mathbf{e} - \boldsymbol{\alpha}_-\mathbf{e}) + \int_0^{\infty} e^{sx}\boldsymbol{\alpha}_+ e^{\mathbf{T}_+ x}\mathbf{t}_+ dx + \int_{-\infty}^{0} e^{sx}\boldsymbol{\alpha}_- e^{\mathbf{T}_- |x|}\mathbf{t}_- dx \\
&= (1 - \boldsymbol{\alpha}_+\mathbf{e} - \boldsymbol{\alpha}_-\mathbf{e}) + \boldsymbol{\alpha}_+(-s\mathbf{I} - \mathbf{T}_+)^{-1}\mathbf{t}_+ + \boldsymbol{\alpha}_-(s\mathbf{I} - \mathbf{T}_-)^{-1}\mathbf{t}_-,
\end{aligned}$$

where both terms $\boldsymbol{\alpha}_+(-s\mathbf{I}-\mathbf{T}_+)^{-1}\mathbf{t}_+$ and $\boldsymbol{\alpha}_-(s\mathbf{I}-\mathbf{T}_-)^{-1}\mathbf{t}_-$ are rational [5]. Thus, $M_X(s)$ is the sum of rational functions in s, and then rational.

On the other hand, let $M_X(s)$ be the moment-generating function of X given by (3.2). We can write $A(s) = A_+(s)A_-(s)$, where $A_+(s)$ is a polynomial that has roots in the positive half-plane and $A_-(s)$ one that has roots in the negative half-plane. Now define $B_+(s)$ and $B_-(s)$ (see appendix) such that

$$B(s) = (1-\boldsymbol{\alpha}_+\mathbf{e}-\boldsymbol{\alpha}_-\mathbf{e})(A_+(s)A_-(s)) + A_+(s)B_-(s) + A_-(s)B_+(s);$$

then the moment-generating function becomes

$$M_X(s) = (1-\boldsymbol{\alpha}_+\mathbf{e}-\boldsymbol{\alpha}_-\mathbf{e}) + \frac{B_+(s)}{A_+(s)} + \frac{B_-(s)}{A_-(s)}, \tag{3.4}$$

where the functions related to $\frac{B_+(s)}{A_+(s)}$ and $\frac{B_-(s)}{A_-(s)}$ are nonnegative, having support on the positive and negative reals, respectively.

If we are dealing with a case where there are no positive (negative) roots, then we define $A_+(s) = 1$ and $B_+(s) = 0$ ($A_-(s) = 1$ and $B_-(s) = 0$).

Then using Lemma 2.1 from Asmussen and Bladt [5] with the appropriate notation, we get that $M_X(s) = (1-\boldsymbol{\alpha}_+\mathbf{e}-\boldsymbol{\alpha}_-\mathbf{e}) + \boldsymbol{\alpha}_+(-s\mathbf{I}-\mathbf{T}_+)^{-1}\mathbf{t}_+ + \boldsymbol{\alpha}_-(s\mathbf{I}-\mathbf{T}_-)^{-1}\mathbf{t}_-$, which represents the moment-generating function of a random variable with density given by (3.3). □

Definition 3.1. We say that X is univariate bilateral matrix-exponentially or simply bilateral matrix-exponentially (BME) distributed, if it has a rational moment-generating function.

We write $X \sim \mathrm{BME}(\boldsymbol{\alpha}_+,\mathbf{T}_+,\mathbf{t}_+,\boldsymbol{\alpha}_-,\mathbf{T}_-,\mathbf{t}_-)$ when X has a density given by (3.3).

Remark 3.1. We have seen that if $X \sim BME(\boldsymbol{\alpha}_+,\mathbf{T}_+,\mathbf{t}_+,\boldsymbol{\alpha}_-,\mathbf{T}_-,\mathbf{t}_-)$, then its moment-generating function can be written as (3.4), where the degree of the polynomial $A_+(s) = \det(-s\mathbf{I}-\mathbf{T}_+)$ is the dimension of $\mathbf{T}_+$, m_+ say, and in the same way if the degree of $A_-(s) = \det(s\mathbf{I}-\mathbf{T}_-)$ is m_-, then

$$\begin{aligned} M_X(s) &= (1-\boldsymbol{\alpha}_+\mathbf{e}-\boldsymbol{\alpha}_-\mathbf{e}) + \frac{B_+(s)}{A_+(s)} + \frac{B_-(s)}{A_-(s)} \\ &= \frac{(1-\boldsymbol{\alpha}_+\mathbf{e}-\boldsymbol{\alpha}_-\mathbf{e})(A_+(s)A_-(s))}{A_+(s)A_-(s)} + \frac{B_+(s)A_-(s)+B_-(s)A_+(s))}{A_+(s)A_-(s)} \end{aligned}$$

has degree $m_+ + m_-$. If $B_+(s)$ and $A_+(s)$ [$B_-(s)$ and $A_-(s)$] have no common factors, then $\mathbf{T}_+$ ($\mathbf{T}_-$) has the lowest dimension possible. When both $\mathbf{T}_+$ and $\mathbf{T}_-$ have the lowest dimension, we say that the representation is of minimal order. The number $m = m_+ + m_-$ is the order of the distribution. We will analyze this issue in more detail in the section "Order of Bilateral Matrix-Exponential Distributions."

A Generalization of Phase-Type Distributions

Let $\tau \sim \mathrm{PH}(\boldsymbol{\alpha}, \mathbf{T})$. We can interpret τ as resulting from a simple reward structure on a finite Markov jump process $\{J(t)\}_{t\geq 0}$. If the reward rate is 1 in each state, then the total reward is PH distributed.

More generally, we may assign a real-valued constant $r(i)$, referred to as the reward rate for state i. Define the reward function

$$W(t) = \int_0^t r(J(s))\mathrm{d}s, \tag{3.5}$$

which is the accumulated reward earned by the process $J(s)$ up to time t.

If the rewards are strictly positive, then the random variable $X = W(\tau)$ is PH distributed, i.e.,

$$X = W(\tau) \sim \mathrm{PH}(\boldsymbol{\alpha}, \boldsymbol{\Delta}(\mathbf{r})^{-1}\mathbf{T}),$$

where $\boldsymbol{\Delta}(\mathbf{r})$ is the diagonal matrix composed of the reward rates of the transient states $\mathbf{r} = (r(1), \ldots, r(m))'$. Its moment-generating function is given by

$$M_X(s) = \alpha_{m+1} + \boldsymbol{\alpha}(s\mathbf{T}^{-1}\boldsymbol{\Delta}(\mathbf{r}) + \mathbf{I})^{-1}\mathbf{e}; \tag{3.6}$$

see [1].

When the reward vector $\mathbf{r}$ is a nonzero real vector, we obtain the class of bilateral PH distributions, which was introduced by Ahn and Ramaswami [1].

Definition 3.2. [1] Let $X = W(\tau)$ be the total accumulated reward until absorption. Then X is said to be bilaterally PH distributed with initial probability vector $\boldsymbol{\alpha}$, subgenerator $\mathbf{T}$, and reward matrix $\boldsymbol{\Delta}(\mathbf{r})$. We denote this by $X \sim \mathrm{BPH}^*(\boldsymbol{\alpha}, \mathbf{T}, \boldsymbol{\Delta}(\mathbf{r}))$.

It is clear from the construction of the BPH* class that it has an atom at zero if and only if $\alpha_{m+1} = 1 - \boldsymbol{\alpha}\mathbf{e} > 0$.

Note that the moment-generating function (3.6) is rational (Theorem 3.1), i.e., X is BME distributed with representation given by Theorem 4.1 of [1]. The expression is also valid when $r(i) = 0$ for some i [9].

Kulkarni [12] used a construction similar to (3.5) when defining a class (MPH*) of multivariate PH distributions. For $j = 1, \ldots, k$ let $\mathbf{r}_j = (r_j(1), \ldots, r_j(m))'$ be k nonnegative m-column reward vectors, and define $\mathbf{R} = (\mathbf{r}_1, \ldots, \mathbf{r}_k)$, the $(m \times k)$-dimensional reward matrix. Considering the following random variables

$$X_j = \int_0^\tau r_j(J(t))\mathrm{d}t, \quad 1 \leq j \leq k,$$

the vector $\mathbf{X} = (X_1, \ldots, X_k)$ is said to have MPH* distribution with representation $(\boldsymbol{\alpha}, \mathbf{T}, \mathbf{R})$, and we write $\mathbf{X} \sim \mathrm{MPH}^*(\boldsymbol{\alpha}, \mathbf{T}, \mathbf{R})$. From Theorem 2.3.2 of [9], we then have that

$$\langle \mathbf{X}, \mathbf{a} \rangle \sim \mathrm{PH}(\boldsymbol{\alpha}, \boldsymbol{\Delta}(\mathbf{Ra})^{-1}\mathbf{T})$$

for all k-dimensional column vectors $\mathbf{a}$ such that $\mathbf{Ra} > \mathbf{0}$. In addition, the moment-generating function of $\langle \mathbf{X}, \mathbf{a} \rangle$ is given by

$$M_{\langle \mathbf{X},\mathbf{a}\rangle}(s) = \alpha_{m+1} + \boldsymbol{\alpha}(s\mathbf{T}^{-1}\boldsymbol{\Delta}(\mathbf{Ra}) + \mathbf{I})^{-1}\mathbf{e}. \tag{3.7}$$

The joint transform of $\mathbf{X}$ is obtained with $s = 1$ as a function of $\mathbf{a}$.

By allowing the reward rates $r_j(i)$ to be real we say that $\mathbf{X} = (X_1, \dots, X_k)$ is multivariate bilateral PH (denoted by MBPH*) distributed with representation $(\boldsymbol{\alpha}, \mathbf{T}, \mathbf{R})$, and we write

$$\mathbf{X} \sim \text{MBPH}^*(\boldsymbol{\alpha}, \mathbf{T}, \mathbf{R}).$$

Formula (3.7) remains valid for the bilateral case as well. Indeed, if we write $\mathbf{X} = \mathbf{X}^+ - \mathbf{X}^-$, where $\mathbf{X}^+$ and $\mathbf{X}^-$ are the rewards earned with only nonnegative, respectively negative, reward rates, then $(\mathbf{X}^+, \mathbf{X}^-) \sim \text{MPH}^*(\boldsymbol{\alpha}, \mathbf{T}, (\mathbf{R}^+, \mathbf{R}^-))$, where $\mathbf{R}^+$ contains all nonnegative rewards and $\mathbf{R}^-$ all negative rewards. Thus

$$\begin{aligned}
M_{\mathbf{X}}(s) &= \mathbb{E}\left(\mathrm{e}^{\langle \mathbf{s}, \mathbf{X}^+ - \mathbf{X}^- \rangle}\right) \\
&= \mathbb{E}\left(\mathrm{e}^{\langle (\mathbf{s}, -\mathbf{s}), (\mathbf{X}^+, \mathbf{X}^-) \rangle}\right) \\
&= \boldsymbol{\alpha}\left(\mathbf{T}^{-1}\boldsymbol{\Delta}\left((\mathbf{R}^+, \mathbf{R}^-)\begin{pmatrix}\mathbf{s} \\ -\mathbf{s}\end{pmatrix}\right) + \mathbf{I}\right)^{-1}\mathbf{e} \\
&= \boldsymbol{\alpha}\left(\mathbf{T}^{-1}\boldsymbol{\Delta}\left((\mathbf{R}^+ - \mathbf{R}^-)\mathbf{s}\right) + \mathbf{I}\right)^{-1}\mathbf{e}.
\end{aligned}$$

Theorem 3.2. $\mathbf{X} = (X_1, \dots, X_k) \sim$ *MBPH*$^*(\boldsymbol{\alpha}, \mathbf{T}, \mathbf{R})$ *if and only if* $\langle \mathbf{X}, \mathbf{a} \rangle \sim$ *BPH** $(\boldsymbol{\alpha}, \mathbf{T}, \boldsymbol{\Delta}(\mathbf{Ra}))$ *for all k-dimensional real vector* $\mathbf{a}$.

Proof. The proof is similar to that of [9]. □

To generalize to the MBPH* class, we present the following definition.

Definition 3.3. For $\mathbf{X} = (X_1, \dots, X_k)$ let MBME* be the class of distributions such that the moment-generating function of $\mathbf{X}$ at $\mathbf{s} \in \mathbb{R}^k$ is given by

$$M_{\mathbf{X}}(\mathbf{s}) = \mathbb{E}(\mathrm{e}^{\langle \mathbf{X}, \mathbf{s} \rangle}) = \alpha_{m+1} + \boldsymbol{\alpha}(\mathbf{T}^{-1}\boldsymbol{\Delta}(\mathbf{Rs}) + \mathbf{I})^{-1}\mathbf{e}; \tag{3.8}$$

then we say that the vector $\mathbf{X}$ is MBME* with representation $(\boldsymbol{\alpha}, \mathbf{T}, \mathbf{R})$. If $\mathbf{X}$ is nonnegative, then we say that it has an MME* distribution [9].

The following theorem gives an explicit formula for calculating cross-moments of the components of an MBME*-distributed random variable. Bladt and Nielsen [9] have proved a similar result for a class that generalizes MPH* distributions.

Theorem 3.3. *The cross-moments* $\mathbb{E}\left(\prod_{i=1}^{k} X_i^{a_i}\right)$, *where* $\mathbf{X} = (X_1, \ldots, X_k) \sim MBME^*$ $(\boldsymbol{\alpha}, \mathbf{T}, \mathbf{R})$ *and* $a_i \in \mathbb{N}$, *are given by*

$$\boldsymbol{\alpha} \sum_{l=1}^{a!} \left(\prod_{i=1}^{a} (-T^{-1}) \boldsymbol{\Delta}(\mathbf{r}_{\sigma_l(i)}) \right) \mathbf{e},$$

where $a = \sum_{i=1}^{k} a_i$, $\boldsymbol{r}_i$ *is the ith column of* $\mathbf{R}$, *and* $\sigma_1, \ldots, \sigma_{a!}$ *are the ordered permutations of a-tuples of derivatives, within* $\sigma_l(i)$ *being the value among* $1, \ldots, k$ *at the ith position of the permutation* σ_l.

Proof. We can obtain the cross-moments by

$$\mathbb{E}\left(\prod_{i=1}^{k} X_i^{a_i}\right) = \left.\frac{\mathrm{d}^a M_{\mathbf{X}}(\mathbf{s})}{\mathrm{d}s_1^{a_1} \mathrm{d}s_2^{a_2} \ldots \mathrm{d}s_k^{a_k}}\right|_{\mathbf{s}=\mathbf{0}},$$

where $M_{\mathbf{X}}(\mathbf{s})$ is given in (3.8). Since

$$\frac{\mathrm{d}}{\mathrm{d}s_i} \left(\mathbf{T}^{-1}\boldsymbol{\Delta}(\mathbf{Rs}) + \mathbf{I}\right)^{-1} = \left(\mathbf{T}^{-1}\boldsymbol{\Delta}(\mathbf{Rs}) + \mathbf{I}\right)^{-1} \left(-\mathbf{T}^{-1}\right) \boldsymbol{\Delta}(\mathbf{r}_i) \left(\mathbf{T}^{-1}\boldsymbol{\Delta}(\mathbf{Rs}) + \mathbf{I}\right)^{-1},$$

then by induction and substituting $\mathbf{s} = \mathbf{0}$, we get the result. □

For more details of the proof we refer the reader to Nielsen et al. [15].

An application in which bilateral PH (or ME) distributions occur naturally is in a queueing model with PH (ME) renewal arrivals and PH (ME) service times. The corresponding Lindley process is then based on a random walk with bilateral PH (ME) increments, and we may calculate the ladder-height distributions [2, 5], maximum, and stationary waiting times in the usual way.

Order of Bilateral Matrix-Exponential Distributions

The *order* of the ME representation $(\boldsymbol{\alpha}, \mathbf{T}, \mathbf{t})$ is given by the dimension of $\mathbf{T}$, and the smallest order among all equivalent representations is called the *degree* [18] or the order of the distribution [11]. A representation whose order is equal to the degree is said to be of *minimal order*.

Asmussen and Bladt [5] identified some necessary and sufficient conditions for an ME representation to be minimal and developed a method for computing a minimal ME representation.

In this section, we will establish a relationship between a moment-generating function and the minimal BME representation using Hankel matrices.

For $j > 0$ the noncentralized moments of $X \sim \mathrm{BME}(\boldsymbol{\alpha}_+, \mathbf{T}_+, \mathbf{t}_+, \boldsymbol{\alpha}_-, \mathbf{T}_-, \mathbf{t}_-)$ are given by

$$M_j = \mathbb{E}(X^j) = M_j^+ + M_j^-,$$

where $M_j^+ = j!\boldsymbol{\alpha}_+(-\mathbf{T}_+)^{-(j+1)}\mathbf{t}_+$ and $M_j^- = (-1)^j j!\boldsymbol{\alpha}_-(-\mathbf{T}_-)^{-(j+1)}\mathbf{t}_-$. The reduced moments are given by

$$\mu_j = \frac{M_j}{j!} = \frac{M_j^+}{j!} + \frac{M_j^-}{j!} =: \mu_j^+ + \mu_j^-, \tag{3.9}$$

where $\mu_j^+ > 0$, for all j, and $\mu_j^- > 0$ if j is even and $\mu_j^- < 0$ if j is odd.

Moreover, the moment-generating function of X is rational and has a power series expansion of the form $M_X(s) = \sum_j \mu_j s^j$, where μ_j is the jth reduced moment. Then by (3.9), we obtain $M_X(s) = \sum_j \mu_j^+ s^j + \sum_j \mu_j^- s^j$.

Let m be the minimal order of the distribution; then its moment-generating function can be written as

$$M_X(s) = \frac{b_m s^m + b_{m-1}s^{m-1} + \cdots + b_1 s + 1}{a_m s^m + a_{m-1}s^{m-1} + \cdots + a_1 s + 1} = \frac{B(s)}{A(s)}, \tag{3.10}$$

which is well defined in a strip containing the imaginary axis. Since $\mu_0 = 1$, we obtain that

$$\frac{B(s)}{A(s)} = 1 + \sum_{j=1}^{\infty} \mu_j s^j, \tag{3.11}$$

and multiplying (3.11) by $A(s)$ and equating coefficients, we obtain that the coefficients corresponding to powers $m+1,\ldots,2m$ of s satisfy the system of equations

$$-\boldsymbol{\mu}_m = \boldsymbol{H}_m \mathbf{a}_m,$$

where $\boldsymbol{\mu}_m = (\mu_{m+1},\ldots,\mu_{2m})'$, $\mathbf{a}_m = (a_m,\ldots,a_1)'$, and $\boldsymbol{H}_m$ is the $(m\times m)$-dimensional Hankel matrix given by

$$\boldsymbol{H}_m = \begin{pmatrix} \mu_1 & \mu_2 & & \mu_m \\ \mu_2 & \mu_3 & & \mu_{m+1} \\ \vdots & \vdots & \ddots & \vdots \\ \mu_m & \mu_{m+1} & & \mu_{2m-1} \end{pmatrix}, \tag{3.12}$$

with the Hankel determinant defined by

$$\phi_m = \det(\boldsymbol{H}_m). \tag{3.13}$$

The equation system must have a unique solution due to the irreducibility of (3.10). Hence $\boldsymbol{H}_m$ must have full rank and $\phi_m \neq 0$. On the other hand, considering the equations corresponding to powers $m+1,\ldots,2m+1$ of s, we obtain that

$$\mathbf{0} = \boldsymbol{H}_{m+1}\mathbf{a}_m^*,$$

where $\mathbf{0}$ is the $(m+1)$-dimensional column vector of zeros and $\mathbf{a}_m^* = (a_m, \ldots, a_1, 1)'$. Note that $\boldsymbol{H}_{m+1}$ has rank m since the determinant of the lower-left $m \times m$ submatrix is different from zero. Hence $\phi_{m+1} = 0$.

By continuation of the argument, we see that $\text{rank}(\boldsymbol{H}_l) = m$ for $l \geq m$, which means that $\phi_l = 0$ for $l > m$. Thus the minimal order of the BME distribution can be checked through the verification of the determinants to be the highest index of the determinant for which it is different from zero. Note that some determinants ϕ_l for $l < m$ could be zero or nonzero. See also Liefvoort [18] and He and Zhang [11].

Example 3.1. Suppose X is a random variable with density given by

$$f_X(x) = p\mathrm{e}^{-x}\mathbf{1}_{\{x>0\}} + (1-p)\mathrm{e}^{x}\mathbf{1}_{\{x<0\}}, \quad p \in (0,1).$$

With the notation presented previously, we have that $m_+ = 1$ and $m_- = 1$. The moment-generating function is given by

$$M_X(s) = \frac{(1-2p)s-1}{s^2-1},$$

and the Hankel determinants are given by

$$\phi_1 = 2p-1, \quad \phi_2 = 4p^2-4p, \quad \phi_l = 0, \text{ for } l > 2.$$

Example 3.2. Suppose X has the following density:

$$f_X(x) = p\left(\frac{2}{3}\mathrm{e}^{-x}(1+\cos(x))\right)\mathbf{1}_{\{x>0\}} + (1-p)\mathrm{e}^{x}\mathbf{1}_{\{x<0\}},$$

with $p \in (0,1)$.

Then we have that $m_+ = 3$ and $m_- = 1$. The moment-generating function of X is given by

$$M_X(s) = \frac{1}{3}\frac{(-7p+3)s^3+(13p-9)s^2+(-10p+12)s-6}{s^4-2s^3+s^2+2s-2},$$

and the Hankel determinants are given by

$$\begin{aligned}
\phi_1 &= (5/3)p-1,\\
\phi_2 &= (9/4)p^2-(13/6)p,\\
\phi_3 &= (307/432)p^3-(103/144)p^2,\\
\phi_4 &= -(25/216)p^4+(25/216)p^3,\\
\phi_l &= 0, \text{ for } l > 4.
\end{aligned}$$

Multivariate Bilateral Matrix-Exponential Distributions

We will define the class of multivariate bilateral ME distributions as a natural extension of the univariate case.

Definition 3.4. A random vector $\mathbf{X} \in \mathbb{R}^k$ of dimension k is multivariate bilateral matrix-exponentially (MVBME) distributed if the joint moment-generating function, $\mathbb{E}(e^{\langle \mathbf{X}, \mathbf{s} \rangle})$, is a multidimensional rational function.

This definition generalizes the class of MBME* since the latter has a rational moment-generating function on a special form.

To prove our main characterization, we proceed by deriving the following two lemmas.

Lemma 3.1. *Assume that $\langle \mathbf{X}, \mathbf{a} \rangle$ has a BME distribution for all $\mathbf{a} \in \mathbb{R}^k \setminus \{\mathbf{0}\}$. Then the (minimal) order $m(\mathbf{a})$ of the univariate BME distribution for $\langle \mathbf{X}, \mathbf{a} \rangle$ is a bounded function of $\mathbf{a}$.*

Proof. Let $\phi_i(\mathbf{a})$ denote the ith-order Hankel determinant [see (3.13)] corresponding to $\langle \mathbf{X}, \mathbf{a} \rangle$, and let $C_i = \{\mathbf{a} \in \mathbb{R}^k \setminus \{\mathbf{0}\} : \phi_j(\mathbf{a}) = 0, j \geq i\}$. For $\mathbf{a}_1 \in \mathbb{R}^k \setminus \{\mathbf{0}\}$ we let $m_1 = m(\mathbf{a}_1)$; then $\phi_i(\mathbf{a}_1) = 0$ for $i > m_1$.

The ith-order Hankel determinant is a sum of monomials of order $i(i+1)$ and, hence, a continuous function. Thus there exists a neighborhood B around $\mathbf{a}_1$ for which $\phi_{m_1}(\mathbf{b}) \neq 0$ for $\mathbf{b} \in B$. Hence the order of the BME distribution of $\langle \mathbf{X}, \mathbf{b} \rangle$ is at least the order of $\langle \mathbf{X}, \mathbf{a} \rangle$ for $\mathbf{b} \in B$.

Since ϕ_{m_1} is a nonvanishing k-dimensional polynomial, then C_{m_1} has k-dimensional Lebesgue measure zero. Suppose there exists $\mathbf{a}_2 \in \mathbb{R}^k \setminus \{\mathbf{0}\}$ such that $m_2 = m(\mathbf{a}_2) > m_1$; then $\mathbf{a}_1 \in C_{m_2}$ and $C_{m_1} \subseteq C_{m_2}$.

If the order of the moment-generating function for $\langle \mathbf{X}, \mathbf{a} \rangle$ is unbounded, then there exists a sequence $\mathbf{a}_i$ with $m_i = m(\mathbf{a}_i)$ such that $m_i \uparrow \infty$, and the set $C = \cup_{i=1}^{\infty} C_{m_i}$ has k-dimensional Lebesgue measure zero, contradicting the assumption of $\langle \mathbf{X}, \mathbf{a} \rangle$ being BME distributed (of finite order). □

The next lemma shows that the rational moment-generating function is of a particularly simple form.

Lemma 3.2. *Assume that $\langle \mathbf{X}, \mathbf{a} \rangle$ has a univariate bilateral ME distribution for all $\mathbf{a} \in \mathbb{R}^k \setminus \{\mathbf{0}\}$, and suppose the order of the distribution of $\langle \mathbf{X}, \mathbf{a} \rangle$ is bounded by some m. Then, we may write the moment-generating function of $\langle \mathbf{X}, \mathbf{a} \rangle$ as*

$$\frac{\tilde{b}_m(\mathbf{a})s^m + \tilde{b}_{m-1}(\mathbf{a})s^{m-1} + \cdots + \tilde{b}_1(\mathbf{a})s + 1}{\tilde{a}_m(\mathbf{a})s^m + \tilde{a}_{m-1}(\mathbf{a})s^{m-1} + \cdots + \tilde{a}_1(\mathbf{a})s + 1},$$

where the terms $\tilde{b}_j(\mathbf{a})$ and $\tilde{a}_j(\mathbf{a})$ are sums of k-dimensional monomials in $\mathbf{a}$ of degree j.

Proof. Since $\langle \mathbf{X}, \mathbf{a} \rangle \sim$ BME, its moment-generating function can be written as

$$\frac{\tilde{b}_m(\mathbf{a})s^m + \tilde{b}_{m-1}(\mathbf{a})s^{m-1} + \cdots + \tilde{b}_1(\mathbf{a})s + 1}{\tilde{a}_m(\mathbf{a})s^m + \tilde{a}_{m-1}(\mathbf{a})s^{m-1} + \cdots + \tilde{a}_1(\mathbf{a})s + 1}, \tag{3.14}$$

where $\tilde{b}_i(\mathbf{a})$ and $\tilde{a}_i(\mathbf{a})$ $[\tilde{a}_m(\mathbf{a}) \neq 0]$ are functions in $\mathbf{a}$.

Let $\tilde{a}_i(\mathbf{a}) = P_i(\mathbf{a}) + E_i(\mathbf{a})$, where $P_i(\mathbf{a})$ is a sum of all, if any, ith-order monomials appearing in the expression for $\tilde{a}_i(\mathbf{a})$, whereas $E_i(\mathbf{a}) = \tilde{a}_i(\mathbf{a}) - P_i(\mathbf{a})$.

Let $\boldsymbol{\mu}_m(\mathbf{a}) = (\mu_{m+1}(\mathbf{a}), \ldots, \mu_{2m}(\mathbf{a}))'$ and let $\boldsymbol{H}_m(\mathbf{a})$ be the Hankel matrix (3.12) which now depends on $\mathbf{a}$.

$$\frac{\tilde{b}_m(\mathbf{a})s^m + \tilde{b}_{m-1}(\mathbf{a})s^{m-1} + \cdots + \tilde{b}_1(\mathbf{a})s + 1}{\tilde{a}_m(\mathbf{a})s^m + \tilde{a}_{m-1}(\mathbf{a})s^{m-1} + \cdots + \tilde{a}_1(\mathbf{a})s + 1} = 1 + \sum_{j=1}^{\infty} \mu_j(\mathbf{a})s^j, \tag{3.15}$$

we obtain the following system of equations:

$$-\boldsymbol{\mu}_m(\mathbf{a}) = \boldsymbol{H}_m(\mathbf{a})\mathbf{P}_m(\mathbf{a}) + \boldsymbol{H}_m(\mathbf{a})\mathbf{E}_m(\mathbf{a}),$$

where $\mathbf{P}_m(\mathbf{a}) = (P_m(\mathbf{a}), \ldots, P_1(\mathbf{a}))'$ and $\mathbf{E}_m(\mathbf{a}) = (E_m(\mathbf{a}), \ldots, E_1(\mathbf{a}))'$.

For $1 \leq j \leq m$, $\mu_{m+j}(\mathbf{a})$ is a sum of monomials of order $m+j$ as the corresponding terms of $\boldsymbol{H}_m(\mathbf{a})\mathbf{P}_m(\mathbf{a})$. Note that we can rewrite $E_j(\mathbf{a})$ as $E_{>j}(\mathbf{a}) + E^j_{\text{irra}}(\mathbf{a}) + E^j_{\text{rat}}(\mathbf{a})$, where $E_{>j}$ represents the sum of monomials with order greater than j, E^j_{rat} is a rational function of lower leading order than j, and E^j_{irra} is a function that cannot be expressed as a rational function. Then we obtain that $\mathbf{E}_m(\mathbf{a}) = \mathbf{E}_{>m}(\mathbf{a}) + \mathbf{E}_{\text{irra}}(\mathbf{a}) + \mathbf{E}_{\text{rat}}(\mathbf{a})$.

It is easy to see that $\boldsymbol{H}_m(\mathbf{a})\mathbf{E}_m(\mathbf{a})$ does not contain monomials of order $m+j$ since:

- $\boldsymbol{H}_m(\mathbf{a})\mathbf{E}_{>m}(\mathbf{a})$ has monomials of order greater than $m+j$ and
- $\boldsymbol{H}_m(\mathbf{a})\mathbf{E}_{\text{irra}}(\mathbf{a})$ cannot contain addends that are rational monomials.

For the rational case, i.e., $\boldsymbol{H}_m(\mathbf{a})\mathbf{E}_{\text{rat}}(\mathbf{a})$, we refer the reader to [9] to see a proof that does not have monomials of order $m+j$. Then, by coefficient matching, we get that

$$\boldsymbol{H}_m(\mathbf{a})\mathbf{E}_m(\mathbf{a}) = \mathbf{0}.$$

This implies that $\mathbf{E}_m(\mathbf{a}) = \mathbf{0}$ since $\boldsymbol{H}_m(\mathbf{a})$ is nonsingular. Hence all $\tilde{a}_i(\mathbf{a})$ are sums of monomials of order i. From (3.15) we can also see that $\tilde{b}_i(\mathbf{a})$ are sums of monomials of order i. □

Our theorem that characterizes the class of MVBME distributions is as follows.

Theorem 3.4. *A vector* $\mathbf{X}$ *follows a multivariate bilateral ME distribution, i.e.,* $\mathbf{X} \sim$ *MVBME, if and only if* $\langle \mathbf{X}, \mathbf{a} \rangle \sim$ *BME for all* $\mathbf{a} \in \mathbb{R}^k \setminus \{\mathbf{0}\}$.

Proof. Let $\mathbf{X} \sim MVBME$; then $\mathbb{E}(e^{\langle \mathbf{X}, s\mathbf{a}\rangle})$ is rational in $s\mathbf{a}$ for $s \in \mathbb{R}$ and $\mathbf{a} \in \mathbb{R}^k \setminus \{\mathbf{0}\}$. Since

$$\mathbb{E}(e^{\langle \mathbf{X}, s\mathbf{a}\rangle}) = \mathbb{E}(e^{s\langle \mathbf{X}, \mathbf{a}\rangle}),$$

$\mathbb{E}(e^{s\langle \mathbf{X}, \mathbf{a}\rangle})$ is rational in s, i.e., $\langle \mathbf{X}, \mathbf{a}\rangle \sim BME$.

On the other hand, suppose that $\langle \mathbf{X}, \mathbf{a}\rangle$ has a rational moment-generating function for all $\mathbf{a} \in \mathbb{R}^k \setminus \{\mathbf{0}\}$. Then we know that the moment-generating function can be expressed in the form of Lemma 3.2. By setting $s = 1$, this rational function coincides with the multidimensional moment-generating function of $\mathbf{X}$ at $\mathbf{a}$. □

Concerning the classes MBME* and MVBME, it is an open and difficult problem as to whether they are equal.

Example 3.3. Wishart distribution.

The Wishart distribution was formulated by John Wishart in 1928 [19]. Let $\mathbf{X}_1 = (X_{i1})_{1\le i\le p}, \mathbf{X}_2 = (X_{i2})_{1\le i\le p}, \ldots, \mathbf{X}_\nu = (X_{i\nu})_{1\le i\le p}$ be p-dimensional random column vectors distributed independently according to the p-dimensional normal distributions $\mathrm{N}_p(\boldsymbol{\mu}_1, \boldsymbol{\Sigma}), \ldots, \mathrm{N}_p(\boldsymbol{\mu}_\nu, \boldsymbol{\Sigma})$, with mean vectors $\boldsymbol{\mu}_1 = (\mu_{i1})_{1\le i\le p}, \ldots, \boldsymbol{\mu}_\nu = (\mu_{i\nu})_{1\le i\le p}$ (respectively) and a common variance–covariance matrix $\boldsymbol{\Sigma}$. The distribution of a $(p \times p)$ symmetric random matrix $\boldsymbol{W} = (w_{ij})_{1\le i,j\le p}$ defined by $w_{ij} = \sum_{t=1}^{\nu} X_{it}X_{jt}$ is the real noncentral Wishart distribution $W_p(\nu, \boldsymbol{\Sigma}, \boldsymbol{\Lambda})$, where $\boldsymbol{\Lambda} = (\lambda_{ij})_{1\le i,j\le p}$ is the mean square matrix defined by $\lambda_{ij} = \sum_{t=1}^{\nu} \mu_{it}\mu_{jt}$. The Wishart distribution for $\boldsymbol{\Lambda} = \mathbf{0}$ is said to be central and is denoted by $W_p(\nu, \boldsymbol{\Sigma})$.

The moment-generating function of the central Wishart distribution [16] is given by

$$M_{\boldsymbol{W}}(\boldsymbol{\Theta}) = \mathbb{E}[e^{\mathrm{tr}(\boldsymbol{\Theta}\boldsymbol{W})}] = \det(\,\mathbf{I} - 2\boldsymbol{\Theta}\boldsymbol{\Sigma})^{-\frac{\nu}{2}}, \tag{3.16}$$

where $\boldsymbol{\Theta} = (\theta_{ij})_{1\le i,j\le p}$ is a symmetric parameter matrix and $\mathrm{tr}(\cdot)$ is the trace of a matrix.

If we define the following vectors in $\mathbb{R}^{p^2}$

$$\mathbf{s} = ((\theta_{i1})_{1\le i\le p}, (\theta_{i2})_{1\le i\le p}, \ldots, (\theta_{ip})_{1\le i\le p}),$$
$$\mathbf{X} = ((w_{i1})_{1\le i\le p}, (w_{i2})_{1\le i\le p}, \ldots, (w_{ip})_{1\le i\le p}),$$

then $\mathbb{E}(e^{\langle \mathbf{X}, \mathbf{s}\rangle})$ is given by (3.16), which is a rational function whenever ν is an even integer number. This means that $\mathbf{X} \sim$ MVBME.

Markov Additive Processes with Absorption

Let $\mathbf{Y} = (Y_1, \ldots, Y_\ell) \sim$ MME*$(\boldsymbol{\alpha}, \mathbf{T}, \mathbf{R})$, where $\mathbf{T}$ is of dimension m. Now we consider a multidimensional reward structure $\mathbf{X} = (X_1, \ldots, X_k)$ such that

$$X_j = \sum_{i=1}^{\ell} B_{ij}, \quad j = 1, \dots, k,$$

where $\mathbf{B}_i = (B_{i1}, \dots, B_{ik}) \sim \mathrm{N}_k(Y_i \mathbf{r}(i), Y_i \boldsymbol{\Sigma}(i))$, with $\mathbf{r}(i) = (r_1(i), \dots, r_k(i))$, and $\boldsymbol{\Sigma}(i)$ is a covariance matrix, $i = 1, \dots, \ell$.

The joint moment-generating function of $\mathbf{X}$ is given by

$$\begin{aligned} M_{\mathbf{X}}(\mathbf{s}) &= \mathbb{E}\left(\mathrm{e}^{\langle \mathbf{X}, \mathbf{s} \rangle}\right) \\ &= \int_0^\infty \cdots \int_0^\infty \prod_{i=1}^{\ell} \exp\left(y_i \mathbf{s}\mathbf{r}(i)' + y_i \frac{1}{2} \mathbf{s}\boldsymbol{\Sigma}(i)\mathbf{s}' \right) \mathrm{d}F(\mathbf{y}), \end{aligned}$$

where F is the joint distribution function of $\mathbf{Y}$, so (3.8) becomes

$$M_{\mathbf{X}}(\mathbf{s}) = \boldsymbol{\alpha} \left(\mathbf{T}^{-1} \boldsymbol{\Delta}(\mathbf{R}\boldsymbol{\theta}) + \mathbf{I}\right)^{-1} \mathbf{e}. \tag{3.17}$$

Here $\boldsymbol{\theta} = (\theta_1, \dots, \theta_\ell)'$, with $\theta_i = \mathbf{s}\mathbf{r}(i)' + \frac{1}{2}\mathbf{s}\boldsymbol{\Sigma}(i)\mathbf{s}'$.

Hence the moment-generating function is rational in $\mathbf{s}$, so $\mathbf{X}$ is MVBME distributed.

Note that all these arguments are also valid for the MPH* class [see (3.7)], where the probabilistic interpretation is easier. In the following analysis we will present an application of this class considering Markov additive processes.

Analysis of Terminal Distributions with Added Multidimensional Brownian Components

Let $J = \{J(t)\}_{t \geq 0}$ and τ be as in section "Background." Then we define the real-valued process $W = \{W(t)\}_{t \geq 0}$ as

$$W(t) = \int_0^t r(J(s))\mathrm{d}s + \int_0^t \sigma(J(s))\mathrm{d}B(s), \tag{3.18}$$

where B is a standard Brownian motion, $r(i)$ is the drift, and $\sigma(i)$ is the diffusion parameter. This is known to be the most general Markov additive process on J with skip-free (continuous) paths [4]. The case of $\sigma^2(i) \equiv 0$ corresponds to a standard fluid flow model leading to W being PH distributed.

Asmussen has proved that $W(\tau)$ has a bilateral PH distribution with representations given by Corollaries 1 and 3 in [4].

For the multivariate case we define

$$\mathbf{W}(t) = \int_0^t \mathbf{r}(J(s))\mathrm{d}s + \int_0^t \boldsymbol{\sigma}(J(s))\mathrm{d}\mathbf{B}(s) \tag{3.19}$$

and $\mathbf{X} = \mathbf{W}(\tau)$, where $\mathbf{B}$ is a k-dimensional standard Brownian motion, $\mathbf{r}(i)$ are k-dimensional drift vectors, and $\boldsymbol{\Sigma}(i) = \boldsymbol{\sigma}(i)\boldsymbol{\sigma}(i)'$ are positive semi–definite diffusion matrices. Then by (3.17), $\mathbf{X}$ belongs to the class of MVBME distributions, with $\mathbf{R} = \mathbf{I}$ and $\ell = m$. This extends the result of [4], though we do not provide an MBME* representation.

Conclusion

In this article we have generalized the class of matrix-exponential distributions to a class of distributions with rational moment-generating functions and support on the whole real line. For this purpose we defined a new class called bilateral ME distributions (distributions with rational moment-generating functions). We also analyzed the multivariate case, whose domain is the real space. Our main characterization of this is based on the one presented in [9] for multivariate ME distributions.

Moreover, we have analyzed and used the theory already written about bilateral PH distributions [1] in order to give a generalization of them for the multivariate case. Indeed, we have applied this to Markov additive processes. We believe that these distributions may find wide application in areas like statistics, finance, and computer science, where general reward rates may have advantages.

Appendix

Existence of B_+ and B_-

In what follows, we will give an analysis of the existence of B_+ and B_- assuming that we do not have an atom at zero.

Suppose that the polynomial $A(s)$ can be written as $A(s) = \prod_{j=1}^{r}(s-\lambda_j)^{\nu_j}$ for some r such as $\sum_{j=1}^{r}\nu_j = \deg(A)$ and whose poles are given by λ_j. Then for

$$A_k(s) = \prod_{j \neq k}(s-\lambda_j)^{\nu_j} = \frac{A(s)}{(s-\lambda_k)^{\nu_k}}, \quad k = 1, \dots, r,$$

we obtain that

$$\frac{B(s)}{A(s)} = \sum_{j=1}^{r} \frac{C_j(s)}{(s-\lambda_j)^{\nu_j}}, \tag{3.20}$$

where the polynomial $C_j(s)$ is the Taylor polynomial of $\frac{B(s)}{A_j(s)}$ of order $\nu_j - 1$ at the point λ_j, i.e.,

$$C_j(s) := \sum_{k=0}^{\nu_j-1} \frac{1}{k!} \left(\frac{B(s)}{A_j(s)} \right)^k \lambda_j (s-\lambda_j)^k.$$

Taylor's theorem (in the real or complex case) provides a proof of the existence and uniqueness of the partial fraction decomposition and a characterization of the coefficients. If we define

$$A_+(s) := \prod_{j=1}^{r} (s-\lambda_j)^{\nu_j} \mathbf{1}_{\{\lambda_j>0\}}, \quad A_-(s) := \prod_{j=1}^{r} (s-\lambda_j)^{\nu_j} \mathbf{1}_{\{\lambda_j<0\}},$$

then from (3.20) we obtain

$$\begin{aligned} \frac{B(s)}{A(s)} &= \sum_{j=1}^{r} \frac{C_j(s)}{(s-\lambda_j)^{\nu_j}} \mathbf{1}_{\{\lambda_j>0\}} + \sum_{j=1}^{r} \frac{C_j(s)}{(s-\lambda_j)^{\nu_j}} \mathbf{1}_{\{\lambda_j<0\}} \\ &= \frac{B_+(s)}{A_+(s)} + \frac{B_-(s)}{A_-(s)}, \end{aligned}$$

where

$$\begin{aligned} B_+(s) &:= \sum_{j=1}^{r} C_j(s) \mathbf{1}_{\{\lambda_j>0\}} \prod_{k\neq j}^{r} (s-\lambda_k)^{\nu_k} \mathbf{1}_{\{\lambda_k>0\}}, \\ B_-(s) &:= \sum_{j=1}^{r} C_j(s) \mathbf{1}_{\{\lambda_j<0\}} \prod_{k\neq j}^{r} (s-\lambda_k)^{\nu_k} \mathbf{1}_{\{\lambda_k<0\}}. \end{aligned}$$

Acknowledgements Luz Judith Rodriguez Esparza and Bo Friis Nielsen would like to thank the Villum Kann Rasmussen Foundation and the Danish Council for Strategic Research for their support through MTlab a VKR centre of excellence and the UNITE project under Grant 2140-08-0011. Mogens Bladt acknowledges the support from the Mexican Research Council, Conacyt, Grant 48538.

References

1. Ahn, S., Ramaswami, V.: Bilateral phase-type distributions. Stoch. Models **21**, 239–259 (2005)
2. Asmussen, S.: Phase-type representations in random walk and queueing problems. Ann. Probab. **20**, 772–789 (1992)
3. Asmussen, S.: Applied Probability and Queues. Springer, New York (2003)
4. Asmussen, S.: Terminal distributions of skipfree Markov additive processes with absorption. Technical Report 14, MaPhySto (2004)
5. Asmussen, S., Bladt, M.: Renewal theory and queueing algorithms for matrix-exponential distributions. In: Chakravarthy, S., Attahiru, S.A. (eds.) Matrix-Analytic Methods in Stochastic Models, CRC, Boca Raton, FL, CRC Press, pp. 313–341 (1996)
6. Assaf, D., Langberg, N.A., Savits, T.H., Shaked, M.: Multivariate phase-type distributions. Oper. Res. **32**(3), 688–702 (1984)

7. Bladt, M., Neuts, M. F.: Matrix-exponential distributions: calculus and interpretations via flows. Stoch. Models **19**, 113–124 (2003)
8. Bladt, M., Nielsen, B.F.: Multivariate matrix-exponential distributions. In: Bini, D., Meini, B., Ramaswami, V., Remiche, M.-A., Taylor, P. (eds.) Numerical Methods for Structured Markov Chains, Dagstuhl Seminar Proceedings, Dagstuhl, Germany, 07461 (2008)
9. Bladt, M., Nielsen, B.F.: Multivariate matrix-exponential distributions. Stoch. Models **26**, 1–26 (2010)
10. Bodrog, L., Hovarth, A., Telek, M.: Moment characterization of matrix-exponential and Markovian arrival processes. Ann. Oper. Res. **160**, 51–68 (2008)
11. He, Q., Zhang, H.: On matrix-exponential distributions. Adv. Appl. Probab. **39**, 271–292 (2007)
12. Kulkarni, V.G.: A new class of multivariate phase-type distributions. Oper. Res. **37**, 151–158 (1989)
13. Neuts, M.F.: Probability distributions of phase-type. In: Liber Amicorum Prof. Emeritus H. Florin, Department of mathematics, University of Louvain, Belgium pp. 173–206 (1975)
14. Neuts, M.F.: Matrix-Geometric Solutions in Stochastic Models, vol. 2. Johns Hopkins University Press, Baltimore (1981)
15. Nielsen, B.F., Nielson, F., Nielson, H.R.: Model checking multivariate state rewards. QEST **17**, 7–16 (2010)
16. Numata, Y., Kuriki, S.: On formulas for moments of the Wishart distributions as weighted generating functions of matchings. In: Discrete Mathematics and Theoretical Computer Science, San Francisco, USA, pp. 821–832 (2010)
17. Shanthikumar, J.G.: Bilateral phase type distributions. Naval Res. Log. Q. **32**, 119–136 (1985)
18. Van de Liefvoort, A.: The moment problem for continuous distributions. Technical Report, University of Missouri, Kansas City (1990)
19. Wishart, J.: The generalised product moment distribution in samples from a normal multivariate population. Biometrika **20A**, 32–52 (1928)

Chapter 4
AutoCAT: Automated Product-Form Solution of Stochastic Models

Giuliano Casale and Peter G. Harrison

Introduction

Performance modeling often involves the abstraction of the various components of a system under study and their mutual interactions as a Markov process. Although there exist several high-level formalisms for specifying particular classes of Markov processes, such as queueing networks or stochastic Petri nets, the state space explosion problem typically limits our ability to compute metrics related to the long-term behavior of the system. A notable exception is the class of product-form models, in which the equilibrium probability of a state is a scaled product of the marginal state probabilities of the Markov processes that represent the individual components of the system. Foremost examples of models enjoying a product form include open and closed queueing networks with single and multiple service classes [6, 29], possibly supporting various forms of blocking [4] and different arrival types [17, 18, 20], stochastic Petri nets [3], Markovian process algebras [24], and stochastic automata networks [19].

We introduce AutoCAT, an optimization-based technique that automatically *constructs* exact or approximate product forms for a large class of performance models. We consider models that may be described as a cooperation (i.e., synchronization) of Markov processes over a given set of named actions [40]. This class of processes includes as special cases queueing networks, stochastic Petri nets, stochastic automata, and several other model types that are popular in performance evaluation [27]. Although certain Markov processes enjoy a number of useful properties for determining a product-form solution, such as reversibility [31], quasireversibility [30, 37], and local balance [38], cooperating Markov processes

G. Casale (✉) • P.G. Harrison
Department of Computing, Imperial College London, 180 Queen's Gate, SW7 2AZ London, UK
e-mail: g.casale@imperial.ac.uk; pgh@doc.ic.ac.uk

G. Latouche et al. (eds.), *Matrix-Analytic Methods in Stochastic Models*, Springer Proceedings in Mathematics & Statistics 27, DOI 10.1007/978-1-4614-4909-6_4,

additionally benefit from their compositional structure, which is conducive to recursive analysis and the reversed compound agent theorem (RCAT) in particular [22, 23].

As we discuss in the section titled "Preliminaries," RCAT defines a set of sufficient conditions for cooperating Markov processes to enjoy a product-form solution. To the best of our knowledge, RCAT is the most general formalism available to construct product forms by means of simple conditions that do not require direct solution of the joint probability distribution of the model at equilibrium. We leverage this result to show that RCAT product-form conditions are equivalent to a nonlinear optimization problem with nonconvex quadratic constraints. Nonconvex global optimization is $\mathcal{NP}$-hard in general [11], and thus we derive efficient linear programming (LP) relaxations that are solved sequentially to find the exact product form of a model when one exists. Since the length of such a sequence depends on the tightness of the LP relaxations, we define a hierarchy of increasingly tighter linear programs, based on a potential theory for Markov processes [12], convexification techniques [36], and a set of linear constraints that we derive from the RCAT product-form conditions.

Most importantly, this procedure is extended to the *approximate* analysis of non-product-form Markov processes, which arise in the vast majority of practical systems. It is applied first in "toy" examples to illustrate the method and then in case studies to validate its main features and to assess its numerical tractability and accuracy. Among these models, it is shown that such approximations may be useful to investigate closed queueing network models with phase-type (PH) distributed service times [8]. Recently, it was shown in [14, 15] that such models may be approximated quite accurately by approximate product-form solutions. Here we provide an example illustrating that the AUTOCAT approximation may provide improved accuracy with respect to the methods of Casale and Harrison [15] and Casale et al. [14].

Our method provides one of the first available non-application-specific algorithms for product-form analysis; moreover, at the same time, it constructs automatically workable approximations for the equilibrium probabilities of interacting Markov processes without a product form. A preliminary constructive tool of this type was proposed by Argent-Katwala [2], where a symbolic solver was proposed for product forms based on the sufficient conditions of RCAT. This tool constructed product forms for Markov processes composed from others for which a product form was already known, but it was not able to *detect* whether a given Markov process admits a product-form solution. Moreover, the cost of symbolic linear algebra inevitably makes the technique applicable only to simple processes. Buchholz [9, 10] defines the first general-purpose automatic technique for identifying exact and approximate product-form solutions in stochastic models. The methodology minimizes a residual error norm using an optimization technique based on efficient quadratic programming. Using the stochastic automata network (SAN) formalism, Buchholz's method uses a Kronecker representation of the cooperations to avoid generating the joint state space. Then, an iterative technique searches for a local optimum that is used to compute an approximate product form.

In Balsamo et al. [34] and Marin and Bulo [5] propose INAP, a fixed-point method to estimate reversed rates in RCAT product forms. The main benefit of the INAP algorithm is computational efficiency, which enables the analysis of large models.

To summarize, our main contributions are as follows:

- We present an algorithm to automatically decide whether a given Markov model has a product-form solution and, if so, to compute it without solving the underlying global balance equations. Specifically, in the section "Does a Product Form Exist?," we introduce a formulation of the problem in the form of a quadratically constrained optimization, and we obtain efficient linear relaxations in the section "Linearization Methodology."
- In the section "Exact Product-Form Construction," we show that this algorithm guarantees, within the boundaries of the numerical tolerance of the optimizer, that a product-form solution will be found if it exists.
- Next, approximation techniques stemming from our methodology are developed in the section "Automated Approximations." Such approximations can be applied to a wide class of performance models that are represented as cooperations of Markov processes.

Our methodology is validated with small examples and case studies in the section "Examples and Case Studies," which testify to the effectiveness of the approach on performance models of practical interest. These include stochastic Petri nets and closed queueing networks with PH distributed service times.

Preliminaries

We consider a collection of M Markov processes that cooperate over a set of A actions. Each cooperating process might represent, for example, a queue, a stochastic automaton, a Petri net, or an agent in a stochastic process algebra. Process k is defined on a set of $N_k \geq 1$ states such that the joint state space of the Markov process comprising the cooperation has up to $N_{\text{prod}} = \prod_k N_k$ states. Process indices are $k, m = 1, \ldots, M$, $m \neq k$, action indices are $a, b, c = 1, \ldots, A$, and (marginal) state indices for process k are $n_k, n'_k = 1, \ldots, N_k$. An action a labels a synchronizing transition in a pairwise cooperation between two processes k and $m \neq k$, which can only take place in both processes simultaneously.

We follow the convention of defining *active* and *passive* roles for each action a in the pair of processes it synchronizes. The set of active (respectively passive) actions for process k is denoted by $\mathcal{A}_k$ (respectively $\mathcal{P}_k$).

Consider an action a such that $a \in \mathcal{A}_k$ and $a \in \mathcal{P}_m$, i.e., which is active in k and passive in $m \neq k$. Further, assume that when action a is enabled, it triggers with rate μ_a state transitions $n_k \to n'_k$ and $n_m \to n'_m$ in processes k and m, respectively. We summarize this information in *rate matrices* $\boldsymbol{A}_a$ and $\boldsymbol{P}_a$ of orders N_k and N_m, respectively. That is, we set the values $\boldsymbol{A}_a[n_k, n'_k] = \mu_a$ and $\boldsymbol{P}_a[n_m, n'_m] = p_m$ for each pair (n_k, n'_k) and (n_m, n'_m) where a is enabled, where p_m is the probability of the transition $n_m \to n'_m$ in the passive process when action a takes place, and $\boldsymbol{M}[i,j]$ stands for the element at row i and column j of matrix $\boldsymbol{M}$. Note that the rate of the passive action is unspecified; it is assigned subsequently according to the equilibrium behavior of the active process (i.e., process k here) [22]. Observe also that the rates of transitions $n_k \to n_k$ lie on the diagonal of $\boldsymbol{A}_a$. Such rates define *hidden transitions*, which do not alter the local state of the active process but can affect the local state of the passive process m. Finally, we account for local state jumps that are not due to cooperations, which we call *local transitions*. The rates of all local transitions for process k are stored in the $N_k \times N_k$ matrix $\boldsymbol{L}_k$.

Product-Form Solutions

We assume the joint process underlying the cooperation to be ergodic, and the goal of our analysis is to determine a product-form expression for the model's joint state probability function at equilibrium. Unless otherwise stated, we always refer to the RCAT product form defined in [25]; note that this is a superset of product forms that can be obtained by quasireversibility [35]. Our goal is to find marginal probability vectors $\pi_k(n_k)$ for each cooperating process k such that the equilibrium solution of the model enjoys the product-form expression

$$\alpha(n_1,\ldots,n_k,\ldots,n_M) = G^{-1}\pi_1(n_1)\pi_2(n_2)\cdots\pi_M(n_M), \tag{4.1}$$

where G is a normalizing constant and $\alpha(n_1,\ldots,n_k,\ldots,n_M)$ is the joint state probability function. Under the RCAT methodology, finding a product-form solution such as (4.1) requires one to analyze each process k in "isolation," i.e., to study its transitions over the marginal state space $\mathcal{S}_k = \{n_k \,|\, 0 \le n_k < N_k\}$. If process k cooperates passively on one or more actions $b \in \mathcal{P}_k$, then their (passive) rates of occurrence in isolation are undefined. Thus, they cannot be solved for the marginal probabilities $\pi_k(n_k)$ before such rates are assigned. This is because the rate of a passive action in process k may depend on the state of the cooperating process m, as we elaborate subsequently. The RCAT theorem introduced in [22,23] establishes that, if we can define a generator matrix $\boldsymbol{Q}_k$ on $\mathcal{S}_k$ satisfying conditions RC1, RC2, and RC3 stated at the end of this section, then the equilibrium vectors $\boldsymbol{\pi}_k$ satisfying $\boldsymbol{\pi}_k\boldsymbol{Q}_k = 0$ and $\boldsymbol{\pi}_k\mathbf{1} = 1$ for $1 \le k \le M$ provide a product-form solution (4.1). Specifically, if the three sufficient conditions of RCAT are met, then a certain outgoing rate x_b, called a *reversed rate*, is associated with each passive action $b \in \mathcal{P}_k$ in each state n_k where b is enabled. We point to [22] for a probabilistic interpretation of x_b as a rate in a time-reversed Markov process. The RCAT conditions together with the reversed rates then allow the generators $\boldsymbol{Q}_k$ of each Markov component process to be defined uniquely as follows:

$$\boldsymbol{Q}_k \equiv \boldsymbol{Q}_k(\boldsymbol{x}) = \boldsymbol{L}_k + \sum_{a \in \mathcal{A}_k} \boldsymbol{A}_a + \sum_{b \in \mathcal{P}_k} x_b \boldsymbol{P}_b - \boldsymbol{\Delta}_k(\boldsymbol{x}), \tag{4.2}$$

where $\boldsymbol{x} = (x_1, \ldots, x_a, \ldots, x_A)^T > \mathbf{0}$ is the vector of reversed rates and $\boldsymbol{\Delta}_k(\boldsymbol{x})$ is the diagonal matrix ensuring that $\boldsymbol{Q}_k \mathbf{1} = \mathbf{0}$. From $\boldsymbol{Q}_k$, we can compute the product-form solution of the cooperation based on (4.1). This provides a major computational advantage over a direct solution of the joint process.

RCAT Sufficient Conditions

The original formulation of RCAT was expressed using the stochastic process algebra PEPA [22,23]. We provide here a reformulation of RCAT's conditions using matrix expressions that are simpler to integrate in optimization programs.

- *RCAT Condition 1 (RC1).* Passive actions are always enabled, i.e., $\boldsymbol{P}_a \mathbf{1} \geq \mathbf{1}$ for all $a \in \mathcal{P}_k, 1 \leq k \leq M$.
- *RCAT Condition 2 (RC2).* For $a \in \mathcal{A}_k$ each state in process k has an incoming transition due to active action a, i.e., $\boldsymbol{A}_a^T \mathbf{1} > \mathbf{0}$ for all $a \in \mathcal{A}_k, 1 \leq k \leq M$.
- *RCAT Condition 3 (RC3).* There exists a vector of reversed rates

  $$\boldsymbol{x} = (x_1, \ldots, x_a, \ldots, x_A)^T > \mathbf{0}$$

  such that the generators $\boldsymbol{Q}_k$ have equilibrium vectors $\boldsymbol{\pi}_k$ that satisfy the following *rate equations*: $\boldsymbol{\pi}_k \boldsymbol{A}_a = x_a \boldsymbol{\pi}_k$ for all $a \in \mathcal{A}_k, 1 \leq k \leq M$. (Note that we use the generalized expression introduced in [35] in place of the original condition in [22], although the two forms are equivalent.)

If the preceding conditions are met, then the vectors $\boldsymbol{\pi}_k$ immediately define a product-form solution (4.1). We stress, however, that verifying RC3 is much more challenging than RC1 and RC2 as it is necessary in practice to *find* a reversed rate vector $\boldsymbol{x}$ that satisfies the rate equations.

Does a Product Form Exist?

Let us now turn to the problem of finding an algorithm that automatically constructs a RCAT product form if one exists. We assume initially that every component process has a finite state space ($N_k < \infty \; \forall k$), the generalization to countably infinite state spaces being simple, as discussed in the appendix, "Infinite Processes." According to RC3, to construct an RCAT product form we need to find a solution $\boldsymbol{x} > 0$ of the *e*xact *n*onlinear *s*ystem of equations

$$\begin{aligned}
\text{ENS}: \qquad \boldsymbol{\pi}_k \boldsymbol{A}_a &= x_a \boldsymbol{\pi}_k, & a \in \mathcal{A}_k, 1 \leq k \leq M,\\
\boldsymbol{\pi}_k \boldsymbol{Q}_k(\boldsymbol{x}) &= \mathbf{0}, & 1 \leq k \leq M,\\
\boldsymbol{\pi}_k \mathbf{1} &= 1, & 1 \leq k \leq M.
\end{aligned}$$

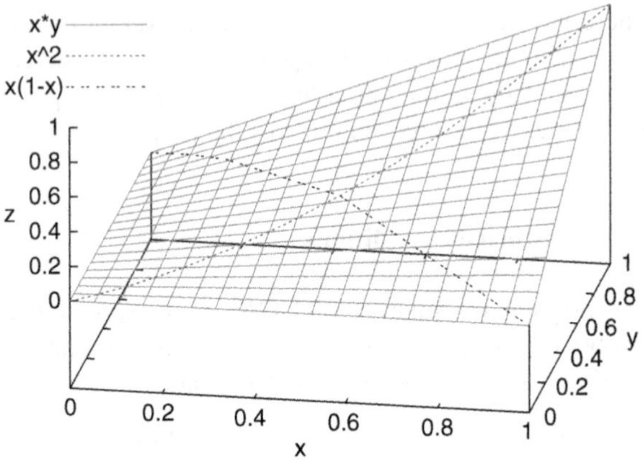

Fig. 4.1 The bilinear surface $z = xy$ is nonconvex since it includes both convex (e.g., $z = x^2$) and concave (e.g., $z = x(1-x)$) functions

This defines a nonconvex feasible region due to the bilinear products $x_a\boldsymbol{\pi}_k$ and $x_b\boldsymbol{\pi}_k$ in the first two sets of constraints. Such a region is illustrated in Fig. 4.1. We now provide the following characterization.

Proposition 4.1. *Consider the vectors* $\boldsymbol{x}^{L,0} = (x_1^{L,0}, x_2^{L,0}, \ldots, x_A^{L,0})^T$ *and* $\boldsymbol{x}^{U,0} = (x_1^{U,0}, x_2^{U,0}, \ldots, x_A^{U,0})^T$ *defined by the values*

$$x_a^{L,0} = \min_{i \in \mathcal{I}^+} \sum_j \boldsymbol{A}_a[i,j], x_a^{U,0} = \max_i \sum_j \boldsymbol{A}_a[i,j],$$

where $\mathcal{I}^+ = \{i \mid \sum_j \boldsymbol{A}_a[i,j] > 0\}$. *Then any feasible solution of* ENS *satisfies the necessary condition* $\boldsymbol{x}^{L,0} \leq \boldsymbol{x} \leq \boldsymbol{x}^{U,0}$.

Proof. The statement follows directly from RC3 since $x_a = x_a\boldsymbol{\pi}_k\mathbf{1} = \boldsymbol{\pi}_k\boldsymbol{A}_a\mathbf{1}$. Thus, $x_a = \boldsymbol{\pi}_k\boldsymbol{A}_a\mathbf{1} \geq \boldsymbol{\pi}_k(x_a^{L,0}\mathbf{1}) = x_a^{L,0}$ and $x_a = \boldsymbol{\pi}_k\boldsymbol{A}_a\mathbf{1} \leq \boldsymbol{\pi}_k(x_a^{U,0}\mathbf{1}) = x_a^{U,0}$. □

Let us also note that $\boldsymbol{x}$ satisfies ENS if and only if it is a global minimum for the *q*uadratically *c*onstrained *p*rogram

$$\begin{aligned}
&\mathsf{QCP}: \quad f_{\text{qcp}} = \min \textstyle\sum_a (\boldsymbol{s}_a^+ + \boldsymbol{s}_a^-) \\
&\boldsymbol{\pi}_k\boldsymbol{A}_a - x_a\boldsymbol{\pi}_k = \boldsymbol{s}_a^+ - \boldsymbol{s}_a^- && a \in \mathcal{A}_k, 1 \leq k \leq M, \\
&\boldsymbol{\pi}_k\boldsymbol{Q}_k(\boldsymbol{x}) = \mathbf{0} && 1 \leq k \leq M, \\
&\boldsymbol{\pi}_k\mathbf{1} = 1 && 1 \leq k \leq M, \\
&\boldsymbol{s}_a^+ \geq \mathbf{0}, \boldsymbol{s}_a^- \geq \mathbf{0} && a \in \mathcal{A}_k, \\
&\boldsymbol{x}^{L,0} \leq \boldsymbol{x} \leq \boldsymbol{x}^{U,0},
\end{aligned}$$

which has $O(A + N_{\text{sum}})$ variables and $O(AMN_{\max})$ constraints, where $N_{\text{sum}} = \sum_k N_k$ and $N_{\max} = \max_k N_k$. Here $\boldsymbol{s}_a^+$ and $\boldsymbol{s}_a^-$ are slack variables that guarantee the feasibility of all constraints in the early stages of the nonlinear optimization where the solver may be unable to determine a feasible assignment of $\boldsymbol{x}$ in ENS. By construction, $f_{\rm qcp} \geq 0$. Furthermore, RC3 holds if and only if $f_{\rm qcp} = 0$. Since all other quantities are bounded, we can also find upper and lower bounds on $\boldsymbol{s}_a^+$ and $\boldsymbol{s}_a^-$. As such, QCP is a quadratically constrained program with box constraints, a class of problems that is known to be $\mathcal{NP}$-hard [11]. The difficulty in a direct solution of ENS or QCP is clear even in "toy problems" such as identifying product forms in Jackson networks, i.e., queueing networks with exponential servers. For example, searching for a product form in a Jackson queueing network with two feedback queues one often finds that MATLAB's fmincon function fails to identify in ENS the search direction due to the small magnitudes of the gradients. QCP has better numerical properties than ENS, but it can take up to 5–10 min on commonly available hardware to find the reversed rates $\boldsymbol{x}$ needed to construct the product form (4.1). Thus QCP quickly becomes intractable on models with several queues. This shows that constructing product-form solutions by numerical optimization methods is, in general, a difficult problem. Moreover, it motivates an investigation of convex relaxations of ENS and QCP to derive efficient techniques for automatically constructing product forms. Indeed, automatic product-form analysis is fundamental to generating approximations for non-product-form models, as we show in the section "Automated Approximations."

Linearization Methodology

We now seek to obtain efficient linear programming (LP) relaxations of ENS that overcome the difficulties of solving a nonlinear system directly. To obtain an effective linearization, we first apply, in section "Convex Envelopes," an established convexification technique [36]. A tighter linear relaxation specific to RCAT is then developed in the section "Tightening the Linear Relaxation" and is shown to dramatically improve the quality of the relaxation. Finally, the section "Potential-Theory Constraints" obtains a tighter formulation based on a potential theory for Markov processes.

Convex Envelopes

To obtain a linearization of ENS, we first rewrite the generator matrix of process k as $\boldsymbol{Q}_k(\boldsymbol{x}) = \widetilde{\boldsymbol{T}}_k + \sum_{b \in \mathcal{P}_k} x_b \widetilde{\boldsymbol{P}}_b$, where $\widetilde{\boldsymbol{P}}_b$ is the sum of the $\boldsymbol{P}_b$ matrices and of the component of $\boldsymbol{\Delta}_k(\boldsymbol{x})$ that multiplies x_b. Then we condense the nonlinear components of ENS into the variables $z_{c,k} = x_c \boldsymbol{\pi}_k$ such that we replace the bilinear terms $x_c \boldsymbol{\pi}_k$ in $\boldsymbol{\pi}_k \boldsymbol{Q}_k(\boldsymbol{x})$ by $z_{c,k}$. Consequently, the only nonlinear constraints left are $z_{c,k} = x_c \boldsymbol{\pi}_k$, which we linearize to obtain an LP relaxation of ENS. We first observe that all variables involved in the bilinear product $x_c \boldsymbol{\pi}_k$ are bounded, as a consequence of

Proposition 4.1 and the fact that probabilities are bounded. We can thereby always write the bounds $\boldsymbol{x}^{L,0} \leq \boldsymbol{x} \leq \boldsymbol{x}^{U,0}$ and $\boldsymbol{\pi}_k^L \leq \boldsymbol{\pi}_k \leq \boldsymbol{\pi}_k^U$. Under these assumptions, for all $c \in \mathcal{A}_k$ and $1 \leq k \leq M$, $\boldsymbol{z}_{c,k}$ is always enclosed in the convex envelope proposed by McCormick in [36], which is known to be the tightest linear relaxation for bounded bilinear variables. Adding the constraint $\boldsymbol{z}_{c,k}\mathbf{1} = x_c$ yields a *l*inear *p*rogramming *r*elaxation of ENS:

$$
\begin{aligned}
\mathsf{LPR}(n) : f_{\text{lpr}}^n = \min f(\boldsymbol{x}, \boldsymbol{\pi}_k, \boldsymbol{z}_{c,k}) \qquad & \text{s.t.} \\
\boldsymbol{\pi}_k \boldsymbol{A}_a - \boldsymbol{z}_{a,k} = \mathbf{0}, \qquad & 1 \leq k \leq M, a \in \mathcal{A}_k, \\
\boldsymbol{\pi}_k \widetilde{\boldsymbol{T}}_k + \textstyle\sum_b \boldsymbol{z}_{b,k} \widetilde{\boldsymbol{P}}_b, = \mathbf{0} \qquad & 1 \leq k \leq M, \\
\boldsymbol{z}_{c,k} \geq x_c^{L,n} \boldsymbol{\pi}_k + x_c \boldsymbol{\pi}_k^{L,n} - x_c^{L,n} \boldsymbol{\pi}_k^{L,n}, \qquad & 1 \leq k \leq M, c \in \mathcal{A}_k \cup \mathcal{P}_k, \\
\boldsymbol{z}_{c,k} \leq x_c^{L,n} \boldsymbol{\pi}_k + x_c \boldsymbol{\pi}_k^{U,n} - x_c^{L,n} \boldsymbol{\pi}_k^{U,n}, \qquad & 1 \leq k \leq M, c \in \mathcal{A}_k \cup \mathcal{P}_k, \\
\boldsymbol{z}_{c,k} \leq x_c^{U,n} \boldsymbol{\pi}_k + x_c \boldsymbol{\pi}_k^{L,n} - x_c^{U,n} \boldsymbol{\pi}_k^{L,n}, \qquad & 1 \leq k \leq M, c \in \mathcal{A}_k \cup \mathcal{P}_k, \\
\boldsymbol{z}_{c,k} \geq x_c^{U,n} \boldsymbol{\pi}_k + x_c \boldsymbol{\pi}_k^{U,n} - x_c^{U,n} \boldsymbol{\pi}_k^{U,n}, \qquad & 1 \leq k \leq M, c \in \mathcal{A}_k \cup \mathcal{P}_k, \\
\boldsymbol{z}_{c,k} \mathbf{1} = x_c, \qquad & 1 \leq k \leq M, c \in \mathcal{A}_k \cup \mathcal{P}_k, \\
\boldsymbol{\pi}_k \mathbf{1} = 1, \qquad & 1 \leq k \leq M, \\
\boldsymbol{\pi}_k^{L,n} \leq \boldsymbol{\pi}_k \leq \boldsymbol{\pi}_k^{U,n}, \qquad & 1 \leq k \leq M, \\
\boldsymbol{x}^{L,n} \leq \boldsymbol{x} \leq \boldsymbol{x}^{U,n} \qquad &
\end{aligned}
$$

for an arbitrary *linear* objective function $f_{\text{lpr}}^n = f(\cdot)$, where n is an integer, used in the section "Exact Product-Form Construction" to parameterize a sequence of upper and lower bounding vectors on $\boldsymbol{x}$ and $\boldsymbol{\pi}_k$. The preceding optimization program is an LP that can be solved in polynomial time using interior-point methods. The number of variables and constraints in ENS grows asymptotically as $O(A + N_{\text{sum}})$ and $O(AMN_{\text{max}})$, respectively. In the foregoing linearized version LPR, the number of variables increases as $O(AN_{\text{sum}})$ and the number of constraints remains at $O(AMN_{\text{max}})$ asymptotically.

Rejecting the Existence of Product Forms. When LPR is infeasible, we can conclude that no RCAT product form exists for the model under study. To see this, it is sufficient to observe that LPR is a relaxation of ENS. Thus, all solutions of ENR are feasible points of LPR, but there exist points in LPR that do not solve ENS. Thus, since the feasible region of LPR is larger than that of ENR, we conclude that if LPR is infeasible, then so is ENS. This provides an interesting innovation over existing techniques for determining product-form solutions since none is currently able to exclude the existence of a product form when one cannot be found. (Note that, although we can then conclude that there is no RCAT product form, we cannot exclude the possibility that there is a non- RCAT product form, were such to exist.)

Tightening the Linear Relaxation

We now define our first method for obtaining tighter linearizations of ENS based on specific properties of the RCAT theorem. This is useful because McCormick's bounds are known to be wide in many cases [1].

Applying recursively the rate equations in RC3 v times we may write $\boldsymbol{\pi}_k \boldsymbol{A}_a^{v+1} = x_a^{v+1} \boldsymbol{\pi}_k$, for all $a \in \mathcal{A}_k$, $1 \leq k \leq M$, since RC3 implies that we can exchange scaling by x_a with right multiplication by $\boldsymbol{A}_a$. Summing over all $v \geq 0$ we obtain $\boldsymbol{\pi}_k \boldsymbol{A}_a \boldsymbol{H}_a = x_a(1-x_a)^{-1}\boldsymbol{\pi}_k$, where $\boldsymbol{H}_a = (\boldsymbol{I} - \boldsymbol{A}_a)^{-1}$, and we have assumed, without loss of generality, that the units of measure of the rates are scaled such that $x_a \leq x_a^U < 1$ and $\rho(\boldsymbol{A}_a) < 1$, where $\rho(\boldsymbol{M})$ denotes the spectral radius of a matrix $\boldsymbol{M}$. Rearranging terms and using $z_{a,k} = x_a \boldsymbol{\pi}_k$, we obtain the new linear constraint

$$\boldsymbol{\pi}_k \boldsymbol{A}_a \boldsymbol{H}_a = z_{a,k}(\boldsymbol{I} + \boldsymbol{A}_a \boldsymbol{H}_a) \tag{4.3}$$

for all active actions $a \in \mathcal{A}_k$ and processes k, $1 \leq k \leq M$. This provides an extra set of constraints that can be added to the linear relaxation of ENS to refine (reduce) the feasible region. Note that since $\boldsymbol{A}_a$ is a constant matrix, (4.3) is a linear equation in $\boldsymbol{\pi}_k$ and $z_{a,k}$.

The advantages of the method outlined above become even more apparent when we consider the generator constraint in LPR. For example, if we left-multiply by x_a, we obtain

$$x_a \boldsymbol{Q}_k = z_{a,k} \widetilde{\boldsymbol{T}}_k + \textstyle\sum_b z_{b,k} \boldsymbol{A}_a \widetilde{\boldsymbol{P}}_b = \boldsymbol{0}, \tag{4.4}$$

where we use the fact that the exchange rule holds for $z_{b,k}$, too, since $x_a z_{b,k} = x_a x_b \boldsymbol{\pi}_k = x_b \boldsymbol{\pi}_k \boldsymbol{A}_a = z_{b,k} \boldsymbol{A}_a$ for all $a \in \mathcal{A}_k$, $1 \leq k \leq M$. Equation (4.4) creates a direct linear relationship between the terms $z_{a,k}$ and $z_{b,k}$ for active and passive actions that cannot be inferred directly from LPR since it is based on exact knowledge of the bilinear relation $z_{b,k} = x_b \boldsymbol{\pi}_k$. As we show in an illustrative example at the end of this subsection, the additional constraints (4.3) and (4.4) greatly improve the LP approximation of ENS. Furthermore, following a similar argument, we can generate a hierarchy of linear constraints for $v = 0, 1, \ldots$

$$z_{a,k} \boldsymbol{A}_a^v \widetilde{\boldsymbol{T}}_k + \textstyle\sum_b z_{b,k} \boldsymbol{A}_a^{v+1} \widetilde{\boldsymbol{P}}_b = \boldsymbol{0}, \tag{4.5}$$

together with the condition obtained by summing over $v \geq 0$:

$$z_{a,k} \boldsymbol{H}_a \widetilde{\boldsymbol{T}}_k + \textstyle\sum_b z_{b,k} \boldsymbol{A}_a \boldsymbol{H}_a \widetilde{\boldsymbol{P}}_b = \boldsymbol{0}. \tag{4.6}$$

In summary, we have refined the linearization into the hierarchy of *t*ight *l*inear *p*rogramming *r*elaxations TLPR (n, V), which extends LPR by including constraints (4.5) for $v = 1, \ldots, V$ and (4.6).

We remark that, even though the preceding formulation is much more detailed than LPR, it inevitably requires an increased number of constraints, which now grows as $O(VAMN_{\max})$, while the complexity in terms of number of variables is the same as LPR. Thus, increased accuracy is obtained at a cost of additional computational complexity.

Potential-Theory Constraints

Potential theory is often applied in sensitivity and transient analyses of Markov processes and in Markov decision process theory [12]. We use it to derive tighter linearizations of ENS; to the best of our knowledge, this is the first time that potentials have been applied to product-form theory.

Consider a process with generator matrix $\boldsymbol{Q}_k(\boldsymbol{x})$ and equilibrium probability vector $\boldsymbol{\pi}_k(\boldsymbol{x})$, and define the vector $\boldsymbol{f}(n_k) = (f_1, \ldots, f_i, \ldots, f_{N_k})^T$, where $f_i = 1$ if $i = n_k$ and $f_i = 0$ otherwise. The linear metric $\eta_k(\boldsymbol{x}) = \boldsymbol{\pi}_k(\boldsymbol{x})\boldsymbol{f}(n_k) = \pi_k(n_k)$ is then the marginal probability of state n_k in process k, given $\boldsymbol{x}$. Potential theory provides compact formulas for studying the changes in the values of linear functions such as $\eta_k(\boldsymbol{x})$ under arbitrarily large perturbations of the generator matrix $\boldsymbol{Q}_k(\boldsymbol{x})$. Let $\boldsymbol{x}_0 > 0$ be an arbitrary reference point for $\boldsymbol{x}$ such that $\boldsymbol{Q}_k(\boldsymbol{x}_0)$ is a valid generator matrix with equilibrium vector $\boldsymbol{\pi}_k^0$ and $\eta_k(\boldsymbol{x}_0) = \pi_k^0(n_k)$. Then it is straightforward to show that the difference between $\eta_k(\boldsymbol{x})$ and $\eta_k(\boldsymbol{x}_0)$ is (as in [12])

$$\eta_k(\boldsymbol{x}) - \eta_k(\boldsymbol{x}_0) = \boldsymbol{\pi}_k(\boldsymbol{x})(\boldsymbol{Q}_k(\boldsymbol{x}) - \boldsymbol{Q}_k(\boldsymbol{x}_0))\boldsymbol{g}(\boldsymbol{x}_0, n_k), \tag{4.7}$$

where $\boldsymbol{g}(\boldsymbol{x}, n_k) = (-\boldsymbol{Q}_k(\boldsymbol{x}) + \boldsymbol{1}\boldsymbol{\pi}_k(\boldsymbol{x}))^{-1}\boldsymbol{f}(n_k)$ is the so-called *zero potential* of the function $\eta_k(\boldsymbol{x})$. [Notice that $\boldsymbol{\pi}_k(\boldsymbol{x}) = \boldsymbol{\pi}_k(\boldsymbol{x})(-\boldsymbol{Q}_k(\boldsymbol{x}) + \boldsymbol{1}\boldsymbol{\pi}_k(\boldsymbol{x}))$.] For the system under study, we can use (4.2) to rewrite (4.7) as

$$\pi_k(n_k) - \pi_k^0(n_k) = \Sigma_b(z_{b,k}(\boldsymbol{x}) - x_b^0 \boldsymbol{\pi}_k(\boldsymbol{x}))\boldsymbol{P}_b \boldsymbol{g}(\boldsymbol{x}_0, n_k).$$

Defining the *potential matrix* $\boldsymbol{G}(\boldsymbol{x}_0) = [\boldsymbol{g}(\boldsymbol{x}_0, n_1) \quad \boldsymbol{g}(\boldsymbol{x}_0, n_2) \quad \ldots \quad \boldsymbol{g}(\boldsymbol{x}_0, N_k)]$ we obtain a new set of linear constraints

$$\boldsymbol{\pi}_k - \boldsymbol{\pi}_k^0 = \Sigma_b(z_{b,k}(\boldsymbol{x}) - x_b^0 \boldsymbol{\pi}_k(\boldsymbol{x}))\boldsymbol{P}_b \boldsymbol{G}(\boldsymbol{x}_0). \tag{4.8}$$

This provides a further tightening of the linear relaxation of ENS.

Note that zero potentials can be memory consuming to evaluate because the matrix inverse $(-\boldsymbol{Q}_k(\boldsymbol{x}) + \boldsymbol{1}\boldsymbol{\pi}_k(\boldsymbol{x}))^{-1}$ does not preserve the sparsity of $\boldsymbol{Q}_k$. Furthermore, the rank 1 update $\boldsymbol{1}\boldsymbol{\pi}_k$ cannot be performed efficiently since $\boldsymbol{Q}_k$ is a singular matrix and updating techniques such as the Sherman–Morrison formula do not apply [39]. We address this computational issue by the algorithm shown in Fig. 4.2, which modifies the classical Jacobi iteration [41] to take advantage of the rank 1 structure of the term $\boldsymbol{1}\boldsymbol{\pi}_k$. That is, at each iteration, we isolate a vector $\boldsymbol{h}$ from the residual matrix in such a way that the matrix $\boldsymbol{1}\boldsymbol{\pi}_k$ is never explicitly computed. Thus, only vectors of the same order of $\boldsymbol{\pi}_k$ are stored in memory, and

Input: $\boldsymbol{Q}(\boldsymbol{x}_0)$, $\boldsymbol{\pi}(\boldsymbol{x}_0)$, $\boldsymbol{f}(n_k)$, ϵ_{tol}; **Output**: $\boldsymbol{g}(\boldsymbol{x}_0, n_k)$
$k = 0$, $\boldsymbol{g}^{(0)} = \boldsymbol{\pi}(\boldsymbol{x}_0)$, $\boldsymbol{R} = diag(-\boldsymbol{Q}(\boldsymbol{x}_0) + \boldsymbol{1}\boldsymbol{\pi}(\boldsymbol{x}_0)) + \boldsymbol{Q}(\boldsymbol{x}_0)$
do $k = k+1$
$\boldsymbol{g}^{(k)} = (diag(-\boldsymbol{Q}(\boldsymbol{x}_0) + \boldsymbol{1}\boldsymbol{\pi}(\boldsymbol{x}_0)))^{-1}(\boldsymbol{R}\boldsymbol{g}^{(k-1)} - \boldsymbol{1}\boldsymbol{\pi}_k(\boldsymbol{x}_0)\boldsymbol{g}^{(k-1)}) + \boldsymbol{f}(n_k))$
while $||\boldsymbol{g}^{(k)}(\boldsymbol{x}_0, n_k) - \boldsymbol{g}^{(k-1)}(\boldsymbol{x}_0, n_k)||_2 > \epsilon_{tol}||\boldsymbol{g}^{(k)}(\boldsymbol{x}_0, n_k)||_2$
return $\boldsymbol{g}^{(k)}(\boldsymbol{x}_0, n_k)$

Fig. 4.2 Memory-efficient computation of potentials; $diag(\boldsymbol{M})$ defines a diagonal matrix from the diagonal of $\boldsymbol{M}$

also $\boldsymbol{Q}_k$ remains in sparse form. In this way, each potential can always be computed efficiently with respect to storage requirements and in the worst case has asymptotic computational cost $O(JN_k^2)$, J being the number of Jacobi iterations. The potential matrix $\boldsymbol{G}$ is therefore computed in $O(JN_k^3)$ steps and, for the fixed reference point $\boldsymbol{x}_0$, needs to be evaluated only once, requiring a computation time that is usually small compared to the time required to solve the linear optimization programs. Moreover, even for the largest models, it is always possible to consider constraints arising from a subset of the columns of $\boldsymbol{G}$, again posing a tradeoff between computational costs and accuracy. Finally, using the exchange rule discussed in the section "Tightening the Linear Relaxation" we again obtain the hierarchy of constraints

$$\boldsymbol{\pi}_k\boldsymbol{A}_a^v - x_a^v\boldsymbol{\pi}_k^0 = \Sigma_b(z_{b,k}(\boldsymbol{x})\boldsymbol{A}_a^v - x_b^0\boldsymbol{\pi}_k(\boldsymbol{x})\boldsymbol{A}_a^v)\boldsymbol{P}_b\boldsymbol{G}(\boldsymbol{x}_0), \tag{4.9}$$

and the asymptotic condition after simple algebra becomes

$$\boldsymbol{\pi}_k\boldsymbol{A}_a'\boldsymbol{H}_a - x_a\boldsymbol{\pi}_k^0 = \Sigma_b(z_{b,k}(\boldsymbol{x}) - x_b^0\boldsymbol{\pi}_k(\boldsymbol{x}))\boldsymbol{A}_a'\boldsymbol{H}_a\boldsymbol{P}_b\boldsymbol{G}(\boldsymbol{x}_0), \tag{4.10}$$

where $\boldsymbol{A}_a' \stackrel{def}{=} \boldsymbol{A}_a - \boldsymbol{A}_a^2$. Summarizing, we can add (4.8)–(4.10) to LPR to generate tighter relaxations. We denote the resulting *z*ero-*p*otential *r*elaxation as ZPR(n,V). Note that ZPR(n,V) has the same asymptotic complexity of TLPR(n,V).

Exact Product-Form Construction

We now consider a technique for finding the solution of ENS that solves a sequence of the linear relaxations defined in the previous section, i.e., LPR(n), TLPR(n), or ZPR(n). Since the approach is identical for all relaxations, we limit the discussion to LPR(n).

The iterative algorithm defines a sequence of progressively tighter bounds $\boldsymbol{x}^{L,n}$ and $\boldsymbol{x}^{U,n}$ on the reversed rates $\boldsymbol{x}$ such that for sufficiently large n, LPR(n) determines a feasible solution $\boldsymbol{x}$ of ENS if one exists. Initial conditions are $\boldsymbol{x}^{L,0} \stackrel{\text{def}}{=} \boldsymbol{x}^L$, $\boldsymbol{x}^{U,0} \stackrel{\text{def}}{=} \boldsymbol{x}^U$, where $\boldsymbol{x}^L$ and $\boldsymbol{x}^U$ are the bounds defined in Proposition 4.1. We have the following result.

Proposition 4.2. *For each $n = 1, 2, \ldots$, consider a sequence of 2A linear relaxations of* ENS*, the first A with objective function $f_{\mathrm{lpr}}^{n,c} = \max x_c$ and the remaining A with objective function $g_{\mathrm{lpr}}^{n,c} = \min x_c$, for action c, $1 \le c \le A$ and bounds $\boldsymbol{x}^{L,n}$, $\boldsymbol{x}^{U,n}$, as previously.*

- *If $f_{\mathrm{lpr}}^{n,c} = x_c^{U,n}$ or $g_{\mathrm{lpr}}^{n,c} = x_c^{L,n}$, then the cth component of the linear relaxation solution $\boldsymbol{x}$ satisfies the bilinear constraint $z_{c,k} = x_c \boldsymbol{\pi}_k$ that is necessary for a solution of* ENS*.*
- *Otherwise, $f_{\mathrm{lpr}}^{n,c} < x_c^{U,n}$ and the bounds at iteration $n+1$ may be refined to*

$$x_c^{U,n+1} \stackrel{\mathrm{def}}{=} f_{\mathrm{lpr}}^{n,c}, \qquad x_c^{L,n+1} \stackrel{\mathrm{def}}{=} g_{\mathrm{lpr}}^{n,c}$$

for all actions c, $1 \le c \le A$, that define a feasible region for LPR(n+1) *that is strictly tighter than for* LPR(n)*.*

Proof. Consider the case $g_{\mathrm{lpr}}^{n} = x_c^{L,n}$, so that $\mathsf{LPR}(n)$ makes the assignment $x_c = x_c^{L,n}$. Then the first two McCormick constraints become

$$z_{c,k} \ge x_c^{L,n} \boldsymbol{\pi}_k + x_c^{L,n} \boldsymbol{\pi}_k^{L,n} - x_c^{L,n} \boldsymbol{\pi}_k^{L,n},$$

$$z_{c,k} \le x_c^{L,n} \boldsymbol{\pi}_k + x_c^{L,n} \boldsymbol{\pi}_k^{U,n} - x_c^{L,n} \boldsymbol{\pi}_k^{U,n},$$

which readily imply the bilinear relation $z_{c,k} = x_c^{L,n} \boldsymbol{\pi}_k = x_c \boldsymbol{\pi}_k$. A similar proof holds for $f_{\mathrm{lpr}}^{n} = x_c^{U,n}$.

Otherwise, if $f_{\mathrm{lpr}}^{n} > x_c^{L,n}$, then the feasible region of $\mathsf{LPR}(\mathsf{n}+1)$ does not include any point outside LPR(n) and excludes the points $x_c = x_c^{L,n}$. Hence it is strictly tighter than the feasible region for LPR(n). □

The preceding result guarantees that, if the sequence of linear relaxations yields feasible solutions, then the bounding box for $\mathsf{LPR}(n)$ defined by

$$[x_1^{L,n}, x_1^{U,n}] \times [x_2^{L,n}, x_2^{U,n}] \times \cdots \times [x_A^{L,n}, x_A^{U,n}]$$

can only decrease its volume or keep it constant as n increases. The volume must therefore converge as n increases, and for sufficiently large n, $x_c^{U,n} \approx x_c^{U,n+1}$ and $x_c^{L,n} \approx x_c^{L,n+1}$ in each dimension c. However, this implies that the outcome of iteration $n+1$ needs to be $f_{\mathrm{lpr}}^{n+1,c} = x_c^{U,n}$ and $g_{\mathrm{lpr}}^{n+1,c} = x_c^{U,n}$, which gives $z_c = x_c \boldsymbol{\pi}_k$ by the first case of Proposition 4.2. Thus, for sufficiently large n, the border of the bounding box intersects points that are feasible for ENS, at least along one dimension c. This yields several possible outcomes for the sequence of linear relaxations:

- The constraints in the linear relaxations are infeasible. As observed earlier, this allows us to conclude that no feasible RCAT product form exists for the model under study.

- One or more solutions $\boldsymbol{x}$ of the $2A$ linear relaxations are also feasible solutions of ENS. This allows us to construct directly a product-form solution by (4.1).
- No solution $\boldsymbol{x}$ of the $2A$ linear relaxations is feasible for ENS for all dimensions $c = 1, \ldots, A$. We have never encountered such a case in product-form detection for stochastic models; however, it can be resolved by a standard branch-and-bound method and reapplying the iteration on each partition of the feasible region.

Summarizing, a sequence of linear relaxations is sufficient to identify a product-form solution if one exists. No guarantee on the maximum number of linear programs to be solved can be given since the problem is $\mathcal{NP}$-hard in general; however, we show in the section "Examples and Case Studies" that this is typically small. In the section "Automated Approximations," we further illustrate how this sequence of linear programs can be modified to identify an approximate product form for a cooperation of Markov processes.

Practical Implementation

Pure Cooperations. If $\boldsymbol{x} = \mathbf{0}$ is a valid solution of ENS, then we call the model a *pure cooperation*. This is because $\boldsymbol{x} = \mathbf{0}$ implies that $\boldsymbol{\pi}_k = \mathbf{0}$, which in turn requires all entries of $\boldsymbol{L}_k$ to be zero for all processes. Hence, the model's rates are solely those of cooperations. Pure cooperations represent a very large class of models of practical interest, e.g., closed queueing networks with exponential or hyperexponential service, but their product-form analysis is harder due to the existence of infinite solutions $\boldsymbol{x}$.

Suppose a model is a pure cooperation and consider a graph defined by the $M \times M$ incidence matrix $\boldsymbol{G}$ such that $\boldsymbol{G}[i,j] = 1$ if and only if process j is passive in a cooperation with the active process i, 0 otherwise. Then, if $r = \text{rank}(\boldsymbol{G}) < M$, the model has $M - r$ degrees of freedom in assigning the values of the $\boldsymbol{x}$ vector. Thus, for these models there exists a continuous solution surface in ENS rather than a single feasible solution. As we show in the section "Closed Stochastic Model," this creates difficulties for existing product-form analysis techniques. However, we show that our method finds the correct solution $\boldsymbol{x} > 0$ under the condition that only objective functions of the type $\max x_c^U$ are used in the linear relaxations. This is because the search algorithm would otherwise converge, due to a lack of a strong lower bound, to the unreliable solution $\boldsymbol{x} = \mathbf{0}$.

Numerical properties. As the area of the bounding boxes decreases, the linear relaxations can be increasingly challenging to solve due to the presence of many hundreds of constraints on a small area and to the numerical scale of the equilibrium probabilities, which can become very small when N_k is several tens or hundreds of states. In such conditions, and without a careful specification of the linear programs, the solver may erroneously return that the program is infeasible, whereas a feasible solution does exist. However, a number of strategies can prevent such problems. First, it is often beneficial to reformulate equality constraints as "soft" constraints, e.g., for a small $\varepsilon_{tol} > 0$

$$\boldsymbol{\pi}_k \boldsymbol{Q}_k = \mathbf{0} \qquad \Rightarrow \qquad -\varepsilon_{tol}\boldsymbol{\pi}_k \leq \boldsymbol{\pi}_k \boldsymbol{Q}_k \leq \varepsilon_{tol}\boldsymbol{\pi}_k$$

that differentiates tolerances depending on the value of each individual term in $\boldsymbol{\pi}_k$. Another useful strategy consists of tuning the numerical tolerances of the LP solver. For instance, in IBM ILOG CPLEX's primal and dual simplex algorithms, this may be achieved by setting the Markowitz numerical tolerance to a large value such as 0.5. In addition, if the relaxation used is TLPR or ZPR, then it is often beneficial for a numerically challenging model to revert to the LPR formulation, which is less constraining. Finally, for models where the feasible region is sufficiently small, one could solve QCP directly without much effort and with the benefit of removing the extra constraints introduced by the linear relaxations. In our implementation, such corrections are done at runtime through a set of retrial runs upon detection that a LP is infeasible.

Automated Approximations

Using the preceding LP-based method, a non-product-form solution may be approximated using a product-form. The particular approximation we propose differs depending on which condition out of RC1, RC2, and RC3 is violated. Two approximations are now elaborated.

Rate Approximation

RC3 becomes infeasible when the solver cannot find a single reversed rate x_a that satisfies the condition for some actions a. Assuming that a solution $\boldsymbol{x}$ defines a valid RCAT product-form, for each process k we can always define a Markov-modulated point process $\mathcal{M}_{a,k}$ associated with the activation of the action $a \in \mathcal{A}_k$ in the Markov process with generator matrix $\mathbf{Q}_k$. Let the random variable $X_{a,k}$ denote the interarrival time between two consecutive activations of action a in $\mathcal{M}_{a,k}$ and define the rate $\lambda_{a,k}$ to be the reciprocal of the mean interarrival time $E[X_{a,k}]$. Then we approximate

$$\lambda_{a,k} = \frac{1}{E[X_{a,k}]} \approx \boldsymbol{\pi}_k \boldsymbol{A}_a \mathbf{1} = x_a \boldsymbol{\pi}_k \mathbf{1} = x_a. \tag{4.11}$$

The principle of rate approximation is to assume (inexactly) that (4.11) is a sufficient condition for a product-form solution. Let $\widetilde{\boldsymbol{x}} = (\widetilde{x}_1, \widetilde{x}_2, \ldots, \widetilde{x}_A)$ be an approximate solution that can be found by the *a*pproximate ENS program, which is defined by replacing RC3 with (4.11) in ENS and its relaxations.

We note that $\widetilde{\boldsymbol{x}}$ includes the exact solution $\widetilde{\boldsymbol{x}} = \boldsymbol{x}$ when it exists; thus AENS is a relaxation of ENS. Note that using the approach introduced in the section "Tightening the Linear Relaxation," one may further tighten the relaxation using a quadratic constraint

$$\boldsymbol{\pi}_k \mathbf{A}_a^2 \mathbf{1} = \widetilde{x}_a^2, \quad \forall a \in \mathcal{A}_k, \tag{4.12}$$

which provides a more accurate approximation of x_a by $\widetilde{x}_a$ but involves relaxation of a convex, and thus efficiently solvable, quadratically constrained program. Such an extension is left for future work.

The foregoing approximation can be applied to all programs introduced in the preceding sections, e.g., for LPR we define the rate approximation ALPR.

Structural Approximation

Example cases where RC1 is violated are models with blocking, where a cooperating process is not allowed to synchronize passively owing to capacity constraints, e.g., a queue with a finite-size buffer. Similarly, violations of RC2 are exemplified by models with priorities, where a low-priority action is disabled until higher-priority tasks are completed. Structural approximation iteratively updates the rate matrices $\boldsymbol{A}_a$ and $\boldsymbol{P}_b$ in order to account for the blocked or disabled status of certain transitions. Then, the search algorithm presented in the section "Exact Product-Form Construction" is run normally, if needed using rate approximation to address any violations of RC3 introduced by the updates. The updating process is detailed in the pseudocode reported in the appendix "Structural Approximation Pseudocode." First, we correct the blocked (respectively disabled) transitions in $\boldsymbol{P}_b$ (respectively $\boldsymbol{A}_a$) by hidden transitions in the synchronizing process that do change the state of the passive (respectively active) process. For $\boldsymbol{A}_a$ such hidden transitions need to be set to the reversed rate of action a in order to satisfy RC3. Next, a local iteration is done to scale the rates of the active (respectively passive) process to account approximately for the probability that the event could not occur in the passive (respectively active) process prior to the updates. Note that the particular way in which $\boldsymbol{A}_a$ and $\boldsymbol{P}_b$ are updated may be customized to reflect how the particular class of models under study handles the specific types of blocking or job priorities. For example, the pseudocode applies to the case of blocking followed by retrials; variants are discussed in the section "Models with Resource Constraints."

Example

Consider two small processes k and m with $N_k = 2$ and $N_m = 3$ states. Suppose there is a single action $a = 1$, $a \in \mathcal{A}_k$, $a \in \mathcal{P}_m$. The rate and local transition matrices are

$$\boldsymbol{L}_k = \begin{bmatrix} 0 & 0 \\ 10 & 0 \end{bmatrix}, \quad \boldsymbol{A}_a = \begin{bmatrix} 10 & 15 \\ 0 & 0 \end{bmatrix}, \boldsymbol{L}_m = \begin{bmatrix} 0 & 1 & 0 \\ 0 & 0 & 0 \\ 0 & 0 & 0 \end{bmatrix}, \quad \boldsymbol{P}_a = \begin{bmatrix} 0 & 0 & 0 \\ 0 & 0 & 1 \\ 1 & 0 & 0 \end{bmatrix}.$$

Process k has a high transition rate between its two states and the $1 \to 2$ one requires synchronization with process m. However, when process m is in state 1, no passive action is enabled (all zeros in the first row of $\boldsymbol{P}_a$). Hence, k is prevented from transiting from state 1 to state 2.

The structural approximation sets $\boldsymbol{P}_a(1,1) = 1$ and corrects the rates in process k to account for the blocking effects. In the resulting model, $\boldsymbol{L}_k$ and $\boldsymbol{L}_m$ are unaffected; instead

$$\boldsymbol{A}_a^{(1)} = \begin{bmatrix} 10 + \alpha_{a,1}15 & (1-\alpha_{a,1})15 \\ 0 & 0 \end{bmatrix}, \boldsymbol{P}_a^{(1)} = \begin{bmatrix} 1 & 0 & 0 \\ 0 & 0 & 1 \\ 1 & 0 & 0 \end{bmatrix},$$

where $\alpha_{a,1} = \boldsymbol{\pi}_m^{(0)}[1]$ and $\boldsymbol{\pi}_m^{(0)}$ is the equilibrium probability distribution for process m in the model for iteration $n = 0$ having the rate matrices $\boldsymbol{A}_a^{(0)} = \boldsymbol{A}_a$ and $\boldsymbol{P}_a^{(0)} = \boldsymbol{P}_a^{(1)}$. In this way, we have adjusted the active rates in such a way that, for the fraction of time where process m is in state 1, process k has the rate of action a's transitions to another state proportionally reduced. For this example, it is found that the fraction of the joint probability mass incorrectly placed by the product-form approximation converges after four iterations to 5.9%, while it is 45% if we just add $\boldsymbol{P}_a^{(1)}(1,1) = 1$ and do not apply corrections to $\boldsymbol{A}_a^{(1)}$.

Examples and Case Studies

Example: LPR *and* TLPR

We use a small example to illustrate and compare typical levels of tightness obtained by TLPR and LPR. The results are shown in Fig. 4.3, where the 2-norm for the current optimal solution with respect to the RC3 formula is evaluated for LPR(n) and TLPR$(n,1)$. The algorithm, described in the section "Exact Product-Form Construction," increases the lower bound on the reversed rates in each iteration. The model is composed of $M = 2$ agents that interact over $A = 2$ actions a and b with $\mathcal{A}_1 = \{a\}, \mathcal{P}_1 = \{b\}, \mathcal{A}_2 = \{b\}$, and $\mathcal{P}_2 = \{a\}$. Process 1 has $N_1 = 2$ states, and process 2 is defined by $N_1 = 4$ states. Rates of active actions and local transitions are given in Table 4.1. The passive rate matrices have $\boldsymbol{P}_a(1,4) = \boldsymbol{P}_a(2,1) = \boldsymbol{P}_a(3,2) = \boldsymbol{P}_a(4,3) = 1$ and $\boldsymbol{P}_b(2,1) = 1$.

For this example, the LP solver finds a product form in both cases, with reversed rates $x_a = 0.659039$ and $x_b = 0.646361$. Linear programs here and in the rest of the paper are generated from MATLAB using YALMIP [33] and solved by IBM ILOG

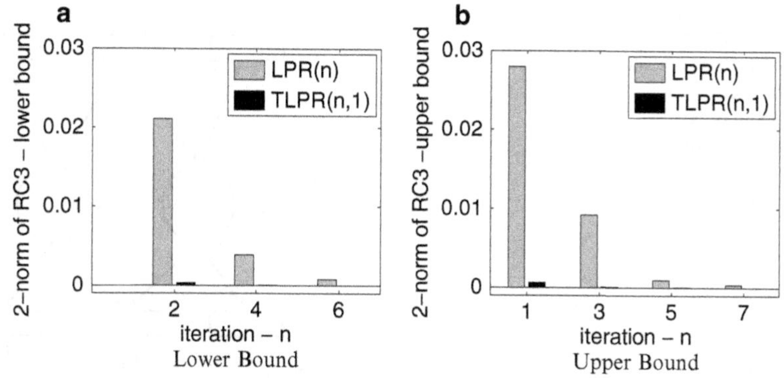

Fig. 4.3 Example showing increased tightness of TLPR compared to McCormick's convexification in LPR. The metric is the 2-norm of the error on RC3 at the current iteration of the search algorithm. Note that TLPR finds the product form at iteration 4, while LPR takes seven iterations

Table 4.1 Two processes cooperating on $A = 2$ action types

Element	Value	Element	Value	Element	Value
$L_1(1,2)$	1.000000	$A_a(1,1)$	0.312700	$A_b(2,1)$	0.758394
$L_1(2,1)$	0.092800	$A_a(1,2)$	0.012900	$A_b(2,2)$	0.000096
$L_2(1,2)$	0.624292	$A_a(2,1)$	0.384000	$A_b(3,2)$	0.684848
$L_2(2,3)$	0.867884	$A_a(2,2)$	0.644700	$A_b(3,3)$	0.521905
$L_2(3,4)$	0.823686	$A_b(1,1)$	0.180881	$A_b(4,3)$	0.073012
$L_2(4,1)$	0.999997	$A_b(1,4)$	0.574032	$A_b(4,4)$	0.064987

CPLEX's parallel barrier method with 16 software threads [28]. CPU time is 28 ms for LPR(n) (20 variables, 82 constraints and bounds) and 51 ms for TLPR$(n,1)$ (20 variables, 106 constraints and bounds).

The case studies in the sections "Closed Stochastic Model" and "A G-Network Model" focus on exact product-form solutions and are used to evaluate the proposed methodology against state-of-the-art techniques, namely, Buchholz's method [9] and INAP [34]. Conversely, the section "Models with Resource Constraints" illustrates the accuracy of rate and structural approximations on two models with resource constraints.

Example: ZPR

The example in this subsection illustrates certain benefits of the zero-potential relaxation over LPR and TLPR. Consider the toy model studied in [34, Fig. 5] composed of $M = 2$ processes $m = 1$ and $k = 2$ defined over $N_m = 4$ and $N_k = 3$ states. The processes cooperate on actions $a \in \mathcal{A}_k$ and $b \in \mathcal{A}_m$ and are defined by the rate and transition matrices given in the appendix "ZPR Example Model." On this model, all relaxations find a product-form solution associated to the reversed

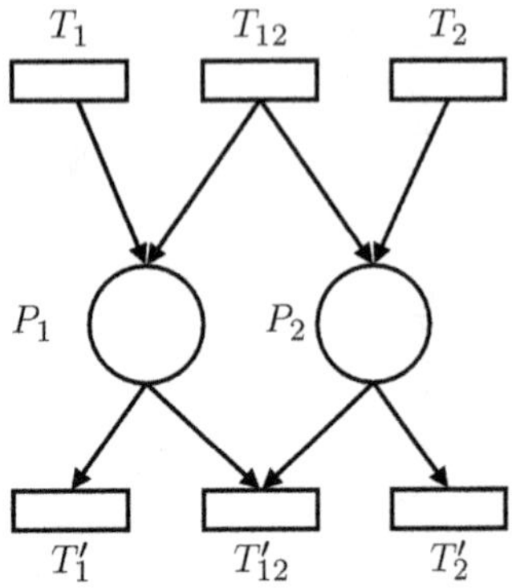

Fig. 4.4 Petri net process with six transitions and two places. The process abstracts a system where some operations may be synchronized between servers, e.g., a parallel storage system

rates $\boldsymbol{x} = (0.70, 1.90)$. LPR requires 14 linear programs to converge to such a solution with $\varepsilon_{tol} = 10^{-4}$ tolerance. Conversely, ZPR obtains the same solution in just six linear programs. Noticeably, at the first iteration ZPR achieves a 2-norm for the residual of RC3 that is achieved by LPR after only five linear programs. This provides a qualitative idea of the benefits of ZPR over LPR. Compared to TLPR, instead, ZPR offers similar accuracy, including in this example where TLPR completes after five linear programs. However, we have found ZPR to be numerically more robust than TLPR on several instances.

Closed Stochastic Model

Next, a challenging model of a closed network comprising three queues, indexed by $k = 1,2,3$, that cooperate with a parallel system modeled by the stochastic Petri net shown in Fig. 4.4, indexed by $k = 4$. This Petri net abstracts a generic parallel system where some operations are synchronized between two servers, e.g., mirrored disk drives. The special structure of this model has been shown recently to admit a product form for certain values of the transition rates [26]. The places P_1 and P_2 receive tokens, representing disk requests, from transitions T_1, T_{12}, and T_2. Such transitions are passive, meaning that they are activated by other components. The other transitions are active and fire after exponentially distributed times when all their input places have at least one token. The rates of the underlying exponential distributions are $\sigma_1 = 0.4$ for T_1', $\sigma_2 = 0.1$ for T_2', and $\sigma_{12} = 0.33$ for T_{12}'. Place P_1 receives jobs passively from transition T_1' and actively outputs into T_1 at the rate $\mu_1 = 0.5$ (actions 4 and 1, respectively); similarly, place P_2 receives jobs from T_2 and feeds T_2' at the rate $\mu_2 = 0.6$ (actions 3 and 6). Similarly, the queue $k = 3$ receives from T_{12} and outputs to T_{12}' at the rate $\mu_3 = 0.9$ (actions 2 and 5). Thus, $N_k = +\infty$, $k = 1,\dots,M$, and the model is a cooperation of $M = 4$ infinite processes on $A = 6$ actions. In the RCAT methodology, any cooperating process is considered in isolation with all its (possibly infinite) states, even if part of a closed model.

This is consistent with the fact that the specific population in the model affects the computation of the normalizing constant, but not the structure of the product-form solution for a joint state [38].

In addition, note that the model is a pure cooperation, due to the lack of local transitions, having the dependency graph

$$\boldsymbol{G} = \begin{pmatrix} 0\,0\,0\,1 \\ 0\,0\,0\,1 \\ 0\,0\,0\,1 \\ 1\,1\,1\,0 \end{pmatrix}.$$

Since $\boldsymbol{G}$ has a rank $r = 2$, there are $M - r = 2$ degrees of freedom in assigning the reversed rate vector $\boldsymbol{x} = (x_1, x_2, \ldots, x_6)$. Specifically, it is shown in [26] that the following necessary conditions hold for a product-form solution: $x_1 = x_4, x_2 = x_5, x_3 = x_6, x_5 = \sigma_{12} x_4 x_6 (\sigma_1 \sigma_2)^{-1}$. We apply our method and the INAP algorithm in [34] to determine a product-form solution of type (4.1). INAP is a simple fixed-point iteration that starting from a random guess of vector $\boldsymbol{x}$ progressively refines it until finding a product-form solution. For an action a the refinement step averages the value of the reversed rates of action a in all states of the active process. Buchholz's method in [9] cannot be used on the present example because it does not apply to closed models. For both INAP and our method we truncate the queue state space to $N_k = 75$ states, the Petri net to $N_4 = 100$ states. Thus, the product-form solution we obtain is valid for closed models with up to $N = 75$ circulating jobs.

Numerical Results. The best performing relaxation on this example is $\mathsf{TLPR}(n,1)$, which returns, after 35.82 s, a solution

$$\boldsymbol{x}^{\text{tlpr}} = (0.4023, 0.3323, 0.1004, 0.4014, 0.3315, 0.1003)$$

that matches the RC3 conditions with a tolerance of 10^{-3}. Since the tolerance of the solver is $\varepsilon_{tol} = 10^{-4}$, we regard this as an acceptable value considering that $\mathsf{TLPR}(n,1)$ describes a tight feasible region that may require the LP to apply numerical perturbations. Note that this is a standard feature of modern state-of-the-art LP solvers. $\mathsf{LPR}(n,1)$ provides a more accurate solution, $\boldsymbol{x}^{lpr} = (0.4008, 0.3305, 0.1001, 0.4001, 0.3300, 0.1001)$, but requires 234.879 s of CPU time to converge and 124 linear programs. INAP seems instead to suffer a significant loss of accuracy with this parameterization and does not converge. The returned solution after 48.64 s and 15,000 iterations is

$$\boldsymbol{x}^{\text{inap}} = (0.3651, 1.0464, 0.6566, 0.3236, 1.0411, 0.4307),$$

which is still quite far from the correct solution, especially concerning the necessary condition $x_3 = x_6$. We have further investigated this problem and observed that, in contrast to our algorithm, INAP ignores ergodicity constraints; hence most of the mass in this example is placed in states near the truncation border. This appears to be the reason for the failed convergence.

Table 4.2 Reversed rates returned by LPR for a G-network. The indexing is identical to that in [5]

$x_1 = 1.1615$	$x_2 = 1.7424$	$x_3 = 2.3230$	$x_4 = 0.5806$
$x_5 = 0.1162$	$x_6 = 0.2324$	$x_7 = 0.4646$	$x_8 = 0.2324$
$x_9 = 0.5228$	$x_{10} = 0.8712$	$x_{11} = 0.3486$	$x_{12} = 0.6970$
$x_{13} = 1.6262$	$x_{14} = 0.7317$	$x_{15} = 0.2559$	$x_{16} = 0.0852$
$x_{17} = 0.4268$	$x_{18} = 0.0852$	$x_{19} = 1.9355$	$x_{20} = 0.1215$
$x_{21} = 0.2430$	$x_{22} = 0.0241$	$x_{23} = 0.4709$	$x_{24} = 0.1178$

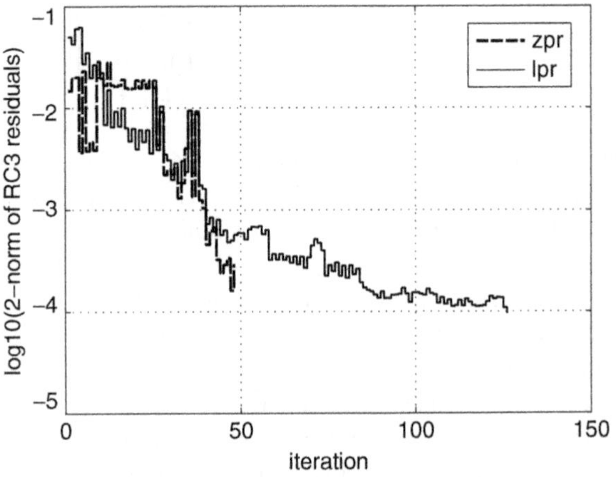

Fig. 4.5 Convergence speed of LPR and ZPR$(n,1)$. An iteration corresponds to the solution of a linear program

A G-Network Model

We next consider a generalized, open queueing network, where customers are of positive and negative types, i.e., a G-network [20]. These models enjoy a product-form solution, but this is not generally available in closed form and requires numerical techniques to determine it. Hence, G-networks provide a useful benchmark to compare different approaches for automated product-form analysis.

The queue parameterization used in this case study is the one given in [5] for a large model with $M = 10$ queues and $A = 24$ actions. Model parameters are given in the appendix "G-Network Case Study." The infinite state space is truncated such that each queue has $N_k = 75$ states. The size of the joint state space for the truncated model is $5.63 \cdot 10^{18}$ states, which is infeasible to solve numerically in the joint process.

Numerical Results. For this model, ENS and QCP fail almost immediately, reporting that the magnitude of the gradient is too small. Conversely, LPR returns the solution in Table 4.2. Quite interestingly, ZPR returns a different set of reversed rates, but these are found to generate the same equilibrium distributions $\boldsymbol{\pi}_k$ for all processes within the numerical tolerance $\varepsilon_{tol} = 10^{-4}$. Thus, this case study again confirms that our approach also provides valid answers in models with multiple solutions. A comparison of the convergence speed of LPR and ZPR is given in Fig. 4.5; TLPR fails in this case due to numerical issues since the feasible region is very tight. We have investigated the problem further and found that the barrier method is responsible for such instabilities and that switching to the simplex algorithm solves the problem and provides the same product-form solution as LPR.

We now compare our technique against Buchholz's method, applied in finding product forms of type (4.1). Buchholz's method involves a quadratic optimization technique that minimizes the residual norm with respect to a product-form solution for the model. This is done without explicitly computing the joint state distribution; hence it is efficient computationally. Comparison with the method proposed here is interesting since Buchholz's method seeks local optima instead of the global optima searched for by AUTOCAT. We have verified that, on small- to medium-scale models, the method is efficient in finding product forms. However, the local optimization approach for large models does not guarantee that a product-form solution will be found when one is known to exist. In particular, we used random initial points and found that, even though the residual errors are similar to those of the optimum solution of LPR, the specific local optimum returned by Buchholz's method can differ substantially in terms of the global product-form probability distribution. In particular, for some local optima, the marginal probability distribution at a queue is not geometric and the error on performance indices can be very large. This confirms the importance of using global optimization methods, such as that proposed in this chapter, for product-form analysis, especially in large-scale models. Furthermore, we believe that including RC3 in Buchholz's method would help to ensure the geometric structure of the marginal distribution.

Models with Resource Constraints

Finally, we consider an automated approximation of performance models with resource constraints. We have considered an open queueing network composed of $M = 5$ exponential, first-come first-served queues with finite buffer sizes described by the vector $(B_1, B_2, \ldots, B_M) = (7, 2, +\infty, 3, 10)$. Routing probabilities and model parameters are given in the appendix "Loss and BBS Models." In particular, arrival rates are chosen such that the equilibrium of the network differs dramatically from that of the corresponding infinite capacity model, where the first queues would be fully saturated. To explore the accuracy of rate and structural approximations, we have considered two opposite blocking types: *blocking before service* (BBS) [4], where a job is blocked before entering the server if its target queue is full, and the classical *loss policy*, where a job reaching a full station leaves the network. Such policies apply homogeneously to all queues. In both models, we study as the target performance metric the mean queue-length vector $\boldsymbol{n} = (n_1, \ldots, n_M)$ because such values are typically harder to approximate than utilizations as they depend more strongly on the entire marginal probability distributions of the queues.

The BBS model requires structural approximation to improve the accuracy of the initial ALPR rate approximation. To adapt the $\boldsymbol{A}_a$ corrections to this specific blocking policy, it is sufficient to delete the term $\alpha_{c,n}\Delta(\boldsymbol{A}_c^{(0)}\mathbf{1})$ from the updating in the structural approximation pseudocode, implying that jobs are not executed while the target station is busy. For this case study, the absolute values of the queue lengths obtained by simulation are $\boldsymbol{n} = (5.946, 1.262, 0.327, 1.1631, 1.653)$. The estimates returned by structural approximation converge after the fifth iteration to $\boldsymbol{n}^{sa(5)} = (5.9580, 1.3117, 0.2871, 1.0631, 1.3559)$ with an error on the bottleneck queue of just 0.20%.

For the loss model, we found that queue lengths are estimated accurately by the ALPR rate approximation alone, after adding hidden transitions to the $\boldsymbol{P}_b$ matrices to correct RC1. In particular, $\boldsymbol{n}^{ra} = (5.0792, 0.9599, 0.2688, 0.5050, 0.5273)$, where the result of the simulation is $\boldsymbol{n} = (5.4877, 1.0642, 0.2766, 0.5248, 0.5536)$, which has an average relative gap of 5.72%.This confirms the quality of the rate approximation in the loss case. Note that both in this case and in the BBS model computational costs are less than 5 min.

We have also tried to apply Buchholz's method to these examples, but as with the model of the section "A G-Network Model," the technique converges to a local optimum that differs from the simulated equilibrium behavior. Conversely, INAP does not apply to approximate analysis.

Closed Phase-Type Queueing Network

We now describe an example of approximate analysis of closed queueing networks. For illustration purposes, we focus on a machine repairman model comprising a single-server first-come, first-served queue in tandem with an infinite-server station. The same methodology can be used for larger models. The infinite-server station has exponentially distributed service times with rate $\mu_2 = 20$ jobs per second. The queue has PH service times (we refer the reader to [8] for an introduction to PH models). The distribution chosen has two states and representation $(\boldsymbol{\alpha}, \boldsymbol{T})$ with initial vector $\alpha = (1, 0)$ and PH subgenerator

$$\boldsymbol{T} = \begin{bmatrix} -1.2705 & 0.0118 \\ 0.0457 & -0.0457 \end{bmatrix}.$$

This PH model generates hyperexponential service times with mean 0.9996, squared coefficient of variation 9.9846, and skewness 19.6193. With this parameterization, the model is solved for a population $N = 15$ jobs by direct evaluation of the underlying Markov chain obtaining a throughput $X_{\text{ex}} = 0.6303$ jobs per second. This is lower than the throughput $X_{\text{pf}} = 0.6701$ jobs per second provided by a corresponding product-form model where the PH service time distribution is replaced by an exponential distribution.

We then approximate the solution of this model by AUTOCAT and study its relative accuracy. To cope with the lack of explicit constraints to find feasible reversed rates different from the degenerate ones $\boldsymbol{x} = (x_1, x_2) = \boldsymbol{0}$, we use the following iterative method. Initially, we set $x_1 = X^{\mathrm{pf}}$. Based on this educated guess, we run our approximation method based on the LPR formulation to find an approximate value for x_2. This allows the model to be solved after computing numerically the normalizing constant of the equilibrium probabilities and readily provides an estimate $X^{(1)}$ for the network throughput. In the following iteration we assign $x_1 = X^{(1)}$ and reoptimize to find a new value of x_2 and corresponding throughput $X^{(2)}$. This iterative scheme is reapplied until convergence is achieved.[1]

For the model under study, this approximation provides a sequence of solutions $X^{(1)}_{lpr} = 1.0004, X^{(2)}_{lpr} = 0.6291, X^{(3)}_{lpr} = 0.6089, X^{(4)}_{lpr} = 0.6107$, and $X^{(5)}_{lpr} = 0.6105$ jobs per second, for a total of 40 solver iterations. The last solution provides a relative error on the exact one of -3.14% compared to the 6.31% error of the product-form approximation, thus reducing the approximation error by about 50%.

We have also compared accuracy with a recent iterative approximation technique for closed networks, inspired by RCAT and proposed in [14, 15]. This technique involves replacing each $-/PH/1$ queueing station by a load-dependent station such that the state probability distribution for a model with M queues is

$$\Pr(n_1, n_2, \ldots, n_M) = G^{-1} \prod_{i=1}^{M} F_i(n_i),$$

where n_i is the number of jobs in queue i, G is a normalization constant, and

$$F_i(n_i) = \begin{cases} 1 - \rho_i, & n_i = 0, \\ \rho_i (1 - \eta_i) \eta_i^{n_i}, & n_i > 0, \end{cases}$$

where η_i is the largest eigenvalue of the rate matrix for the quasi-birth-and-death process obtained by studying the ith station as an open $PH/PH/1$ queue with appropriate input process and utilization ρ_i. We point to [14] for further details on this construction; here we simply stress that this particular approximation differs from the AutoCAT one by using only the slowest decay rate of the queue-length marginal probabilities for such a $PH/PH/1$ queue, whereas in this chapter we developed more general approximations that do not resort to asymptotic arguments to simplify the model and that may be applied also to stochastic systems other than queueing networks.

A comparison with the method proposed in [14, 15] reveals that the throughput returned by the approximation is $X = 0.5843$ jobs per second with a relative error of -7.30%. While classes of models exist where it can be shown that this method is far more accurate than the product-form one [14], this example convincingly illustrates a case where the AutoCAT approximation is the most accurate available.

[1]Note that all test cases did converge, but no rigorous convergence proof is available.

Conclusion

We have introduced an optimization-based approach to product-form analysis of performance models that can be described as a cooperation of Markov processes, e.g., queueing networks and stochastic Petri nets. Our methodology consists of solving a sequence of linear programming relaxations for a nonconvex optimization program that captures a set of sufficient conditions for a product form. The main limitation of our methodology is that we cannot represent cooperations involving actions that synchronize over more than two processes. However, multiple cooperations are useful only in specialized models, e.g., queueing networks with catastrophes [18]. We believe that such extension is possible, although it may require a sequence of independent product-form search problems to be solved. Hence, the computational costs of such solutions should be evaluated for models of practical interest.

Finally, we plan to study the effects of integrating new constraints into the linear programs, such as costs or bounds on the variables that may help in determining a particular reversed rate vector among a set of multiple feasible solutions. For instance, for models that enjoy bounds on their steady state that may be expressed as linear programs, e.g., stochastic Petri nets [32], this could enable the generation of exact or approximate product forms that are guaranteed to be within the known theoretical bounds.

Appendix

Infinite Processes

Numerical optimization techniques generally require matrices of finite size. In both ENS and its relaxations, we therefore used exact or approximate aggregations to truncate the state spaces of any infinite processes. Let $C+1$ be the maximum acceptable matrix order. Then we decompose the generator matrix and its equilibrium probability vector of an infinite process k as

$$\boldsymbol{Q}_k = \begin{bmatrix} \boldsymbol{Q}_k^{C,C} & \boldsymbol{Q}_k^{C,\infty} \\ \boldsymbol{Q}_k^{\infty,C} & \boldsymbol{Q}_k^{\infty,\infty} \end{bmatrix}, \quad \boldsymbol{\pi}_k = \left[\boldsymbol{\pi}_k^C, \boldsymbol{\pi}_k^\infty\right],$$

where $\boldsymbol{Q}_k^{C_1,C_2}$ is a $C_1 \times C_2$ matrix. Similar partitionings are also applied to the transition matrix $\boldsymbol{L}_k$ and to the rate matrices $\boldsymbol{A}_a$ and $\boldsymbol{P}_b$, $a \in \mathcal{A}_k, b \in \mathcal{P}_k$. We define the truncation such that the total probability mass in the first C states is 1 relative to the numerical tolerance of the optimizer, i.e., $\boldsymbol{\pi}_k^\infty \mathbf{1} < \varepsilon_{tol}$. Notice that the latter condition can also be used to determine the ergodicity of the infinite process. Furthermore, from condition RC1 (respectively RC2) we need to account for the cases where passive (respectively active) actions associated with the first C states are only enabled in $\boldsymbol{P}_k^{C,\infty}$ (respectively only incoming from $\boldsymbol{A}_k^{\infty,C}$). Such problems are easily handled by adding one fictitious state to the truncated set $\{1,2,\ldots,C\}$. For example, for $\boldsymbol{A}_a$ and $\boldsymbol{P}_b$ we consider the truncated matrices

$$\boldsymbol{A}_a = \begin{bmatrix} \boldsymbol{A}_a^{C,C} & \boldsymbol{A}_a^{C,\infty}\mathbf{1} \\ \mathbf{1}^T\boldsymbol{A}_a^{\infty,C} & 0 \end{bmatrix}, \quad \boldsymbol{P}_b = \begin{bmatrix} \boldsymbol{P}_b^{C,C} & \boldsymbol{P}_b^{C,\infty}\mathbf{1} \\ \mathbf{1}^T\boldsymbol{P}_b^{\infty,C} & 0 \end{bmatrix},$$

where $\mathbf{1}$ is now an infinite column of 1s. Note that the fictitious state is excluded from the validation of conditions RC1 and RC2; thus the value of the diagonal rate on the last row is irrelevant with respect to finding a product form.

Finally, we comment on the choice of the parameter C for a given process k. Since this determines the number of states N_k for the truncated process, an optimal choice of this value can provide substantial computational savings. Let us first note that starting from a small C, it is easy to integrate additional constraints or potential vectors in the linear formulations for a value $C' > C$. Recall that we propose in the rest of the paper a sequence of linear programs in order to obtain a feasible solution $\boldsymbol{x}$. Then, if a linear program is infeasible, this can be due either to a lack of a product form or to a truncation where C is too small. The latter case can be readily diagnosed by adding slack variables, as in QCP, to the ergodicity condition and verifying if this is sufficient to restore feasibility. In such a case, the C value is updated to the smallest value such that feasibility is restored in the main linear program.

ZPR Example Model

$$\boldsymbol{A}_a = \begin{bmatrix} 0 & 0 & 0 \\ 0 & 0 & 0.2170 \\ 2.9105 & 2.2575 & 0 \end{bmatrix} \qquad \boldsymbol{P}_a = \begin{bmatrix} 0 & 1 & 0 & 0 \\ 0 & 1 & 0 & 0 \\ 1 & 0 & 0 & 0 \\ 1 & 0 & 0 & 0 \end{bmatrix}$$

$$\boldsymbol{A}_b = \begin{bmatrix} 0 & 0 & 0 & 0 \\ 5.65 & 0 & 0.52 & 2.13 \\ 0 & 7.00 & 0 & 0 \\ 0 & 0 & 0 & 0 \end{bmatrix} \qquad \boldsymbol{P}_b = \begin{bmatrix} 0 & 1 & 0 \\ 0 & 1 & 0 \\ 1 & 0 & 0 \end{bmatrix}$$

$$\boldsymbol{L}_m = \begin{bmatrix} 0 & 8 & 0 & 3 \\ 6.15 & 0 & 8.28 & 7.67 \\ 15 & 9.70 & 0 & 0 \\ 16 & 0 & 0 & 0 \end{bmatrix} \qquad \boldsymbol{L}_k = \begin{bmatrix} 0 & 0 & 0 \\ 0 & 0 & 3.78 \\ 3.09 & 2.74 & 0 \end{bmatrix}$$

Structural Approximation Pseudocode

Input: RLX$\in$ {ALPR, ATLPR, AZPR}, $\boldsymbol{L}_k$, $\boldsymbol{A}_a$, $\boldsymbol{P}_b$, $\mathcal{A}_k$, $\mathcal{P}_k$, $\forall k$, a;
Output: $\boldsymbol{x}$, $\boldsymbol{\pi}_k$, $\boldsymbol{Q}_k$, $\forall k$ ignore RC1 and RC2, get approximate product-form solution $\boldsymbol{x}^{(0)}$ by RLX
for $k = 1,\ldots,M$ `/* correct RC1 and RC2 */`
  for all $a \in \mathcal{A}_k$ **do** $\boldsymbol{A}_a(j,j) = x_a^{(0)}, \forall j \in \mathcal{J}_b$, $\mathcal{J}_b = \{j \mid \sum_i \boldsymbol{A}_a[i,j] = 0\}$
 end for all
  for all $b \in \mathcal{P}_k$ **do** $\boldsymbol{P}_b(i,i) = 1$, $\forall i \in \mathcal{I}_b$, $\mathcal{I}_b = \{i \mid \sum_j \boldsymbol{P}_b[i,j] = 0\}$
 end for all
$\alpha_{c,0} = 1$; $\boldsymbol{A}_c^{(0)} = \alpha_{c,0}\boldsymbol{A}_c$, $c = 1,2,\ldots,A$;
$\beta_{c,0} = 1$; $\boldsymbol{P}_c^{(0)} = \beta_{c,0}\boldsymbol{P}_c$, $c = 1,2,\ldots,A$;
while current iteration number $n \geq 1$ is less than the maximum number of iterations
  get by RLX an approximate product-form solution $\boldsymbol{x}^{(n)}$ for $\boldsymbol{L}_k$, $\boldsymbol{A}_a^{(n)}$, $\boldsymbol{P}_b^{(n)}$
  for $c = 1,\ldots,A$, where $c \in \mathcal{A}_k$ and $c \in \mathcal{P}_m$
`/* update blocking probabilities */`
   $\alpha_{c,n} = \sum_{i \in \mathcal{I}_c} \boldsymbol{\pi}_m(\boldsymbol{x}^{(n)})[i]$; $\boldsymbol{A}_c^{(n)} = (1-\alpha_{c,n})\boldsymbol{A}_c^{(0)} + \alpha_{c,n}\Delta(\boldsymbol{A}_c^{(0)}\boldsymbol{1})$
   $\beta_{c,n} = \sum_{j \in \mathcal{J}_c} \boldsymbol{\pi}_k(\boldsymbol{x}^{(n)})[j]$; $\boldsymbol{P}_c^{(n)} = (1-\beta_{c,n})\boldsymbol{P}_c^{(0)} + \beta_{c,n}\Delta(\boldsymbol{P}_c^{(0)}\boldsymbol{1})$
  end for
  if $\max_c(||\alpha_{c,n} - \alpha_{c,n-1}||_2, ||\beta_{c,n} - \beta_{c,n-1}||_2) \leq \varepsilon_{tol}$ **return** $\boldsymbol{x}^{(n)}$, $\boldsymbol{\pi}_k(\boldsymbol{x}^{(n)})$, $\boldsymbol{Q}_k(\boldsymbol{x}^{(n)})$
end while
return $\boldsymbol{x}^{(n)}$, $\boldsymbol{\pi}_k(\boldsymbol{x}^{(n)})$, $\boldsymbol{Q}_k(\boldsymbol{x}^{(n)})$

G-Network Case Study

We report the parameters for the G-network given in [5]. The network consists of $M = 10$ queues with exponentially distributed service times having rates $\mu_1 = 4.5$ and $\mu_i = 4.0 + (0.1)i$ for $i \in [2,10]$. The external arrival rate defines a Poisson process with rate $\lambda = 5.0$. The routing matrix for (positive) customers has in row i and column j the probability $r_{i,j}^{+}$ of a (positive) customer being routed to queue j, as a positive customer, upon leaving queue i. In this case study, this routing matrix is given by

$$R^{+} = [r^{+}_{i,j}] = \begin{bmatrix} 0 & 0.2 & 0.3 & 0.4 & 0 & 0 & 0 & 0 & 0 & 0 \\ 0.1 & 0 & 0 & 0 & 0.2 & 0 & 0 & 0.2 & 0 & 0 \\ 0 & 0 & 0 & 0 & 0.3 & 0.5 & 0.2 & 0 & 0 & 0 \\ 0.3 & 0 & 0 & 0 & 0 & 0 & 0.7 & 0 & 0 & 0 \\ 0 & 0 & 0 & 0 & 0 & 0 & 0 & 1 & 0 & 0 \\ 0 & 0 & 0 & 0 & 0 & 0 & 0 & 0.3 & 0 & 0.5 \\ 0 & 0 & 0 & 0 & 0 & 0 & 0 & 0 & 0 & 1 \\ 0 & 0 & 0 & 0 & 0 & 0 & 0 & 0 & 0.2 & 0 \\ 0 & 0 & 0 & 0 & 0 & 0 & 0 & 0 & 0 & 0 \\ 0 & 0 & 0 & 0 & 0 & 0 & 0 & 0 & 0 & 0 \end{bmatrix}.$$

Conversely, the probability $r^{-}_{i,j}$ of a customer leaving queue i and becoming a negative signal upon arrival at queue j is

$$R^{-} = [r^{-}_{i,j}] = \begin{bmatrix} 0 & 0 & 0 & 0 & 0.1 & 0 & 0 & 0 & 0 & 0 \\ 0 & 0 & 0 & 0 & 0 & 0.4 & 0 & 0 & 0 & 0 \\ 0 & 0 & 0 & 0 & 0 & 0 & 0 & 0 & 0 & 0 \\ 0 & 0 & 0 & 0 & 0 & 0 & 0 & 0 & 0 & 0 \\ 0 & 0 & 0 & 0 & 0 & 0 & 0 & 0 & 0 & 0 \\ 0 & 0 & 0 & 0 & 0 & 0 & 0 & 0.1 & 0 & 0.1 \\ 0 & 0 & 0 & 0 & 0 & 0 & 0 & 0 & 0 & 0 \\ 0 & 0 & 0 & 0 & 0.1 & 0 & 0 & 0 & 0 & 0 \\ 0.1 & 0 & 0 & 0 & 0 & 0 & 0 & 0 & 0 & 0 \\ 0 & 0 & 0 & 0 & 0 & 0 & 0.2 & 0 & 0.05 & 0 \end{bmatrix}.$$

Loss and BBS Models

The model is composed of $M = 5$ queues that cooperate on a set of $A = 12$ actions, one for each possible job movement from and inside the network. The routing probabilities $R[k, j]$ from queue k to queue j are as follows:

$$R = \begin{bmatrix} 0.16 & 0 & 0.04 & 0.50 & 0.30 \\ 0.08 & 0.29 & 0.02 & 0.08 & 0.52 \\ 0 & 0 & 0.78 & 0 & 0 \\ 0.29 & 0.24 & 0 & 0.25 & 0.22 \\ 0 & 0.49 & 0 & 0.20 & 0 \end{bmatrix}.$$

Service times are exponential at all queues with rates $mu_k = k$, $k =, 1 \ldots, 5$. For a queue k, the probability of departing from the network is $r_{k,0} = 1 - \Sigma_{j=1}^{5} R[k, j]$. The Poisson arrival rates from the outside world are given by the vector

$$\boldsymbol{\lambda} = (0.6600, 0.1500, 0.0750, 0.1650, 0.4500).$$

References

1. Al-Khayyal, F.A., Falk, J.E.: Jointly constrained biconvex programming. Math. Oper. Res. **8**(2), 273–286 (1983)
2. Argent-Katwala, A.: Automated product-forms with Meercat. In: Proceedings of SMCTOOLS, October 2006
3. Balbo, G., Bruell, S.C., Sereno, M.: Product form solution for generalized stochastic Petri nets. IEEE TSE **28**(10), 915–932 (2002)
4. Balsamo, S., Onvural, R.O., De Nitto Personé, V.: Analysis of Queueing Networks with Blocking. Kluwer, Norwell, MA (2001)
5. Balsamo, S., Dei Rossi, G., Marin, A.: A numerical algorithm for the solution of product-form models with infinite state spaces. In Computer Performance Engineering (A. Aldini, M. Bernardo, L. Bononi, V. Cortellessa, Eds.) LNCS 6342, Springer 2010. (7th Europ. Performance Engineering Workshop EPEW 2010, Bertinoro (Fc), Italy, (2010)
6. Baskett, F., Chandy, K.M., Muntz, R.R., Palacios, F.G.: Open, closed, and mixed networks of queues with different classes of customers. J. ACM **22**(2), 248–260 (1975)
7. Bertsimas, D., Tsitsiklis, J.: Introduction to Linear Optimization. Athena Scientific, Nashua, NH (1997)
8. Bolch, G., Greiner, S., de Meer, H., Trivedi, K.S.: Queueing Networks and Markov Chains. Wiley, New York (1998)
9. Buchholz, P.: Product form approximations for communicating Markov processes. In: Proceedings of QEST, pp. 135–144. IEEE, New York (2008)
10. Buchholz, P.: Product form approximations for communicating Markov processes. Perform. Eval. **67**(9), 797–815 (2010)
11. Burer, S., Letchford, A.N.: On nonconvex quadratic programming with box constraints. SIAM J. Optim. **20**(2), 1073–1089 (2009)
12. Cao, X.R.: The relations among potentials, perturbation analysis, and Markov decision processes. Discr. Event Dyn. Sys. **8**(1), 71–87 (1998)
13. Cao, X.R., Ma, D.J.: Performance sensitivity formulae, algorithms and estimates for closed queueing networks with exponential servers. Perform. Eval. **26**, 181–199 (1996)
14. Casale, G., Harrison, P.G.: A class of tractable models for run-time performance evaluation. In: Proceedings of ACM/SPEC ICPE (2012)
15. Casale, G., Harrison, P.G., Vigliotti, M.G.: Product-form approximation of queueing networks with phase-type service. ACM Perf. Eval. Rev. **39**(4) (2012)
16. de Souza e Silva, E., Ochoa, P.M.: State space exploration in Markov models. In: Proceedings of ACM SIGMETRICS, pp. 152–166 (1992)
17. Dijk, N.: Queueing Networks and Product Forms: A Systems Approach. Wiley, Chichester (1993)
18. Fourneau, J.M., Quessette, F.: Computing the steady-state distribution of G-networks with synchronized partial flushing. In: Proceedings of ISCIS, pp. 887–896. Springer, Berlin (2006)
19. Fourneau, J.M., Plateau, B., Stewart, W.: Product form for stochastic automata networks. In: Proceedings of ValueTools, pp. 1–10 (2007)
20. Gelenbe, E.: Product-form queueing networks with negative and positive customers. J. App. Probab. **28**(3), 656–663 (1991)
21. GNU GLPK 4.8. http://www.gnu.org/software/glpk/
22. Harrison, P.G.: Turning back time in Markovian process algebra. Theor. Comput. Sci **290**(3), 1947–1986 (2003)
23. Harrison, P.G.: Reversed processes, product forms and a non-product form. Lin. Algebra Appl. **386**, 359–381 (2004)
24. Harrison, P.G., Hillston, J.: Exploiting quasi-reversible structures in Markovian process algebra models. Comp. J. **38**(7), 510–520 (1995)
25. Harrison, P.G., Lee, T.: Separable equilibrium state probabilities via time reversal in Markovian process algebra. Theor. Comput. Sci **346**, 161–182 (2005)

26. Harrison, P.G., Llado, C.: A PMIF with Petri net building blocks. In: Proceedings of ICPE (2011)
27. Hillston, J.: A compositional approach to performance modelling. Ph.D. Thesis, University of Edinburgh (1994)
28. IBM ILOG CPLEX 12.0 User's Manual, 2010
29. Jackson, J.R.J.: Jobshop-like queueing systems. Manage. Sci. **10**(1), 131–142 (1963)
30. Kelly, F.P.: Networks of queues with customers of different types. J. Appl. Probab. **12**(3), 542–554 (1975)
31. Kelly, F.P.: Reversibility and Stochastic Networks. Wiley, New York (1979)
32. Liu, Z.: Performance analysis of stochastic timed Petri nets using linear programming approach. IEEE TSE **11**(24), 1014–1030 (1998)
33. Löfberg, J.: YALMIP: A toolbox for modeling and optimization in MATLAB. In: Proceedings of CACSD (2004)
34. Marin, A., Bulò, S.R.: A general algorithm to compute the steady-state solution of product-form cooperating Markov chains. In: Proceedings of MASCOTS, pp. 1–10 (2009)
35. Marin, A., Vigliotti, M.G.: A general result for deriving product-form solutions in Markovian models. In: Proceedings of ICPE, pp. 165–176 (2010)
36. McCormick, G.P.: Computability of global solutions to factorable nonconvex programs. Math. Prog. **10**, 146–175 (1976)
37. Muntz, R.R.: Poisson departure processes and queueing networks. Tech. Rep. RC 4145, IBM T.J. Watson Research Center, Yorktown Heights, NY (1972)
38. Nelson, R.D.: The mathematics of product form queuing networks. ACM Comp. Surv. **25**(3), 339–369 (1993)
39. Nocedal, J., Wright, S.J.: Numerical Optimization. Springer, Berlin (1999)
40. Plateau, B.: On the stochastic structure of parallelism and synchronization models for distributed algorithms. SIGMETRICS 147–154 (1985)
41. Saad, Y.: Iterative Methods for Sparse Linear Systems. SIAM, Philadelphia (2000)

Chapter 5
Markovian Trees Subject to Catastrophes: Would They Survive Forever?

Sophie Hautphenne, Guy Latouche, and Giang T. Nguyen

Introduction

It is easy to recognize if a simple branching process has a chance of surviving forever: if individuals act independently, then the key parameter is the total number of children that each of them has, on average. We know that if the expected number of children is greater than one, then the process has a strictly positive probability of remaining alive for all times; otherwise, excluding the degenerate case where each individual has one and only one child, extinction occurs almost surely [13]. In the first case, we say that the process is supercritical.

If the branching process is subject to an external environment that affects all individuals simultaneously, then the situation becomes more complex. The probability of extinction is conditionally linked to the asymptotic growth rate of the population given the history of the environment. This gives precise criteria that, unfortunately, are, in general, not easily evaluated.

We consider in this chapter multitype Markovian branching processes subject to catastrophic events. In this case, determining whether the process is supercritical is akin to computing the maximal Lyapunov exponent of a sequence of random matrices, a notoriously difficult problem [16, 25]. We show that there is a simple characterization in the case where all individuals have the same probability of surviving a catastrophe, and we determine upper and lower bounds in the case where survival depends on the type of individual.

S. Hautphenne (✉)
Department of Mathematics and Statistics, University of Melbourne,
Melbourne, Victoria 3010, Australia
e-mail: sophiemh@unimelb.edu.au

G. Latouche • G.T. Nguyen
Département d'informatique, Université libre de Bruxelles, Blvd du Triomphe,
CP 212, 1050 Brussels, Belgium
e-mail: latouche@ulb.ac.be; giang.nguyen@adelaide.edu.au

G. Latouche et al. (eds.), *Matrix-Analytic Methods in Stochastic Models*, Springer Proceedings in Mathematics & Statistics 27, DOI 10.1007/978-1-4614-4909-6_5,

After giving some background material in the next section, we characterize our bounds in the section "A Duality Approach." To do so, we define a dual process which involves a single individual. In addition, we derive in the section "Markovian Catastrophes" we drive explicit expressions for the bounds when catastrophes are Markovian. We give a few numerical examples in the section "Numerical Illustration" that indicate that the upper bound is usually closer to the actual value than the lower bound.

Background

Branching processes in a random environment have a long history: Smith and Wilkinson [22], Athreya and Karlin [4], and Kingman [16] are among the important early researchers; two references most relevant for our purposes are by Tanny [23, 24]. We use here the *Markovian tree* description of Bean et al. [6] to define the branching process.

Each individual is characterized by an $m \times m$ matrix D, a sequence of $m \times m^{k+1}$ matrices B_k, $k \geq 1$, and an $m \times 1$ vector $\boldsymbol{d}$. The order m is the number of *types*, which corresponds to the physiological states of an individual, categories in a population, or ..., depending on the viewpoint. The matrices D, $\{B_k\}_{k\geq 1}$ and the vector $\boldsymbol{d}$ are defined as follows: D_{ij}, for $i \neq j$, is the instantaneous transition rate at which an individual of type i changes to type j without producing an offspring, $(B_k)_{i;j_1 j_2 \cdots j_k j}$ is the rate at which an individual of type i gives birth to k children *and* simultaneously changes to type j, the k children starting their lives with types j_1, $j_2, \ldots, j_k$ respectively, and d_i is the rate at which an individual of type i dies.

The diagonal elements of D are strictly negative, and $|D_{ii}|$ is the parameter of the exponential distribution of the sojourn time of type i before an event occurs: a change of type, the birth of children, or the death of the individual. The matrices and vector satisfy $D\mathbf{1} + \sum_{k\geq 1}(B_k \mathbf{1}) + d = 0$, where $\mathbf{1}$ denotes a column vector of which all elements are equal to one, the size being clear by the context, and 0 is a vector of elements all equal to zero.

We assume that every individual eventually dies, which is expressed algebraically by the requirement that the matrix $D + \sum_{k\geq 1} B_k (I \otimes \mathbf{1}^{(k)})$ be the generator of a transient Markov process, where $\mathbf{1}^{(k)}$ stands for the kth-fold Kronecker product of the m-vector $\mathbf{1}$ with itself: $\mathbf{1}^{(0)} = 1$, and $\mathbf{1}^{(k)} = \mathbf{1}^{(k-1)} \otimes \mathbf{1}$, for $k \geq 1$. We also require that every type be accessible from any type, which implies that the matrix

$$\Omega = D + \sum_{k\geq 1} B_k \sum_{i=0}^{k} (\mathbf{1}^{(i)} \otimes I \otimes \mathbf{1}^{(k-i)})$$

is irreducible.

The matrix Ω plays an important role [5]: in the absence of catastrophe, the branching process is supercritical, and extinction occurs with a probability strictly less than one if and only if the eigenvalue λ of maximal real part of Ω is strictly positive. Furthermore, $\exp(\Omega t)$ is the matrix of expected population size at time t: its (i,j)th component is the expected number of individuals of type j alive at time t, given that the population initially consists of one individual of type i.

We superimpose on this a process $\{\tau_n : n \in \mathbb{Z}\}$ of *catastrophe* epochs. All the individuals alive at a time of catastrophe are subject to an event with two outcomes – to die or to survive – and we assume that the probability that an individual of type i survives is $\delta_i > 0$, independently of the fate of the other individuals, of the time of catastrophe, and of the effect of previous events. The intervals between catastrophes are denoted by $\{\xi_n : n \in \mathbb{Z}\}$ with $\xi_n = \tau_n - \tau_{n-1}$ and are assumed to form an ergodic stationary process with finite mean.

We denote by $\{Z_n, n \geq 0\}$ the population process embedded at the epochs of catastrophes: Z_n is a vector of size m, and its component $Z_{n,i}$ is the number of individuals of type i that are alive at time τ_n, immediately after the catastrophe. We define $\Delta_\delta = \mathrm{diag}(\delta)$, and we note that $(\mathrm{e}^{\Omega\xi}\Delta_\delta)_{ij}$ is the matrix of the expected number of survivors of type j after a catastrophe that occurs ξ units of time after the beginning of the process, given that there was one individual of type i alive at time zero.

We define the *conditional* extinction probability given the successive epochs of catastrophes,

$$q_i(\xi) = \mathrm{P}[\lim_{n\to\infty} Z_n = 0 | Z_0 = e_i, \xi],$$

where $\xi = (\xi_1, \xi_2, \ldots)$ and e_i is a vector with all components equal to zero, except for the ith one, which is equal to one. The two theorems below give criteria to determine if the process is supercritical or not.

If $m = 1$, then all individuals have the same type, and the following property immediately follows from Tanny [23, Theorem 5.5, Corollary 6.3].

Theorem 5.1. *If $m = 1$, then $\Omega = D + \sum_{k\geq 1}(k+1)B_k$ and $\Delta_\delta = \delta$ are both scalars and*

(i) If $\mathrm{e}^{\Omega \mathrm{E}[\xi]}\delta \leq 1$, then $\mathrm{P}[q(\xi) = 1] = 1$;
(ii) If $\mathrm{e}^{\Omega \mathrm{E}[\xi]}\delta > 1$, then

$$\lim_{n\to\infty} 1/n \log(Z_n) = \Omega\mathrm{E}[\xi] + \log(\delta) \qquad a.s.$$

and $\mathrm{P}[q(\xi) < 1] = 1$,

where ξ is a random variable with the common distribution of the ξ_n. □

The dichotomy is very simple: one applies a catastrophe after an interval of time of expected duration $\mathrm{E}[\xi]$; if the expected number of survivors is at most one, then extinction occurs a.s.; otherwise the branching process a.s. has a strictly positive probability of surviving forever.

If $m > 1$, then Theorem 5.2 follows from Tanny [24, Theorem 9.10]; the proof that the assumptions in [24] are satisfied is purely technical and not very enlightening, it is given in the appendix.

Theorem 5.2. *There exists a constant* ω *such that*

$$\omega = \lim_{n\to\infty} 1/n \log ||e^{\Omega\xi_1}\Delta_\delta \cdots e^{\Omega\xi_n}\Delta_\delta|| \qquad a.s., \tag{5.1}$$

independently of the matrix norm, and

(i) If $\omega \le 0$*, then* $\mathrm{P}[q(\xi) = 1] = 1$*;*
(ii) If $\omega > 0$*, then* $\mathrm{P}[q(\xi) < 1] = 1$ *and*

$$\mathrm{P}[\lim_{n\to\infty} 1/n \log ||Z_n|| = \omega | Z_0 = e_i, \xi] = 1 - q_i(\xi) \qquad a.s.,$$

for $1 \le i \le m$*.* □

We see that ω is a key quantity: extinction occurs with probability one if $\omega \le 0$, and with probability strictly less than one if $\omega > 1$.

The limit in (5.1) may take different forms, and one also has

$$\omega = \lim_{n\to\infty} 1/n \mathrm{E} \log ||e^{\Omega\xi_1}\Delta_\delta \cdots e^{\Omega\xi_n}\Delta_\delta|| \tag{5.2}$$

$$= \lim_{n\to\infty} 1/n \log (e^{\Omega\xi_1}\Delta_\delta \cdots e^{\Omega\xi_n}\Delta_\delta)_{ij} \qquad \text{a.s.} \tag{5.3}$$

$$= \lim_{n\to\infty} 1/n \mathrm{E} \log (e^{\Omega\xi_1}\Delta_\delta \cdots e^{\Omega\xi_n}\Delta_\delta)_{ij} \tag{5.4}$$

for all i and j, as shown in Athreya and Karlin [4] and Kingman [16].

In attempting to determine ω, the situation is complicated by the fact that a catastrophe may have a stronger or weaker effect, depending on the value of ξ, because the survival probability may depend on the type, and the mix of population evolves over time.

The parameter ω may be likened to the maximal Lyapunov exponent of the set of matrices $\{e^{\Omega x}\Delta_\delta : x \ge 0\}$. Given a set $\mathcal{A}$ of real matrices A_i and a probability distribution P on $\mathcal{A}$, the maximal Lyapunov exponent ρ for $\mathcal{A}$ and P is defined to be

$$\rho(\mathcal{A}, P) = \lim_{n\to\infty} 1/n \mathrm{E} \log ||A_1 \ldots A_n||, \tag{5.5}$$

where A_n, $n \ge 1$, are independent and identically distributed random matrices on $\mathcal{A}$ with the distribution P. The limit exists and does not depend on the choice of matrix norm [10, 20]. Lyapunov exponents are hard to compute [16, 25], except under special circumstances, such as in Key [14], where the matrices in the family are assumed to be simultaneously diagonalizable, or in Lima and Rahibe [17], where $\mathcal{A}$ contains two matrices of order 2 only, one of which is singular. For a thorough survey on the basics of Lyapunov exponents, we refer the reader to Watkins [26].

Here, the exponential matrices obviously share a strong common structure, but the factor Δ_δ creates enough of a disturbance that we did not find a simple expression for ω. We focus therefore our attention on finding an upper bound ω_u and a lower bound ω_ℓ for ω.

For nonnegative matrices, under the assumption that $\mathcal{A}$ is a finite set, Gharavi and Anantharam [11] give an upper bound for the maximal Lyapunov exponent in the form of the maximum of a nonlinear concave function. Key [15] gives both upper and lower bounds determined as follows, on the basis of (5.2) and (5.4). Define $\sigma_n = 1/n\mathrm{E}\log||\mathrm{e}^{\Omega\xi_1}\Delta_\delta\cdots\mathrm{e}^{\Omega\xi_n}\Delta_\delta||$ and $\sigma_n^* = 1/n\mathrm{E}\log(\mathrm{e}^{\Omega\xi_1}\Delta_\delta\cdots\mathrm{e}^{\Omega\xi_n}\Delta_\delta)_{jj}$ for some arbitrarily fixed j. One verifies that $\{\sigma_{2^k}\}$ is nonincreasing and that $\{\sigma_{2^k}^*\}$ is nondecreasing, so that these form sequences of upper and lower bounds for ω; they are, unfortunately, not much easier to compute than the Lyapunov exponent itself.

A Duality Approach

Before looking for bounds, however, we determine an explicit expression for ω in Theorem 5.1, under the added assumption that the survival probabilities are independent of the type.

Since we assume that Ω is irreducible, we know that $\mathrm{e}^{\Omega t} > 0$ for all $t > 0$, which implies that $\mathrm{e}^{\Omega t} = \mathrm{e}^{\lambda t}C + o(\mathrm{e}^{\lambda t})$, where λ is the eigenvalue of maximal real part of Ω and C is a finite matrix, independent of t. Furthermore, the process $\{\xi_n\}$ is ergodic, so that $\tau_n = \xi_1 + \cdots + \xi_n$ tends to infinity as n tends to infinity, except in the uninteresting case where the ξ_n are equal to zero a.s. These two observations combine to give the following property.

Theorem 5.1. *If $\{\xi_n\}$ is an ergodic stationary process with finite mean and if $\delta_i = \delta$ for all i, then*

$$\omega = \lambda\mathrm{E}[\xi] + \log\delta. \tag{5.6}$$

Proof. If $\delta_i = \delta$ for all i, then (5.1) becomes

$$\begin{aligned}\omega &= \lim_{n\to\infty} 1/n\log||\delta^n\mathrm{e}^{\Omega\xi_1}\ldots\mathrm{e}^{\Omega\xi_n}||\\ &= \log\delta + \lim_{n\to\infty} 1/n\log||\mathrm{e}^{\Omega(\xi_1+\cdots+\xi_n)}||\\ &= \log\delta + \lim_{n\to\infty} 1/n\log\mathrm{e}^{\lambda(\xi_1+\cdots+\xi_n)},\end{aligned}$$

since $\xi_1 + \cdots + \xi_n$ tends to infinity a.s.,

$$\begin{aligned}&= \log\delta + \lim_{n\to\infty}\lambda(\xi_1+\cdots+\xi_n)/n\\ &= \log\delta + \lambda\mathrm{E}[\xi],\end{aligned}$$

since the ξ_i form an ergodic sequence.

Observe the similarity with Theorem 5.1: the value of ω given by (5.6) is strictly positive if and only if $\mathrm{sp}(\mathrm{e}^{\Omega \mathrm{E}[\xi]}\delta) > 1$, where $\mathrm{sp}(\cdot)$ is the spectral radius, so that the conclusion is based in both cases on the expected numbers of survivors if a catastrophe occurs after an interval of expected length.

Define $\Omega^* = \Omega - \lambda I$ and rewrite ω as

$$\begin{aligned}\omega &= \lim_{n\to\infty} \left\{\lambda \tau_n/n + 1/n \log||\mathrm{e}^{\Omega^*\xi_1}\Delta_\delta \cdots \mathrm{e}^{\Omega^*\xi_n}\Delta_\delta||\right\} \\ &= \lambda \mathrm{E}[\xi] + \lim_{n\to\infty} \left\{1/n \log||\mathrm{e}^{\Omega^*\xi_1}\Delta_\delta \cdots \mathrm{e}^{\Omega^*\xi_n}\Delta_\delta||\right\}. \end{aligned} \tag{5.7}$$

The off-diagonal elements of Ω^* are nonnegative, the matrix has one eigenvalue equal to zero, and all others have a strictly negative real part. Therefore, $-\Omega^*$ is an irreducible M-matrix and its left and right eigenvectors u and v corresponding to the eigenvalue 0, normalized by $u^{\mathrm{T}}1 = 1$ and $u^{\mathrm{T}}v = 1$, are strictly positive. This is a consequence of the Perron–Frobenius theorem for nonnegative matrices; for details of the proof, see Fiedler and Plák [9, Theorem 5.6] and Berman and Plemmons [7, Theorem 4.16].

Now let us define

$$\Theta = \Delta_u^{-1}(\Omega^*)^{\mathrm{T}}\Delta_u, \tag{5.8}$$

where $\Delta_u = \mathrm{diag}(u)$. It is easy to verify that Θ is a generator: it has nonnegative off-diagonal and strictly negative diagonal elements, and the row sums are equal to zero. Furthermore, its stationary probability vector π is given by

$$\pi = \Delta_u v = \Delta_v u. \tag{5.9}$$

Lemma 5.1. *Denote by $\eta_n = \xi_{-n}$ the intervals between events in the time-reversed version of the catastrophe process. One has $\omega = \lambda \mathrm{E}[\xi] + \psi$, where*

$$\psi = \lim_{n\to\infty} 1/n \log(\mathrm{e}^{\Theta\eta_1}\Delta_\delta \cdots \mathrm{e}^{\Theta\eta_n}\Delta_\delta)_{ij} \qquad a.s. \tag{5.10}$$

for all i, j.

Proof. Using (5.3), (5.7), and (5.8) we write $\omega = \lambda \mathrm{E}[\xi] + \psi$, where

$$\begin{aligned}\psi &= \lim_{n\to\infty} 1/n \log(\mathrm{e}^{\Omega^*\xi_1}\Delta_\delta \cdots \mathrm{e}^{\Omega^*\xi_n}\Delta_\delta)_{ji} \\ &= \lim_{n\to\infty} 1/n \log(\Delta_\delta \Delta_u \mathrm{e}^{\Theta\xi_n}\Delta_\delta \cdots \mathrm{e}^{\Theta\xi_1}\Delta_u^{-1})_{ij} \\ &= \lim_{n\to\infty} 1/n \log(\mathrm{e}^{\Theta\xi_n}\Delta_\delta \cdots \mathrm{e}^{\Theta\xi_1}\Delta_\delta)_{ij}\end{aligned}$$

since the extra factors at the beginning and the end of the matrix product do not matter.

The process $\{\xi_n\}$ is stationary, so that the n-tuple $(\eta_1,\ldots,\eta_n)$, which is equal to $(\xi_{-1},\ldots,\xi_{-n})$, has the same distribution as $(\xi_n,\ldots,\xi_1)$, and we may write

$$\psi = \lim_{n\to\infty} 1/n \log(e^{\Theta\eta_1}\Delta_\delta \cdots e^{\Theta\eta_n}\Delta_\delta)_{ij} \tag{5.11}$$

in probability. Since $\{\xi_n\}$ is ergodic, so is $\{\eta_n\}$, on the same probability space, and we may once again apply Tanny [24, Theorem 9.10] to conclude that there exists a constant ψ^* such that

$$\psi^* = \lim_{n\to\infty} 1/n \log(e^{\Theta\eta_1}\Delta_\delta \cdots e^{\Theta\eta_n}\Delta_\delta)_{ij} \qquad \text{a.s.,}$$

which, together with (5.11), proves (5.10). The proof that the assumptions in [24] are satisfied follows analogous arguments to those given for Theorem 5.2 in the appendix.

In this manner, working with $\exp(\Theta\eta)$ instead of $\exp(\Omega\xi)$, we replace the collection of random matrices $e^{\Omega\xi}$ by the collection of random *stochastic* matrices $e^{\Theta\eta}$. The interpretation is that we follow one single particle that evolves according to the generator Θ, instead of the whole population of a branching process, and we denote by $\{\varphi_t\}$ the Markov process with generator Θ. This particle is subject to a process of accidents: it survives with probability δ_i if it is in state i at the time of the accident. We define $\theta_0 = 0$, $\theta_n = \theta_{n-1} + \eta_n$, for $n \geq 1$; that is, $\{\theta_n\}$ is the process of the successive epochs of accidents. We further define S as the first epoch when the particle does not survive an accident, and $\varphi_n = \varphi_{\theta_n}$. We obtain the following bounds.

Theorem 5.3. *If the epochs of catastrophes form an ergodic stationary process, then* $\omega_\ell \leq \omega \leq \omega_u$, *with*

$$\omega_u = \lambda E[\xi] + \lim_{n\to\infty} 1/n \log P[S > \theta_n, \varphi_n = j | \varphi_0 = i], \tag{5.12}$$

and

$$\omega_\ell = \lambda E[\xi] + \sum_{1\leq i\leq m} u_i v_i \log \delta_i, \tag{5.13}$$

where u and v are the eigenvectors of Ω *for the eigenvalue* λ.

Proof. Equation (5.10) may be written as

$$\psi = \lim_{n\to\infty} 1/n \log P[S > \theta_n, \varphi_n = j | \varphi_0 = i, \theta_1, \ldots, \theta_n],$$

so that

$$e^\psi = \lim_{n\to\infty} P[S > \theta_n, \varphi_n = j | \varphi_0 = i, \theta_1, \ldots, \theta_n]^{1/n}. \tag{5.14}$$

By the dominated convergence theorem, this becomes

$$e^\psi = \lim_{n\to\infty} E[P[S > \theta_n, \varphi_n = j | \varphi_0 = i, \theta_1, \ldots, \theta_n]^{1/n}], \tag{5.15}$$

where the expectation is with respect to $\theta_1, \ldots, \theta_n$,

$$\leq \lim_{n\to\infty} \mathrm{E}[\mathrm{P}[S > \theta_n, \varphi_n = j | \varphi_0 = i, \theta_1, \ldots, \theta_n]]^{1/n}$$

by Jensen's inequality (see Ross [21, Proposition 1.7.3]),

$$= \lim_{n\to\infty} \mathrm{P}[S > \theta_n, \varphi_n = j | \varphi_0 = i]^{1/n} \tag{5.16}$$

because $\mathrm{E}[Y] = \mathrm{E}[\mathrm{E}[Y|\mathcal{G}]]$ for any random variable Y and σ-algebra $\mathcal{G}$.

This shows that $\omega \leq \omega_{\mathrm{u}}$. In short, we obtain an upper bound by replacing the conditional probability in (5.14) by its expectation (5.15), which contains less information, and then replacing $\mathrm{E}[X^{1/n}]$ with its upper bound $\mathrm{E}[X]^{1/n}$.

To obtain a lower bound, we start from a conditional expectation given more information:

$$\begin{aligned}\mathrm{P}[S > \theta_n, \varphi_n &= j | \varphi_1, \ldots, \varphi_{n-1}, \eta_n] \\ &= \delta_{\varphi_1} \cdots \delta_{\varphi_{n-1}} (e^{\Theta \eta_n} \Delta_\delta)_{\varphi_{n-1}, j} \\ &= \delta_1^{n_1} \cdots \delta_m^{n_m} (e^{\Theta \eta_n} \Delta_\delta)_{\varphi_{n-1}, j},\end{aligned}$$

where n_i is the number of times that the type is i during the first $n-1$ accidents, $1 \leq i \leq m$. Clearly, the right-hand side of the preceding equation is conditionally independent of φ_0 and of the epochs of accidents, given $\varphi_1, \ldots, \varphi_{n-1}$, and we may write

$$\begin{aligned}&\lim_{n\to\infty} \mathrm{P}[S > \theta_n, \varphi_n = j | \varphi_0 = i, \varphi_1, \ldots, \varphi_{n-1}, \theta_1, \ldots, \theta_n]^{1/n} \\ &\quad = \lim_{n\to\infty} \delta_1^{n_1/n} \cdots \delta_m^{n_m/n} (e^{\Theta \eta_n} \Delta_\delta)_{\varphi_{n-1}, j}^{1/n} \\ &\quad = \delta_1^{\kappa_1} \cdots \delta_m^{\kappa_m},\end{aligned}$$

where κ is the stationary distribution of the Markov process embedded at epochs of accidents. Clearly, $\kappa = \pi$ since the process $\{\theta_n\}$ is independent of the Markovian tree itself.

We now take the expectation with respect to $\varphi_1, \ldots, \varphi_{n-1}$ and follow the same argument as before to obtain

$$\begin{aligned}&\delta_1^{\pi_1} \cdots \delta_m^{\pi_m} \\ &\quad = \lim_{n\to\infty} \mathrm{E}\left[\mathrm{P}[S > \theta_n, \varphi_n = j | \varphi_0 = i, \theta_1, \varphi_1, \ldots, \varphi_{n-1}, \theta_n]^{1/n}\right] \\ &\quad \leq \lim_{n\to\infty} \mathrm{E}\left[\mathrm{P}[S > \theta_n, \varphi_n = j | \varphi_0 = i, \theta_1, \varphi_1, \ldots, \varphi_{n-1}, \theta_n]\right]^{1/n} \\ &\quad = \lim_{n\to\infty} \mathrm{P}[S > \theta_n, \varphi_n = j | \varphi_0 = i, \theta_1, \ldots \theta_n]^{1/n},\end{aligned}$$

so that $e^{\psi} \geq \delta_1^{\pi_1} \cdots \delta_m^{\pi_m}$ by (5.14). Since π is given by (5.9), this concludes the proof.

The next property shows that the two bounds are tight: they are both equal to ω if the survival probability does not depend on the type.

Corollary 5.1. *If* $\delta_i = \delta$ *for all i, then* $\omega_\ell = \omega = \omega_u$.

Proof. It is obvious that $\omega_\ell = \omega$ since $u^T v = 1$, and we only need to focus on proving that $\omega = \omega_u$.

Conditionally given $[\theta_n = u]$,

$$\begin{aligned} P[S > \theta_n, \varphi_n = j | \varphi_0 = i, \theta_n = u] &= \delta^n P[\varphi_u = j | \varphi_0 = i] \\ &= \delta^n (e^{\Theta u})_{ij}. \end{aligned}$$

Thus,

$$\begin{aligned} &\lim_{n\to\infty} 1/n \log P[S > \theta_n, \varphi_n = j | \varphi_0 = i] \\ &\quad = \log\delta + \lim_{n\to\infty} 1/n \log P[\varphi_{\theta_n} = j | \varphi_0 = i]. \end{aligned} \tag{5.17}$$

Since Θ is irreducible, $P[\varphi_{\theta_n} = j | \varphi_0 = i]$ is bounded away from zero for n sufficiently large and the limit on the right-hand side of (5.17) is zero, which proves the claim.

As we show through a few examples in the section "Numerical Illustration," ω_u is closer than ω_ℓ to ω. A clear indication that this should hold is that $\omega_u = \omega \ge \omega_\ell$ if the intervals ξ_n between catastrophes have a constant length equal to $1/\beta$: in that case, it is obvious that $\omega = \log \operatorname{sp}(e^{1/\beta\Omega}\Delta_\delta)$, and we obtain from (5.12) that

$$\begin{aligned} \omega_u &= \lambda/\beta + \lim_{n\to\infty} 1/n \log[(e^{1/\beta\Theta}\Delta_\delta)^n]_{ij} \\ &= \lambda/\beta + \log \operatorname{sp}(e^{1/\beta\Theta}\Delta_\delta) \\ &= \lambda/\beta + \log \operatorname{sp}(e^{1/\beta\Omega^*}\Delta_\delta) \\ &= \omega. \end{aligned}$$

We obtain from (5.13) that

$$\omega_\ell \le \lambda/\beta + \log\left(\sum_{1\le i\le m} u_i v_i \delta_i\right) = \lambda/\beta + \log u^T \Delta_\delta v;$$

we recognize that $u^T \Delta_\delta v$ is equal to the limit of the spectral radius of $e^{1/\beta\Omega^*}\Delta_\delta$ as $1/\beta$ goes to infinity. For finite values of β, however, $\omega_\ell \ne \omega$, and thus $\omega_\ell < \omega_u$, unless the spectral radius of $e^{1/\beta\Omega}\Delta_\delta$ is independent of β, which is not true in general.

Markovian Catastrophes

Theorem 5.3 and its corollary hold under very general conditions. Here, we are more specific and we assume that the process of catastrophes is Markovian. The simplest case is when $\{\tau_n\}$ is a Poisson process with rate $\beta = 1/\mathrm{E}[\xi]$.

Define P as the transition matrix for the process $\{\varphi_t\}$ embedded at these epochs:

$$P = \int_0^\infty \mathrm{e}^{\Theta t} \beta \mathrm{e}^{-\beta t}\,\mathrm{d}t = \beta(\beta I - \Theta)^{-1}. \tag{5.18}$$

The transition matrix for the dual process immediately after accidents is $K = P\Delta_\delta$, and the left-hand side in (5.16) is $\lim_{n\to\infty}((P\Delta_\delta)^n_{ij})^{1/n}$, equal to $\mathrm{sp}(P\Delta_\delta)$. Thus,

$$\omega_\mathrm{u} = \lambda \mathrm{E}[\xi] + \log \mathrm{sp}\{\beta(\beta I - \Theta)^{-1}\Delta_\delta\}, \tag{5.19}$$

which is easily computed.

Lemma 5.1. *If the catastrophes follow a Poisson process of rate β, then the upper bound ω_u is expressed as follows in terms of the original branching process:*

$$\omega_\mathrm{u} = \log \mathrm{sp}\,\mathrm{E}\left[\mathrm{e}^{\Omega\xi}\mathrm{e}^{-\lambda(\xi - \mathrm{E}[\xi])}\Delta_\delta\right]. \tag{5.20}$$

Proof. We readily see that

$$\begin{aligned}
\mathrm{sp}\left\{\beta(\beta I - \Theta)^{-1}\Delta_\delta\right\} &= \mathrm{sp}\left\{\beta\Delta_u^{-1}\left(\beta I - (\Omega^*)^\mathrm{T}\right)^{-1}\Delta_\delta\Delta_u\right\} \\
&= \mathrm{sp}\left\{\beta\Delta_\delta(\beta I - \Omega^*)^{-1}\right\} \\
&= \mathrm{sp}\left\{\beta(\beta I - \Omega^*)^{-1}\Delta_\delta\right\} \\
&= \mathrm{sp}\left\{\int_0^\infty \beta \mathrm{e}^{-\beta t}\mathrm{e}^{\Omega t - \lambda t}\Delta_\delta\,\mathrm{d}t\right\}.
\end{aligned}$$

From this we find that

$$\omega_\mathrm{u} = \log \mathrm{sp}\left\{\int_0^\infty \beta \mathrm{e}^{-\beta t}\mathrm{e}^{\Omega t}\mathrm{e}^{-\lambda(t - \mathrm{E}[\xi])}\Delta_\delta\,\mathrm{d}t\right\},$$

and the lemma is proved.

Remark 5.1. The scalar factor $\mathrm{e}^{-\lambda(\xi - \mathrm{E}[\xi])}$ is a random variable with mean equal to $\beta\mathrm{e}^{\lambda/\beta}/(\lambda+\beta) > 1$, and it is tempting to speculate that its presence is the reason why ω_u is an *upper* bound, while ω itself should be equal to $\log \mathrm{sp}\mathrm{E}[\mathrm{e}^{\Omega\xi}\Delta_\delta]$. This, however, is not true, as we show in the case where all δ_is are equal: then,

$$\begin{aligned}\log \mathrm{sp}\mathrm{E}[\mathrm{e}^{\Omega\xi}\Delta_\delta] &= \log \mathrm{sp}(\beta(\beta I-\Omega)^{-1}\Delta_\delta)\\ &= \log \mathrm{sp}(I-1/\beta\Omega)^{-1}+\log\delta\\ &= \log(1+\lambda\mathrm{E}[\xi])+\log\delta,\end{aligned}$$

which is different from ω by (5.6).

Next, we assume that the epochs of catastrophe form a Markovian arrival process (MAP). These are processes in a random environment $\{\phi(t)\}$ that controls the counter $\{M(t)\}$ of the number of catastrophes in $(0,t)$ and are very versatile [2, 19]. A MAP is characterized by two transition-rate matrices: A_0 gives the phase transition rates without catastrophe and A_1 gives the rates at which catastrophes occur. We choose the distribution α of the phase at time 0 to be the stationary probability vector of $(-A_0)^{-1}A_1$, the phase transition matrix at epochs of catastrophes.

The time-reversed version of the catastrophe process is a MAP with transition matrices $\tilde{A}_i=\Delta_\varepsilon^{-1}A_i^{\mathrm{T}}\Delta_\varepsilon$, $i=0$ and 1, where $\Delta_\varepsilon=\mathrm{diag}(\varepsilon)$ and ε, proportional to $\alpha(-A_0)^{-1}$, is the stationary probability vector of A_0+A_1 [1, 3]. The initial phase for the time-reversed process at a time of catastrophe has the stationary distribution $\tilde{\alpha}$ of $(-\tilde{A}_0)^{-1}\tilde{A}_1$; it is proportional to $\varepsilon\,\mathrm{diag}(A_1\mathbf{1})$.

To thoroughly characterize the time S when the dual process terminates, we need to keep track of the two-dimensional Markov process $\{\varphi_t,\chi_t\}$, where φ_t is the phase of the dual process at time t and χ_t is the phase of the time-reversed MAP process of accidents. Its infinitesimal generator is $Q=Q_0+Q_1$, where $Q_0=\Theta\otimes I+I\otimes\tilde{A}_0$ is the matrix of transition rates without accident and $Q_1=I\otimes\tilde{A}_1$ is the rate matrix for transitions with an accident.

The transition probability matrix at epochs of accidents is $\bar{P}=(-Q_0)^{-1}Q_1$, and the transition matrix immediately after an accident is

$$K=\bar{P}(\Delta_\delta\otimes I)=-(\Theta\otimes I+I\otimes\tilde{A}_0)^{-1}(\Delta_\delta\otimes\tilde{A}_1).$$

With this,

$$\mathrm{P}[S>\theta_n,\varphi_n=j|\varphi_0=i]=[(I\otimes\tilde{\alpha})K^n(I\otimes\mathbf{1})]_{ij},$$

so that

$$\omega_{\mathrm{u}}=\lambda\mathrm{E}[\xi]+\log \mathrm{sp}K. \tag{5.21}$$

It is now a simple matter to prove the following lemma.

Lemma 5.3. *If the catastrophes follow a MAP process with transition matrices A_0 and A_1, then*

$$\omega_{\mathrm{u}}=\lambda\mathrm{E}[\xi]+\log \mathrm{sp}[-(\Omega^*\otimes I+I\otimes A_0)^{-1}(\Delta_\delta\otimes A_1)]. \tag{5.22}$$

Proof. The algebraic argument goes as follows:

$$\begin{aligned}
\mathrm{sp}K &= \mathrm{sp}\left[-(\Theta\otimes I+I\otimes\tilde{A}_0)^{-1}(\Delta_\delta\otimes\tilde{A}_1)\right]\\
&= \mathrm{sp}\left[-(\Delta_u^{-1}(\Omega^*)^{\mathrm{T}}\Delta_u\otimes I+I\otimes\Delta_\varepsilon^{-1}A_0^{\mathrm{T}}\Delta_\varepsilon)^{-1}\right.\\
&\qquad \left.(\Delta_\delta\otimes\Delta_\varepsilon^{-1}A_1^{\mathrm{T}}\Delta_\varepsilon)\right]\\
&= \mathrm{sp}\left[-((\Omega^*)^{\mathrm{T}}\otimes I+I\otimes A_0^{\mathrm{T}})^{-1}(\Delta_\delta\otimes A_1^{\mathrm{T}})\right]\\
&= \mathrm{sp}\left[-(\Delta_\delta\otimes A_1)(\Omega^*\otimes I+I\otimes A_0)^{-1}\right],
\end{aligned}$$

which completes the proof.

We may also express ω_{u} in a manner similar to the right-hand side of (5.20) and write

$$\omega_{\mathrm{u}} = \log\mathrm{sp}\left\{\int_0^\infty \mathrm{e}^{(\Omega\otimes I+I\otimes A_0)t}\mathrm{e}^{-\lambda(t-\mathrm{E}[\xi])}(\Delta_\delta\otimes A_1)\,\mathrm{d}t\right\}.$$

Numerical Illustration

We performed some numerical experimentation to evaluate the quality of the two bounds. We used different examples with one birth at a time but report only two here, because although naturally details vary we reached qualitatively similar conclusions.

Right Whale Model

This first example is inspired from a model for North Atlantic right whales in 1980–1981 [8, page 323]. There are five stages: calf, immature, mature, reproducing female, and postbreeding female, numbered from 1 to 5 in that order. The time unit is 1 year, and the transition matrices are

$$D = \begin{bmatrix} -1 & 0.93 & \cdot & \cdot & \cdot \\ \cdot & -0.15 & 0.12 & \cdot & \cdot \\ \cdot & \cdot & -0.41 & \cdot & \cdot \\ \cdot & \cdot & \cdot & -1 & 0.97 \\ \cdot & \cdot & 0.99 & \cdot & -1 \end{bmatrix},$$

$$D' = \begin{bmatrix} \cdot & \cdot & \cdot & \cdot & \cdot \\ \cdot & \cdot & \cdot & \cdot & \cdot \\ \cdot & \cdot & \cdot & 0.40 & \cdot \\ \cdot & \cdot & \cdot & \cdot & \cdot \\ \cdot & \cdot & \cdot & \cdot & \cdot \end{bmatrix},$$

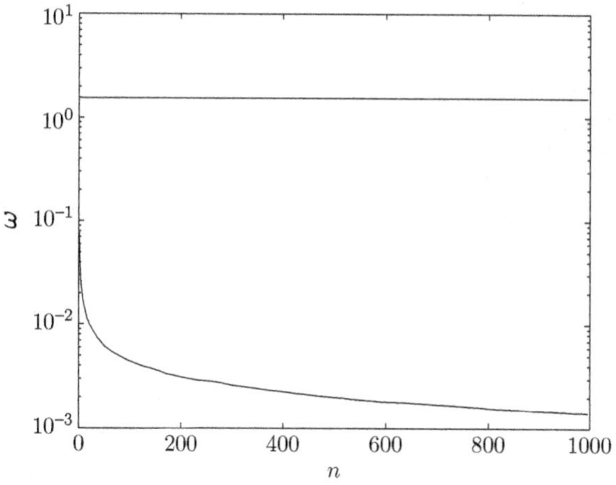

Fig. 5.1 Estimation of the mean of ω_n (*top line*) and its standard deviation (*bottom curve*) for $1 \le n \le 1{,}000$. Right whale example, with $\mathrm{E}[\xi] = 25$ years and $\delta = [0.2, 0.8, 0.2, 0.8, 0.8]$

$B_1 = D' \otimes \begin{bmatrix} 1 & 0 & \cdots & 0 \end{bmatrix}$ and $d = -D\mathbf{1} - B_1\mathbf{1}$. The expected lifetime of a newborn whale is $L = 58.6$ years, and its expected number of children is $C = 11.5$. Catastrophes occur according to a Poisson process, and the expected interval of time between catastrophes is $\mathrm{E}[\xi] = 25$.

Estimation

To compare ω to its bounds, we need to know its value or, more realistically, a sufficiently close approximation. This may be obtained by simulation, as we now explain. We have run simulations of $\xi_1, \xi_2, \ldots, \xi_{1{,}024}$ to obtain samples of $\omega_n = 1/n \log ||e^{\Omega\xi_1}\Delta_\delta \cdots e^{\Omega\xi_n}\Delta_\delta||_\infty$ for $n = 1, \ldots, 1{,}024$, and we have replicated the simulation 1,000 times to estimate the mean and standard deviation of ω_n. The results are shown on Fig. 5.1, where we plot the sample mean $\bar{\omega}_n$ and the sample standard deviation s_n over 1,000 replications.

The first observation is that the sample mean rapidly converges as n increases: we see on the graph that it is nearly constant, and the values reported below help confirm the visual impression.

n	1	10	100	1,000
$\bar{\omega}_n$	1.5502	1.5495	1.5491	1.5491

The standard deviation decreases steadily, as one would expect, given that ω_n converges to a constant. It is noteworthy that the decrease is relatively slow, after an initial sharp drop.

We plot in Fig. 5.2 the mean $\bar{\omega}_n$ (curve in the middle) and the 0.999 confidence interval $\bar{\omega}_n \pm 3.27 s_n/\sqrt{1{,}000}$, and we draw for visual reference a horizontal dashed line at the abscissa $\bar{\omega}_{1{,}000}$.

As an added indication, we estimate the bounds σ_n and σ_n^* defined in the introduction, for $n = 2^{10} = 1{,}024$, by taking sample means over the 1,000 simulations.

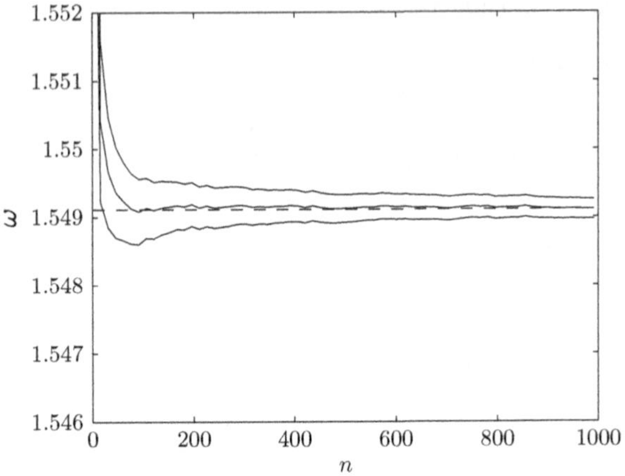

Fig. 5.2 Mean and confidence intervals for 1,000 trajectories and $1 \leq n \leq 1{,}000$. Right whale example, with $\mathrm{E}[\xi] = 25$ years and $\delta = [0.2, 0.8, 0.2, 0.8, 0.8]$

The obtained values are respectively equal to 1.54911 and 1.54836. The lower bound is outside the confidence interval for $n = 1{,}000$, which is $(1.5491 \pm 0.1433\ 10^{-3})$, but it is clear here that ω is determined with four significant digits.

As we are not interested in obtaining a highly precise estimate of ω, it is not necessary to calculate 1,000 independent replications of $\omega_{1,000}$ in all experiments, and we limit ourselves to a smaller number. For this example, ten replications yield the confidence interval (1.5510 ± 0.0016), which still gives us three significant digits. Similar conclusions were reached for other examples, and we shall henceforth denote by ω_{sim} the estimation obtained by the sample mean $\bar{\omega}_{1,000}$ of ten independent replications.

Observations

Since the difficulty in finding an explicit analytic expression for ω is due to the complex interplay between the dynamics of the multitype branching process and the differentiated effect of the catastrophes, we have randomly chosen several different δs in $(0,1)^m$ to have a good mix of cases.

The graph in Fig. 5.3 shows the evolution of the approximation ω_{sim} and of the two bounds as a function of the average survival probability, defined as $\pi\delta$. We also draw, for visual reference, the least-squares polynomial of order 4 fitted to the values of ω_{sim}.

Our first observation is that, as expected, ω is an increasing function of the survival probability and it converges to $\lambda\mathrm{E}[\xi]$ as δ approaches 1; note that $\lambda\mathrm{E}[\xi] = 2.19$ in this case.

The second observation is that ω_{u} is a much better bound than ω_ℓ: at this scale, there is no discernible difference between ω_{u} and ω_{sim}, while ω_ℓ is noticeably smaller. This is not surprising, given that ω_ℓ is independent of the process of catastrophe epochs and only depends on Ω (the population model) and on δ, the

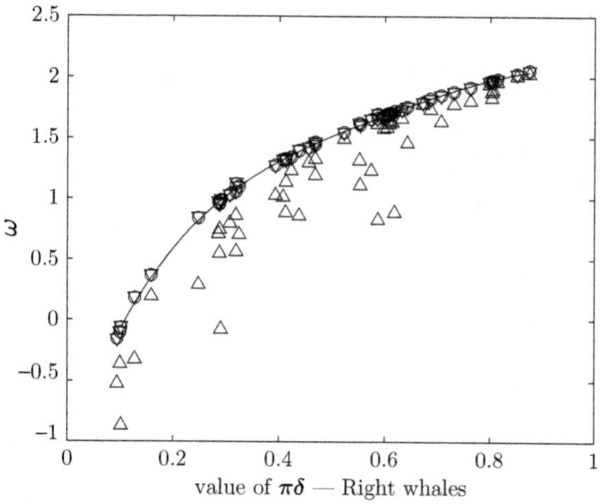

Fig. 5.3 Approximation and bounds against the average survival probability, right whale example. The *circle* symbol represents ω_{sim}, and the *inverted triangle* and the *triangle* represent ω_{u} and ω_{ℓ}, respectively

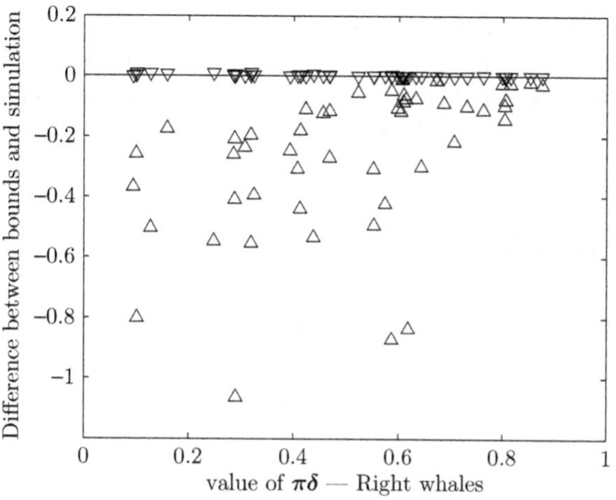

Fig. 5.4 Differences $\omega_{\text{u}} - \omega_{\text{sim}}$ and $\omega_{\ell} - \omega_{\text{sim}}$ as a function of the expected survival probability for the right whale example. The *inverted triangle* and the *triangle* symbols represent the differences $\omega_{\text{u}} - \omega_{\text{sim}}$ and $\omega_{\ell} - \omega_{\text{sim}}$

survival probabilities. More details may be seen in Fig. 5.4, where we show the *differences* between the bounds and the simulation approximation. This very clearly shows the good quality of ω_{u}, and we observe that ω_{ℓ} improves as the survival probabilities get closer to one.

We have also examined the effect of the homogeneity of the δ_is. The reason for our interest is that we saw with Corollary 5.1 that if the δ_i are all equal, then the two bounds are equal, and we expect that if the survival probabilities are not much different, then the bounds might be nearly equal. Our measure of nonhomogeneity is

$$d = \sum_{1 \le i \le m} |\delta_i - \bar{\delta}|,$$

where $\bar{\delta}$ is the arithmetic mean of the δ_i, and the possible values for d range from 0 to $\lfloor m/2 \rfloor$.

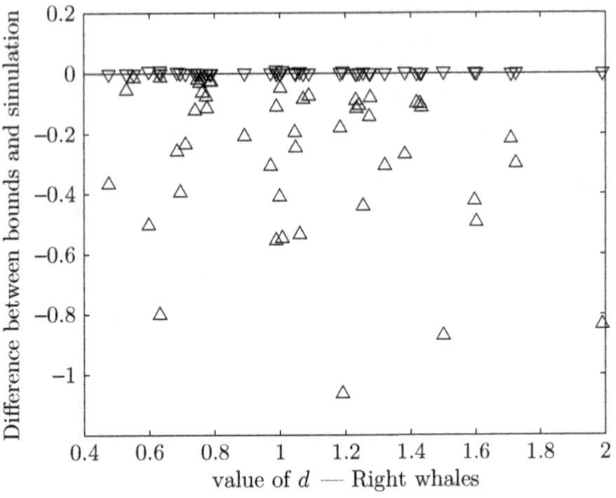

Fig. 5.5 Differences $\omega_u - \omega_{sim}$ and $\omega_\ell - \omega_{sim}$ as a function of the homogeneity measure d for the right whale model. The symbols *inverted triangle* and the *triangle* represent the differences $\omega_u - \omega_{sim}$ and $\omega_\ell - \omega_{sim}$

We show in Fig. 5.5 the same data as in Fig. 5.4, but the points are organized by increasing values of d. We observe that the precision of the bounds is generally better for smaller values of d, despite the fact that the pattern is made fuzzy due to the conflicting influence of the mean $\pi\delta$.

Insect Model

This is a model of insect reproduction: an insect does not reproduce until the end of its life, at which time it produces a geometrically distributed number of eggs; meanwhile, there are ample opportunities for it to die.

Here the parameters are

$$D = \begin{bmatrix} -\alpha_2 & \gamma_0 & & & 0 \\ & -\alpha_2 & \gamma_0 & & \\ & & \ddots & \ddots & \\ & & & -\alpha_2 & \gamma_0 \\ 0 & & & & -\alpha_1 \end{bmatrix},$$

$$D' = \begin{bmatrix} 0 & & & & \\ & 0 & & & \\ & & \ddots & & \\ & & & 0 & \\ & & & & \gamma_1 \end{bmatrix},$$

$d = \begin{bmatrix} \gamma_2 \; \gamma_2 \; \cdots \; \gamma_2 \; \gamma_0 \end{bmatrix}^{\mathrm{T}}$, and $B_1 = D' \otimes \begin{bmatrix} 1 \; 0 \; \cdots \; 0 \end{bmatrix}$, where $\alpha_1 = \gamma_0 + \gamma_1$ and $\alpha_2 = \gamma_0 + \gamma_2$.

The system is characterized by the number m of stages, the expected life $L = m/\gamma_0$, the expected number $E = \gamma_1/\gamma_0$ of eggs laid at the last stage, if the insects survive until then, and the number $C = (\gamma_0/(\gamma_0+\gamma_2))^{m-1}E$ of eggs that would eventually be laid given that the insect is still in its first stage. We have fixed $m = 5, L = 12, E = 100, C = 2$.

We evaluate the upper and lower bounds in cases where catastrophes occur according to one of three MAPs, all with the same expected value $\mathrm{E}[\xi] = 12$ for the intervals between events.

The first MAP is a renewal process with Erlang distributed intervals between renewals, with order 6 and parameter equal to 0.5. The second is a bursty process for which catastrophes are 20 times more frequent when the process is in its first phase. The transition matrices are

$$A_0 = \begin{bmatrix} -21a & a & & & \\ & -2a & a & & \\ & & \ddots & \ddots & \\ & & & & a \\ a & & & & -2a \end{bmatrix},$$

$$A_1 = \begin{bmatrix} 20a & & & & 0 \\ & a & & & \\ & & \ddots & & \\ & & & a & \\ 0 & & & & a \end{bmatrix}$$

of order 6 with $a = 0.02$.

The third MAP is a seesaw process with

$$A_0 = \begin{bmatrix} -2b & b & & & & & \\ & -4b & b & & & & \\ & & -4b & b & & & \\ & & & -8b & b & & \\ & & & & -4b & b & \\ & & & & & -4b & b \\ b & & & & & & -2b \end{bmatrix},$$

$$A_1 = \begin{bmatrix} b & & & & & & 0 \\ & 3b & & & & & \\ & & 3b & & & & \\ & & & 7b & & & \\ & & & & 3b & & \\ & & & & & 3b & \\ 0 & & & & & & b \end{bmatrix},$$

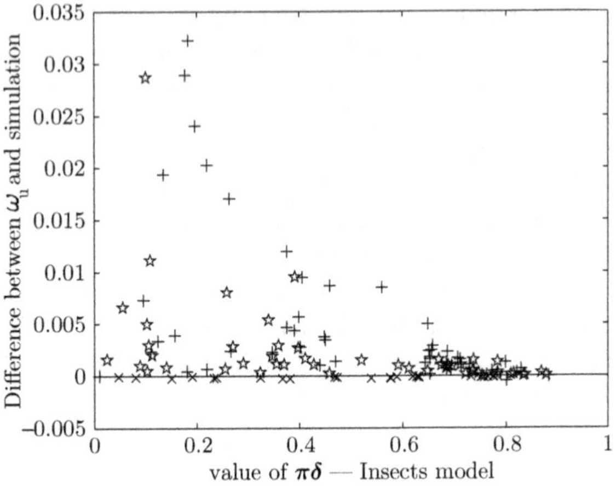

Fig. 5.6 Comparison of ω_{u} for different catastrophes models, as a function of the expected survival probability, for the Insects model. The *times symbol* represent the Erlang renewal process, and the *plus symbol* and the *star symbol* respectively represent the bursty and seesaw MAP processes

and $b = 1/36$. Here, the rate of catastrophes increases and decreases in a periodic fashion.

We give in the table below the correlation coefficient ρ of two successive intervals and the coefficient of variation C.V. $= \sqrt{\mathrm{Var}\,\xi}/\mathrm{E}[\xi]$ for the different models.

MAP	Erlang	Bursty	See-saw
C.V.	0.408	2.20	1.29
ρ	0	0.292	0.128

We show in Fig. 5.6 the difference $\omega_{\mathrm{u}} - \omega_{\mathrm{sim}}$ for the basic model subject to the three MAPs of catastrophes. We immediately notice that ω_{u} can hardly be distinguished from the simulation results when the catastrophes follow a renewal process with Erlang distribution. This is expected since the intervals are very regular, and we saw that $\omega_{\mathrm{u}} = \omega$ if the ξ_ns are constant. Conversely, the discrepancy between the upper bound and the simulation approximation is greatest when the catastrophe process is more irregular and the survival probabilities are small.

Computational Complexity

The numerical evaluation of the two bounds require a few matrix computations only: one needs to determine the spectral norm of Ω and its corresponding eigenvector u, to compute Θ, at a cost of $O(m^3)$ flops. Next, one must perform one matrix inversion and one spectral radius calculation to compute ω_{u} from (5.19) or (5.21); the computational complexity is $O(m^3)$ in the first case and $O(m^3 m'^3)$ in the second, where m' is the number of phases for the MAP.

This may be compared to the $O(NKm^3)$ complexity of obtaining an approximation by simulation, where N is the length of the simulation, K the number of replications, each exponential of a matrix costing 10 to 20 times m^3 flops. See Moler and Van Loan [18] and Golub and Van Loan [12] for further details.

Appendix

To prove Theorem 5.2, we need to verify the three technical assumptions of [24, Theorem 9.10].

The first one is that $\mathrm{E}[\max(0, \log ||M||)] < \infty$, where $M = \mathrm{e}^{\Omega\xi}\Delta_\delta$. Since the statement of the theorem is norm independent, we may choose the ∞-norm without loss of generality and write that

$$\log\delta^{(1)} + \log\max_i(\mathrm{e}^{\Omega\xi}\mathbf{1})_i \le \log||M|| \le \log\delta^{(2)} + \log\max_i(\mathrm{e}^{\Omega\xi}\mathbf{1})_i,$$

where $\delta^{(1)}$ and $\delta^{(2)}$ are respectively the minimum and the maximum among the δ_i. Thus, $\mathrm{E}[\max(0, \log ||M||)] < \infty$ if $\mathrm{E}[\log\max_i(\mathrm{e}^{\Omega\xi}\mathbf{1})_i] < \infty$.

Now, if the eigenvalue λ of maximal real part of Ω is negative, then $\mathrm{e}^{\Omega\xi}$ is bounded and the property holds. If λ is positive, then $\mathrm{e}^{\Omega\xi} = O(\mathrm{e}^{\lambda\xi})$, so that $\log\max_i(\mathrm{e}^{\Omega\xi}\mathbf{1})_i = O(\lambda\xi)$, and the property holds as well.

The second condition is that there exists k such that

$$\mathrm{P}\left[\min_{ij}(\mathrm{e}^{\Omega\xi_1}\Delta_\delta\cdots\mathrm{e}^{\Omega\xi_k}\Delta_\delta)_{ij} > 0\right] = 1.$$

Since we assume that $\delta > 0$, this holds for all k and, in particular, for $k = 1$.

The third condition is that there exists i such that

$$\mathrm{E}\big[|\log(1 - \mathrm{P}[Z_{1i} = 0|Z_0 = e_i, \xi])|\big] < \infty.$$

Assume that $Z_0 = e_i$. The event that there is no birth, death, or change of phase in the interval $(0, \xi_1)$ and that the unique individual survives the catastrophe at time ξ_1 implies that $Z_{1i} = 1$. Therefore,

$$1 - \mathrm{P}[Z_{1i} = 0|Z_0 = e_i, \xi] = \mathrm{P}[Z_{1i} \ge 1|Z_0 = e_i, \xi] \ge \mathrm{e}^{D_{ii}\xi_1}\delta_i$$

and

$$\mathrm{E}\big[|\log(1 - \mathrm{P}[Z_{1i} = 0|Z_0 = e_i, \xi])|\big] \le |D_{ii}|\mathrm{E}[\xi_1] + |\log\delta_i| < \infty.$$

Acknowledgements This work was subsidized by the ARC Grant AUWB-08/13–ULB 5 financed by the Ministère de la Communauté française de Belgique. The first author also gratefully acknowledges the support from the Fonds de la Recherche Scientifique (FRS-FNRS) and from the Australian Research Council, Grant No. DP110101663.

References

1. Andersen, A.T., Neuts, M.F., Nielsen, B.F.: On the time reversal of Markovian arrival processes. Stoch. Models **20**, 237–260 (2004)
2. Asmussen, S., Koole, G.: Marked point processes as limits of Markovian arrival streams. J. Appl. Probab. **30**, 365–372 (1993)
3. Asmussen, S., Ramaswami, V.: Probabilistic interpretations of some duality results for the matrix paradigms in queueing theory. Comm. Stat. Stoch. Models **6**, 715–733 (1990)
4. Athreya, K., Karlin, S.: On branching processes with random environments, i: Extinction probabilities. Ann. Math. Stat. **5**, 1499–1520 (1971)
5. Athreya, K.B., Ney, P.E.: Branching Processes. Springer, New York (1972)
6. Bean, N.G., Kontoleon, N., Taylor, P.G.: Markovian trees: properties and algorithms. Ann. Oper. Res. **160**, 31–50 (2008)
7. Berman, A., Plemmons, R.J.: In: Nonnegative Matrices in the Mathematical Sciences. Classics in Applied Mathematics. SIAM, Philadelphia (1994)
8. Caswell, H.: Applications of Markov chains in demography. In: Langville, A.N., Stewart, W.J. (eds.) MAM 2006: Markov Anniversary Meeting, pp. 319–334. Boson Press, Raleigh (2006)
9. Fiedler, M., Plák, V.: On matrices with non-positive off-diagonal elements and positive principal minor. Czech. Math. J. **12**(3), 382–400 (1962)
10. Furstenberg, H., Kesten, H.: Products of random matrices. Ann. Math. Stat. **31**, 457–469 (1960)
11. Gharavi, R., Anantharam, V.: An upper bound for the largest Lyapunov exponent of a Markovian random matrix product of nonnegative matrices. Theor. Comp. Sci. **332**, 543–557 (2005)
12. Golub, G.H., Van Loan, C.F.: Matrix Computations, 3rd edn. Johns Hopkins University Press, Baltimore (1996)
13. Harris, T.: The Theory of Branching Processes. Dover, New York (1963)
14. Key, E.: Computable examples of the maximal Lyapunov exponent. Probab. Theor. Relat. Fields **75**, 97–107 (1987)
15. Key, E.: Lower bounds for the maximal Lyapunov exponent. J. Theor. Probab. **3**(3), 477–488 (1987)
16. Kingman, J.F.C.: Subadditive ergodic theory. Ann. Probab. **1**, 883–909 (1973)
17. Lima, R., Rahibe, M.: Exact Lyapunov exponent for infinite products of random matrices. J. Phys. A Math. Gen. **27**, 3427–3437 (1994)
18. Moler, C., Van Loan, C.: Nineteen dubious ways to compute the exponential of a matrix, twenty-five years later. SIAM Rev. **45**, 3–49 (2003)
19. Neuts, M.F.: A versatile Markovian point process. J. Appl. Probab. **16**, 764–779 (1979)
20. Oseledec, V.I.: A multiplicative ergodic theorem: Lyapunov characteristic numbers for dynamical systems. Trans. Moscow Math. Soc. **19**, 197–231 (1968)
21. Ross, S.M.: Stochastic Processes, 2nd edn. Wiley, New York (1996)
22. Smith, W., Wilkinson, W.: On branching processes in random environments. Ann. Math. Stat. **40**, 814–827 (1969)
23. Tanny, D.: Limit theorems for branching processes in a random environment. Ann. Probab. **5**, 100–116 (1977)
24. Tanny, D.: On multitype branching processes in a random environment. Adv. Appl. Probab. **13**, 464–497 (1981)
25. Tsitsiklis, J., Blondel, V.: The Lyapunov exponent and joint spectral radius of pairs of matrices are hard – when not impossible – to compute and to approximate. Math. Control Signals Syst. **10**, 31–40 (1997)
26. Watkins, J.C.: Limit theorems for products of random matrices. In: Cohen, J.E., Kesten, H., Newman, C.M. (eds.) Random Matrices and Their Applications. Contemporary Mathematics, vol. 50, pp. 5–22. American Mathematical Society, Providence (1986)

Chapter 6
Majorization and Extremal *PH* Distributions

Qi-Ming He, Hanqin Zhang, and Juan C. Vera

Introduction

Let T be an $m \times m$ invertible matrix with (1) negative diagonal elements, (2) nonnegative off-diagonal elements, and (3) nonpositive row sums, where m is a positive integer. Such a matrix T is called a *PH* generator. Let α be a substochastic vector of order m, i.e., $\alpha \geq 0$ and $\alpha\mathbf{e} \leq 1$, where $\mathbf{e}$ is the column vector of ones. Then (α, T) is called a *PH* representation of a phase-type (*PH*) random variable (distribution) X. In this chapter, we find bounds on the moments of X in terms of the elements of α and T and identify Coxian distributions to be the extremal *PH* distributions in certain subsets of *PH* distributions.

The set of *PH* distributions was introduced by Neuts [13]. Since the set of *PH* distributions is dense in the set of probability distributions on the nonnegative half-line and *PH* representations provide a Markovian structure for stochastic modeling, *PH* distributions have been used widely in the study of queueing, inventory, risk/insurance, manufacturing, and telecommunications models [9, 14]. In almost all applications of *PH* distributions, *PH* representations play a key role. Thus, the study of *PH* representations has attracted great attention from researchers (see [3, 4, 15, 17], and references therein).

Aldous and Shepp [1] find the minimum coefficient of variation of *PH* distributions with a *PH* representation of a fixed order m. They also find that the

Qi-M. He (✉)
University of Waterloo, Waterloo, ON, Canada N2L 3G1
e-mail: q7he@uwaterloo.ca

H. Zhang
University of Singapore, Singapore 119245, Singapore
e-mail: bizzhq@nus.edu.sg

J.C. Vera
Tilburg University, Tilburg, The Netherlands
e-mail: j.c.veralizcano@uvt.nl

G. Latouche et al. (eds.), *Matrix-Analytic Methods in Stochastic Models*, Springer Proceedings in Mathematics & Statistics 27, DOI 10.1007/978-1-4614-4909-6_6,

minimum is attained at *PH* representations of Erlang distributions. Their result is useful in determining the order of *PH* representations needed for fitting probability distributions if their coefficient of variation is known. In [17], a number of open problems related to *PH* representations are brought up and investigated. The results in [17] and in subsequent papers on the open problems (e.g., [5, 18, 19]) reveal the relationship between *PH* representations, density functions, and variances of *PH* distributions. In [17], a lower bound on the density of triangular *PH* distributions is found. In [5], it is shown that not every *PH* representation has an equivalent unicyclic *PH* representation of the same order. In [18], it is shown that, for a *PH* distribution with a *PH* representation of order 2, a minimal-norm representation can be found and the norm coincides with the minimal parameter in Maier's property [10]. While O'Cinneide [17] attempts to show *PH* distributions with a unicyclic *PH* representation as extremal *PH* distributions, this chapter aims to prove that *PH* distributions with some Coxian representations are extremal with respect to the moments of the distribution.

This chapter focuses on the relationship between *PH* representations and the moments of *PH* distributions. In the section "Two Majorization Lemmas," two majorization results are shown for the vector $-T^{-1}\mathbf{e}$. It is worth mentioning that the majorization approach [11] seems quite useful in the study of *PH* distributions and *PH* representations [7, 16]. The majorization results are used to obtain bounds on the mean (i.e., first moment) of *PH* distributions in the section "Bounds on Phase-Type Distributions." All bounds on the expectation are partially independent of the transition structure of the underlying Markov chain associated with the *PH* distribution. Results in the section "Bounds on Phase-Type Distributions" indicate that exponential/Coxian distributions are extreme cases, with respect to the mean, if the vector $-\mathbf{e}'T$ or the sum $-\mathbf{e}'T\mathbf{e}$ is fixed, where $\mathbf{e}'$ is the transpose of the vector $\mathbf{e}$. The section "Extremal Phase-Type Distributions" extends the results in the section "Bounds on Phase-Type Distributions" from the first moment to higher moments. A highlight of the results is the lower bounds on the moments of any *PH* distribution (α, T), i.e., $E[X_k] \geq k!/(-\mathbf{e}'T\mathbf{e})^k$ for all $k \geq 0$, that is independent of the order of the *PH* representation and the transitions within the underlying Markov chain. Results in the section "Extremal Phase-Type Distributions" demonstrate that exponential/Coxian distributions are extremal *PH* distributions with respect to all the moments and the Laplace–Stieltjes transform. All proofs are given in the section "Proofs." The section "Conclusion and Discussion" concludes the chapter with a discussion of the potential applications of the results obtained in this chapter.

Two Majorization Lemmas

For the vector $\mathbf{x} = (x_1, x_2, \ldots, x_m)$, rearrange the elements of $\mathbf{x}$ in ascending order and denote the elements by $x_{[1]} \leq x_{[2]} \leq \cdots \leq x_{[m]}$, where ([1], [2], …, [$m$]) is a permutation of (1, 2, …, m). A vector $\mathbf{x}$ is *weakly submajorized* by vector $\mathbf{y}$, denoted by $\mathbf{x} \prec_w \mathbf{y}$, if $x_{[m]} + x_{[m-1]} + \cdots + x_{[k]} \le y_{[m]} + y_{[m-1]} + \cdots + y_{[k]}$ for $1 \le k \le m$. A vector $\mathbf{x}$ is *weakly supermajorized* by vector $\mathbf{y}$, denoted by $\mathbf{x} \prec^w \mathbf{y}$, if $x_{[1]} + x_{[2]} + \cdots + x_{[k]} \ge y_{[1]} + y_{[2]} + \cdots + y_{[k]}$ for $1 \le k \le m$. A vector $\mathbf{x}$ is *majorized* by $\mathbf{y}$, denoted as $\mathbf{x} \prec \mathbf{y}$, if $\mathbf{xe} = \mathbf{ye}$ and $x_{[1]} + x_{[2]} + \cdots + x_{[k]} \ge y_{[1]} + y_{[2]} + \cdots + y_{[k]}$ for $1 \le k \le m-1$, or, equivalently, $\mathbf{xe} = \mathbf{ye}$, and $x_{[m]} + x_{[m-1]} + \cdots + x_{[k]} \le y_{[m]} + y_{[m-1]} + \cdots + y_{[k]}$ for $2 \le k \le m$. It is easy to see that $\mathbf{x} \prec \mathbf{y}$ if and only if $\mathbf{x} \prec_w \mathbf{y}$ and $\mathbf{x} \prec^w \mathbf{y}$. We refer the reader to Marshall and Olkin [11] for more about majorization.

Consider a *PH* generator T of order m. Define $\mathbf{r} = -\mathbf{e}'T = (r_1, r_2, \ldots, r_m)$. Rearrange the elements of $\mathbf{r}$ in ascending order as $r_{[1]} \le r_{[2]} \le \cdots \le r_{[m]}$. Since T is invertible and $T\mathbf{e} \le 0$, we must have $-\mathbf{e}'T\mathbf{e} = \mathbf{re} > 0$. It is possible that some of $\{r_1, r_2, \ldots, r_m\}$ are negative, but the summation $r_{[j]} + r_{[j+1]} + \cdots + r_{[m]}$ is positive for $1 \le j \le m$. For fixed $\mathbf{r}$, we shall construct two matrices $T_\downarrow^*$ and $T_\uparrow^*$ and find majorization relationships between the vectors $-T^{-1}\mathbf{e}$, $-(T_\downarrow^*)^{-1}\mathbf{e}$, and $-(T_\uparrow^*)^{-1}\mathbf{e}$. Define

$$T_\downarrow^* = \begin{pmatrix} -\sum_{j=1}^{m} r_{[j]} & & & & \\ \sum_{j=2}^{m} r_{[j]} & -\sum_{j=2}^{m} r_{[j]} & & & \\ & \ddots & \ddots & & \\ & & \sum_{j=m-1}^{m} r_{[j]} & -\sum_{j=m-1}^{m} r_{[j]} & \\ & & & r_{[m]} & -r_{[m]} \end{pmatrix}. \tag{6.1}$$

It is easy to see that the matrix $T_\downarrow^*$ is a *PH* generator. In fact, $T_\downarrow^*$ is a Coxian generator for Coxian distributions [6]. Define

$$\begin{aligned} \mathbf{b}_\downarrow^* &= -(T_\downarrow^*)^{-1}\mathbf{e} \\ &= \left(\sum_{i=m}^{m} \left(\sum_{j=m-i+1}^{m} r_{[j]} \right)^{-1}, \ldots, \sum_{i=k}^{m} \left(\sum_{j=m-i+1}^{m} r_{[j]} \right)^{-1}, \ldots, \sum_{i=1}^{m} \left(\sum_{j=m-i+1}^{m} r_{[j]} \right)^{-1} \right)'. \end{aligned} \tag{6.2}$$

It is readily seen that the elements in $\mathbf{b}_\downarrow^*$ are positive and are in ascending order.

Lemma 6.1. *Assume that T is a PH generator of order m. Then $-T^{-1}\mathbf{e}$ is weakly submajorized by $\mathbf{b}_\downarrow^*$ defined in (6.2).*

Next, we define $T_\uparrow^*$ such that $-T^{-1}\mathbf{e}$ is weakly submajorized by $-(T_\uparrow^*)^{-1}\mathbf{e}$ under an additional condition. If $r_{[1]} = \min\{r_1, r_2, \ldots, r_m\} > 0$, then we define

$$T_{\uparrow}^{*} = \begin{pmatrix} -\sum_{j=1}^{m} r_{[j]} & & & & & \\ \sum_{j=1}^{m-1} r_{[j]} & -\sum_{j=1}^{m-1} r_{[j]} & & & & \\ & \ddots & \ddots & & & \\ & & \sum_{j=1}^{2} r_{[j]} & -\sum_{j=1}^{2} r_{[j]} & & \\ & & & r_{[1]} & -r_{[1]} \end{pmatrix}. \tag{6.3}$$

It is easy to see that the matrix $T_{\uparrow}^{*}$ is a *PH* generator. Define

$$\begin{aligned} \mathbf{b}_{\uparrow}^{*} &= -(T_{\uparrow}^{*})^{-1}\mathbf{e} \\ &= \left(\sum_{i=1}^{1} \left(\sum_{j=1}^{m-i+1} r_{[j]} \right)^{-1}, \ldots, \sum_{i=1}^{k} \left(\sum_{j=1}^{m-i+1} r_{[j]} \right)^{-1}, \ldots, \sum_{i=1}^{m} \left(\sum_{j=1}^{m-i+1} r_{[j]} \right)^{-1} \right)'. \end{aligned} \tag{6.4}$$

It is readily seen that the elements in $\mathbf{b}_{\uparrow}^{*}$ are nonnegative and are in ascending order.

Lemma 6.2. *Assume that T is a PH generator of order m, and $r_{[1]} > 0$. Then $-T^{-1}\mathbf{e}$ is weakly submajorized by $\mathbf{b}_{\uparrow}^{*}$ defined in (6.4).*

Bounds on Phase-Type Distributions

Now we focus on a random variable X with a *PH* distribution with *PH* representation (α, T). It is well known that the expectation of *PH* distribution X is given by $E[X] = -\alpha T^{-1}\mathbf{e}$. Since $-\alpha T^{-1}\mathbf{e} = -(\alpha\mathbf{e})(\alpha/(\alpha\mathbf{e}))T^{-1}\mathbf{e}$, without loss of generality, we shall assume α normalized such that $\alpha\mathbf{e} = 1$ in the rest of the paper.

For vector $\mathbf{x}$ let $\mathbf{x}_{\uparrow} = (x_{[1]}, x_{[2]}, \ldots, x_{[m]})$ denote the ascending rearrangement of $\mathbf{x}$, and let $\mathbf{x}_{\downarrow} = (x_{[m]}, x_{[m-1]}, \ldots, x_{[1]})$ denote the descending rearrangement of $\mathbf{x}$. For stochastic vector α, the vectors $\alpha_{\uparrow}$ and $\alpha_{\downarrow}$ are defined accordingly. For any vector $\mathbf{x}$ it is easy to verify $\alpha_{\downarrow}\mathbf{x}_{\uparrow} \le \alpha\mathbf{x} \le \alpha_{\uparrow}\mathbf{x}_{\uparrow}$ [12]. For vectors $\mathbf{x}$ and $\mathbf{y}$, (1) if $\mathbf{x} \prec_w \mathbf{y}$, then we have $\alpha_{\uparrow}\mathbf{x}_{\uparrow} \le \alpha_{\uparrow}\mathbf{y}_{\uparrow}$; (2) if $\mathbf{x} \prec^w \mathbf{y}$, then we have $\alpha_{\downarrow}\mathbf{x}_{\uparrow} \ge \alpha_{\downarrow}\mathbf{y}_{\uparrow}$; and (3) if $\mathbf{x} \prec \mathbf{y}$, then we have $\alpha_{\downarrow}\mathbf{x}_{\uparrow} \ge \alpha_{\downarrow}\mathbf{y}_{\uparrow}$, and $\alpha_{\uparrow}\mathbf{x}_{\uparrow} \le \alpha_{\uparrow}\mathbf{y}_{\uparrow}$ [11].

Now we are ready to state the main results.

Theorem 6.1. *Consider a PH generator T of order m. For any random variable X with a PH distribution with PH representation (α, T) we have*

$$E[X] \geq -\alpha_{\downarrow}(T_{\downarrow}^{*})^{-1}\mathbf{e} \geq -\frac{1}{\mathbf{e}'T\mathbf{e}}, \tag{6.5}$$

where $T_{\downarrow}^{*}$ is defined in (6.1). That is: the mean of the PH distribution (α, T) is greater than or equal to that of the PH distribution $(\alpha_{\downarrow}, T_{\downarrow}^{*})$.

Moreover, if all elements of $\mathbf{r} = \mathbf{e}'(-T)$ are positive, then we have

$$E[X] \leq -\alpha_{\uparrow}(T_{\uparrow}^{*})^{-1}\mathbf{e} \leq \sum_{i=1}^{m}\left(\sum_{j=1}^{m-i+1} r_{[j]}\right)^{-1}, \tag{6.6}$$

where $T_{\uparrow}^{*}$ is defined in (6.3). That is, the mean of the PH distribution (α, T) is less than or equal to that of the PH distribution $(\alpha_{\uparrow}, T_{\uparrow}^{*})$.

Note that the lower bound $-1/(\mathbf{e}'T\mathbf{e})$ in (6.5) is totally independent of the transition structure of the underlying Markov chain (i.e., the transition within T). The upper bound in (6.6) is only partially independent of the transition structure of the underlying Markov chain.

Example 6.1. Consider a *PH* generator

$$T = \begin{pmatrix} -10 & 8 \\ 2 & -2 \end{pmatrix}. \tag{6.7}$$

It is easy to find $\mathbf{e}'(-T) = (8, -6)$, $-T^{-1}\mathbf{e} = (2.5, 3)'$,

$$T_{\downarrow}^{*} = \begin{pmatrix} -2 & 0 \\ 8 & -8 \end{pmatrix}, \tag{6.8}$$

and $-(T_{\downarrow}^{*})^{-1}\mathbf{e} = (0.5, 0.625)'$. For any *PH* distribution (α, T) with $\alpha\mathbf{e} = 1$, by Theorem 6.1, we have $0.5 \leq 0.5\alpha_{[2]} + 0.625\alpha_{[1]} \leq E[X]$.

For this case, the lower bound is not sharp since $2.5 \leq E[X] \leq 3$ for all feasible α with $\alpha\mathbf{e} = 1$. Following He and Zhang [6], the *PH* generator T can be Coxianized, i.e., there is a Coxian generator

$$S = \begin{pmatrix} -6-\sqrt{32} & 0 \\ -6-\sqrt{32} & -6+\sqrt{32} \end{pmatrix} \tag{6.9}$$

such that any *PH* representation (α, T) has an equivalent Coxian representation (β, S), where β is a stochastic vector. The difference between $T_{\downarrow}^{*}$ and S explains why the lower bounds are too small for this case. This example warrants further investigation on the relationship between the matrices $T_{\downarrow}^{*}$ and $T_{\uparrow}^{*}$ and the Coxianization of T. On the other hand, finding bounds on the mean of a *PH* distribution is not the objective of this research. The results on bounds are used for characterizing *PH* distributions and for finding extremal *PH* distributions (see the section "Extremal Phase-Type Distributions").

Example 6.2. Consider a *PH* generator

$$T = \begin{pmatrix} -2 & 1 \\ x & -x \end{pmatrix}, \tag{6.10}$$

where $x > 0$. It is easy to verify $-T^{-1}\mathbf{e} = (1+1/x, 1+2/x)'$. The expectation of (α, T) with $\alpha\mathbf{e} = 1$ goes to positive infinity if x goes to zero. Note that $-\mathbf{e}'T\mathbf{e} = 1$ holds for any positive x. Thus, while there is a lower bound that is totally independent of the transition structure, there may not be such an upper bound.

For some special *PH* generators, lower and upper bounds can be obtained simultaneously.

Theorem 6.2. *Consider a PH generator T of order m and satisfying $-\mathbf{e}'T^{-1}\mathbf{e} = -\mathbf{e}'(T^*_\downarrow)^{-1}\mathbf{e}$. For any PH distributed random variable X with PH representation (α, T) we have*

$$-\frac{1}{\mathbf{e}'T\mathbf{e}} \le -\alpha_\downarrow(T^*_\downarrow)^{-1}\mathbf{e} \le E[X] \le -\alpha_\uparrow(T^*_\downarrow)^{-1}\mathbf{e} \le \sum_{i=1}^{m}\left(\sum_{j=m-i+1}^{m} r_{[j]}\right)^{-1}. \tag{6.11}$$

Consider a PH generator T such that (i) $-\mathbf{e}'T^{-1}\mathbf{e} = -\mathbf{e}(T^*_\uparrow)^{-1}\mathbf{e}$ and (ii) all elements of $\mathbf{r} = \mathbf{e}'(-T)$ are positive. For any PH distributed random variable X with PH representation (α, T) we have

$$-\frac{1}{\mathbf{e}'T\mathbf{e}} \le -\alpha_\downarrow(T^*_\uparrow)^{-1}\mathbf{e} \le E[X] \le -\alpha_\uparrow(T^*_\uparrow)^{-1}\mathbf{e} \le \sum_{i=1}^{m}\left(\sum_{j=1}^{m-i+1} r_{[j]}\right)^{-1}. \tag{6.12}$$

What follows is a special case of Theorem 6.2 that was proved in [7].

Corollary 6.1. *For any PH distribution (α, T) for which T satisfies $\mathbf{e}'T = -\mu\mathbf{e}'$ we have*

$$-\frac{1}{\mu m} \le -\alpha_\downarrow(T^*_\downarrow)^{-1}\mathbf{e} \le E[X] \le -\alpha_\uparrow(T^*_\downarrow)^{-1}\mathbf{e} \le \frac{1}{\mu}\sum_{i=1}^{m}\frac{1}{i}. \tag{6.13}$$

Example 6.3. Consider a *PH* generator

$$T = \begin{pmatrix} -3 & 1 \\ 2 & -2 \end{pmatrix}. \tag{6.14}$$

It is easy to find $\mathbf{e}'(-T) = (1,1)$, $-T^{-1}\mathbf{e} = (3/4, 5/4)'$,

$$T^*_\downarrow = \begin{pmatrix} -2 & 0 \\ 1 & -1 \end{pmatrix}, \tag{6.15}$$

and $-(T_{\downarrow}^{*})^{-1}\mathbf{e} = (0.5, 1.5)'$. For any *PH* distribution (α, T) with $\alpha\mathbf{e} = 1$, by Corollary 6.1, we have $0.5 \le 0.5\alpha_{[2]} + 1.5\alpha_{[1]} \le E[X] \le 0.5\alpha_{[1]} + 1.5\alpha_{[2]} \le 1.5$.

Extremal Phase-Type Distributions

Let $X_{\min}$ be the exponential random variable with parameter λ. Denote by Ω_λ the set of all *PH* distributions with a *PH* representation (α, T) satisfying $\alpha\mathbf{e} = 1$ and $\lambda = -\mathbf{e}'T\mathbf{e}$.

By Theorem 6.1, $E[X_{\min}] = \min\{E[X] : X \in \Omega_\lambda\}$, which implies that $X_{\min}$ is an extremal random variable, with respect to the first moment, in Ω_λ. Note that the result in Theorem 6.1 is independent of the order of the *PH* representation. The result can be generalized to all moments and Laplace-Stieltjes transforms (LSTs) of *PH* distributions.

Corollary 6.2. *For $\lambda > 0$ and $X \in \Omega_\lambda$ we have*

(i) $E[X^k] \ge E[X_{\min}^k] = \dfrac{k!}{(-\mathbf{e}'T\mathbf{e})^k}, \quad k \ge 1;$

(ii) $E[\mathrm{e}^{-sX_{\min}}] \le E[\mathrm{e}^{-sX}]$, $s_{\text{lower}} < s \le 0$, *for some negative number* s_{lower}*; and* $E[\mathrm{e}^{-sX_{\min}}] \ge E[\mathrm{e}^{-sX}]$, $0 \le s < s_{\text{upper}}$, *for some positive number* s_{upper}.

Corollary 6.2 indicates that $X_{\min}$ is an extremal distribution in Ω_λ with respect to the moments and the LST. Define nonnegative random variable $Y_{\min}$ by

$$P\{Y_{\min} \le t\} = \frac{m-1}{m} + \frac{1}{m}(1 - \exp\{-\theta t\}), \quad \text{for } t \ge 0, \tag{6.16}$$

where θ is positive. Then $Y_{\min}$ equals zero, w.p. $(m-1)/m$, and an exponential random variable with parameter θ, w.p. $1/m$. Define

$$\Psi_{m,\theta} = \left\{X : \ X \sim (\alpha, T) \ \text{of order } m, \alpha\mathbf{e} = 1, \ \theta = -\frac{\mathbf{e}'T\mathbf{e}}{m}\right\}, \tag{6.17}$$

where "$\sim$" means equivalency in probability distribution.

Corollary 6.3. *For $\theta > 0$ and $X \sim (\alpha, T) \in \Psi_{m,\theta}$ we have, for $s \ge 0$,*

$$E[\mathrm{e}^{-sX}] \le E[\mathrm{e}^{-sY_{\min}}] = \frac{m-1}{m} + \frac{\theta}{m(s+\theta)}. \tag{6.18}$$

We remark that, while the extremal random variable $X_{\min}$ is in Ω_λ, $Y_{\min}$ is not in $\Psi_{m,\theta}$. Yet the LST of $Y_{\min}$ provides a bound on the LSTs of all *PH* distributions in $\Psi_{m,\theta}$.

Next, let $X_{\max}$ be the exponential random variable with parameter μ. Denote by Φ_μ the set of all *PH* distributions with a *PH* representation (α, T) satisfying $\alpha\mathbf{e} = 1$ and

$$\sum_{i=1}^{m}\left(\sum_{j=1}^{m-i+1} r_{[j]}\right)^{-1} = \frac{1}{\mu}, \tag{6.19}$$

where $\mathbf{r} = -\mathbf{e}'T > 0$ and $m = 1, 2, \ldots$. By Theorem 6.1, $E[X_{\max}] = 1/\mu = \max\{E[X] : X \in \Phi_\mu\}$, which implies that $X_{\max}$ is an extremal random variable, with respect to the first moment, in Φ_μ. The result can be generalized to all moments and LSTs of *PH* distributions.

Corollary 6.4. *For $\mu > 0$ and Φ_μ we have*

(i) $E[X_{\max}^k] \geq E[X^k],\ \ k \geq 1;$

(ii) $E[\mathrm{e}^{-sX_{\max}}] \geq E[\mathrm{e}^{-sX}]$, $s_{\mathrm{lower}} < s \leq 0$, *for some negative number* s_{lower}; *and* $E[\mathrm{e}^{-sX_{\max}}] \leq E[\mathrm{e}^{-sX}]$, $0 \leq s < s_{\mathrm{upper}}$, *for some positive number* s_{upper}.

Define

$$\Theta_m = \{X :\ X \sim (\alpha, T) \text{ of order } m,\ \alpha\mathbf{e} = 1,\ -\mathbf{e}'T > 0\}. \tag{6.20}$$

Corollary 6.5. *For $X \sim (\alpha, T) \in \Theta_m$ we have, for $s \geq 0$,*

$$E[\mathrm{e}^{-sX}] \geq 1 - \sum_{i=1}^{m} \frac{s}{i(s+\delta_i)}, \tag{6.21}$$

where $\delta_i = r_{[1]} + \cdots + r_{[i]}$, $i = 1, 2, \ldots, m$, and $\mathbf{r} = -\mathbf{e}'T$.

Note that $\mathbf{e}'(sI - T) > 0$ for sufficiently large s. For any *PH* distribution, (6.21) holds if s is sufficiently large.

Example 6.4. Consider *PH* generator T defined as

$$T = \begin{pmatrix} -5 & 1 & 1 & 0 & 1 \\ 2 & -15 & 0 & 1 & 5 \\ 0 & 1 & -3 & 1 & 0 \\ 1 & 0 & 0 & -5 & 1 \\ 1 & 1 & 1 & 0 & -8 \end{pmatrix}. \tag{6.22}$$

Note that $-\mathbf{e}'T = (1, 12, 1, 3, 1)$ is positive.

For $X \sim (\alpha, T)$ with $\alpha = (0.2, 0.5, 0.1, 0.1, 0.1)$ we have $E[X_{\min}^k] \leq E[X^k] = k!\alpha(-T^{-1})^k\mathbf{e} \leq E[X_{\max}^k]$ for $k \geq 1$. As shown in Fig. 6.1, the two logarithmic ratios are less than zero for all k, which confirms the inequalities numerically. We further obtain $E[\mathrm{e}^{-sX_{\min}}] \leq E[\mathrm{e}^{-sX}] \leq E[\mathrm{e}^{-sX_{\max}}]$ for $s \leq 0$, for which the expectations exist.

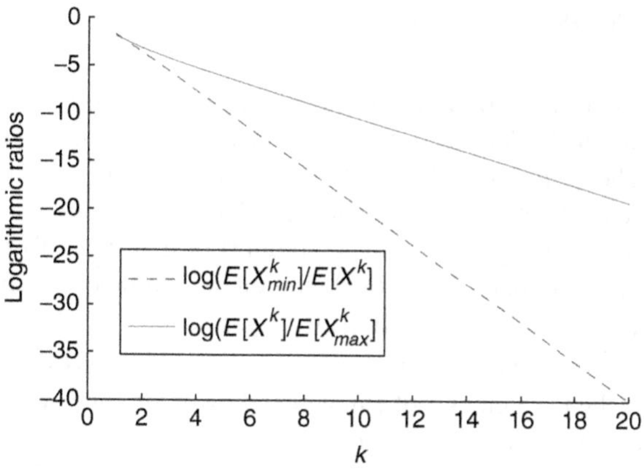

Fig. 6.1 Logarithmic ratios $\log(E[X^k_{\min}]/E[X^k])$ and $\log(E[X^k]/E[X^k_{\max}])$

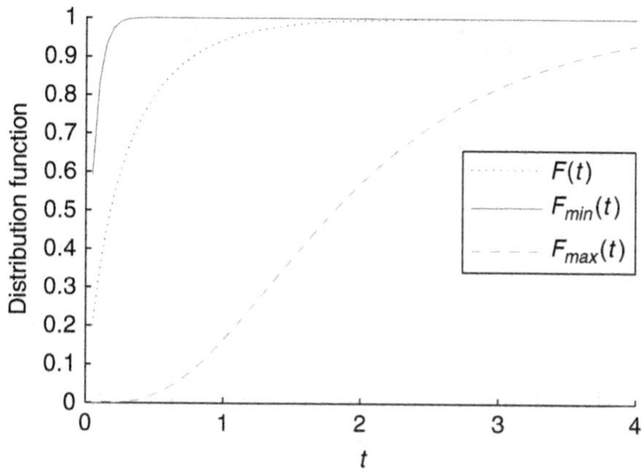

Fig. 6.2 Distribution functions $F_{\max}(t)$, $F(t)$, and $F_{\min}(t)$

For Example 6.4, further numerical results indicate that $(-T^{-1})^k\mathbf{e} \prec^w (-(T^*_\downarrow)^{-1})^k\mathbf{e}$ and $(-T^{-1})^k\mathbf{e} \prec_w (-(T^*_\uparrow)^{-1})^k\mathbf{e}$ for $k \geq 1$. Such results are stronger than those in Corollaries 6.2 and 6.4. If the results are true, then the moments of (α, T) are upper bounded by that of the Coxian distribution $((0,\ldots,0,1), T^*_\uparrow)$, which is different from the distribution function of $X_{\max}$ (which is actually an exponential random variable). Denote by $F_{\min}(t)$, $F(t)$, and $F_{\max}(t)$ the probability distribution functions of the *PH* distributions $((1,0,\ldots,0), T^*_\downarrow)$, (α, T), and $((0,\ldots,0,1), T^*_\uparrow)$, respectively. Numerical results also indicate that $F_{\max}(t) \leq F(t) \leq F_{\min}(t)$ for $t \geq 0$ (Fig. 6.2), which implies that the three probability distributions are stochastically ordered. The result is interesting since $F_{\max}(t)$ is a Coxian (not an exponential) distribution in general. Extensive numerical tests demonstrate that those results may hold for all *PH* distributions with *PH* generators satisfying $-\mathbf{e}'T > 0$.

Proofs

Proof of Lemma 6.1. Denote by $\mathbf{e}(i)$ the column vector with zero everywhere but one in the ith place. Since the matrix $-T$ is an M-matrix, $-T^{-1}$ is nonnegative (Theorem 4.5 [12]). Let $\mathbf{b} = -T^{-1}\mathbf{e}$. Without loss of generality, we assume that elements of $\mathbf{b}$ are in ascending order, i.e., $b_1 \le b_2 \le \cdots \le b_m$, which can be done by permuting the rows and columns of matrix T. To prove that $-T^{-1}\mathbf{e}$ is weakly supermajorized by $\mathbf{b}^*_\downarrow$, by definition, it is sufficient to show that $b_1 + b_2 + \cdots + b_k \ge (\mathbf{b}^*_\downarrow)_1 + (\mathbf{b}^*_\downarrow)_2 + \cdots + (\mathbf{b}^*_\downarrow)_k$, for $1 \le k \le m$.

For fixed $k \le m$ let $\mathbf{z} = (\mathbf{e}(1)' + \mathbf{e}(2)' + \cdots + \mathbf{e}(k)')(-T^{-1})$. Let $z_{n_1} \ge z_{n_2} \ge \cdots \ge z_{n_m}$ be the elements of $\mathbf{z}$ in descending order. Since $\mathbf{ze} = b_1 + b_2 + \cdots + b_k$, our goal is to prove, for $1 \le k \le m$,

$$\mathbf{ze} \ge k\left(\sum_{j=1}^{m} r_{[j]}\right)^{-1} + (k-1)\left(\sum_{j=2}^{m} r_{[j]}\right)^{-1} + \cdots + \left(\sum_{j=k}^{m} r_{[j]}\right)^{-1}. \tag{6.23}$$

Since $\mathbf{z}(-T) = \mathbf{e}(1)' + \mathbf{e}(2)' + \cdots + \mathbf{e}(k)'$, we have, for $1 \le j \le m$,

$$\begin{aligned}
&\mathbf{z}(-T)(\mathbf{e}(n_j) + \mathbf{e}(n_{j+1}) + \cdots + \mathbf{e}(n_m)) \\
&= (\mathbf{e}(1)' + \mathbf{e}(2)' + \cdots + \mathbf{e}(k)')(\mathbf{e}(n_j) + \mathbf{e}(n_{j+1}) + \cdots + \mathbf{e}(n_m)) \\
&\ge \max\{0,\, k - j + 1\}.
\end{aligned} \tag{6.24}$$

By definition, we have $T\mathbf{e} \le 0$ and $T_{i,j} \ge 0$ for $1 \le i \le j \le m$. Then, for any $\{i_1, i_2, \ldots, i_n\} \subset \{1, 2, \ldots, m\}$ and $i_0 \in \{1, 2, \ldots, m\}$, note that

$$\mathbf{e}(i_0)'(-T)(\mathbf{e}(i_1) + \mathbf{e}(i_2) + \cdots + \mathbf{e}(i_n)) = -\sum_{j=1}^{n} T_{(i_0, i_j)}, \tag{6.25}$$

which is nonnegative if $i_0 \in \{i_1, i_2, \ldots, i_n\}$ and nonpositive if $i_0 \notin \{i_1, i_2, \ldots, i_n\}$. For $i < j$ we have $z_{n_i} - z_{n_j} \ge 0$ and $\mathbf{e}(n_i)'(-T)(\mathbf{e}(n_j) + \mathbf{e}(n_{j+1}) + \cdots + \mathbf{e}(n_m)) \le 0$. For $i \ge j$, we have $z_{n_i} - z_{n_j} \le 0$ and $\mathbf{e}(n_i)'(-T)(\mathbf{e}(n_j) + \mathbf{e}(n_{j+1}) + \cdots + \mathbf{e}(n_m)) \ge 0$. Combining the two cases, for $1 \le i, j \le m$ we obtain

$$(z_{n_i} - z_{n_j})\mathbf{e}(n_i)'(-T)(\mathbf{e}(n_j) + \mathbf{e}(n_{j+1}) + \cdots + \mathbf{e}(n_m)) \le 0. \tag{6.26}$$

Equation (6.26) leads to

$$z_{n_i}\mathbf{e}(n_i)'(-T)\left(\sum_{h=j}^{m} \mathbf{e}(n_h)\right) \le z_{n_j}\mathbf{e}(n_i)'(-T)\left(\sum_{h=j}^{m} \mathbf{e}(n_h)\right). \tag{6.27}$$

Summing up over $i = 1, 2, \ldots, m$, in (6.27), yields

$$\begin{aligned}\mathbf{z}(-T)\left(\sum_{h=j}^{m}\mathbf{e}(n_h)\right) &= \sum_{i=1}^{m} z_{n_i}\mathbf{e}(n_i)'(-T)\left(\sum_{h=j}^{m}\mathbf{e}(n_h)\right)\\ &\le \sum_{i=1}^{m} z_{n_j}\mathbf{e}(n_i)'(-T)\left(\sum_{h=j}^{m}\mathbf{e}(n_h)\right)\\ &= z_{n_j}\mathbf{e}'(-T)\left(\sum_{h=j}^{m}\mathbf{e}(n_h)\right)\\ &= z_{n_j}\mathbf{r}\left(\sum_{h=j}^{m}\mathbf{e}(n_h)\right). \end{aligned}\tag{6.28}$$

We also have

$$\mathbf{r}\left(\sum_{h=j}^{m}\mathbf{e}(n_h)\right) = \sum_{h=j}^{m} r_{n_h} \le \sum_{h=j}^{m} r_{[h]}. \tag{6.29}$$

Combining (6.24), (6.28), and (6.29) we obtain

$$z_{n_j} \ge \max\{0, k-j+1\}\left(\sum_{h=j}^{m} r_{[h]}\right)^{-1}. \tag{6.30}$$

Adding over $j = 1, 2, \ldots, m$, (6.23) follows. This completes the proof of Lemma 6.1.

Proof of Lemma 6.2. This proof is similar to that of Lemma 6.1, but some details are different. Let $\mathbf{b} = -T^{-1}\mathbf{e}$. Without loss of generality, we assume that elements of $\mathbf{b}$ are in descending order, i.e., $b_1 \ge b_2 \ge \cdots \ge b_m$. To prove that $-T^{-1}\mathbf{e}$ is weakly submajorized by $\mathbf{b}_{\uparrow}^*$, it is sufficient to show that $b_1 + b_2 + \cdots + b_k \le (\mathbf{b}_{\uparrow}^*)_m + (\mathbf{b}_{\uparrow}^*)_{(m-1)} + \cdots + (\mathbf{b}_{\uparrow}^*)_{(m-k+1)}$, for $1 \le k \le m$.

For fixed $k \le m$ let $\mathbf{z} = (\mathbf{e}(1)' + \mathbf{e}(2)' + \cdots + \mathbf{e}(k)')(-T^{-1})$. Let $z_{n_1} \le z_{n_2} \le \cdots \le z_{n_m}$ be the elements of $\mathbf{z}$ in ascending order, where $(n_1, n_2, \ldots, n_m)$ is a permutation of $(1, 2, \ldots, m)$. Since $\mathbf{ze} = b_1 + b_2 + \cdots + b_k$, our goal is to prove

$$\mathbf{ze} \le \sum_{i=1}^{m}\min\{k, m-i+1\}\left(\sum_{j=1}^{m-i+1} r_{[j]}\right)^{-1}. \tag{6.31}$$

Since $\mathbf{z}(-T) = \mathbf{e}(1)' + \mathbf{e}(2)' + \cdots + \mathbf{e}(k)'$, we have, for $1 \le j \le m$,

$$\mathbf{z}(-T)(\mathbf{e}(n_j) + \mathbf{e}(n_{j+1}) + \cdots + \mathbf{e}(n_m)) \le \min\{k, m-j+1\}. \tag{6.32}$$

Similar to (6.26), we can show, for $1 \le i, j \le m$,

$$(z_{n_i} - z_{n_j})\mathbf{e}(n_i)'(-T)(\mathbf{e}(n_j) + \mathbf{e}(n_{j+1}) + \cdots + \mathbf{e}(n_m)) \ge 0, \tag{6.33}$$

which leads to

$$\begin{aligned}\mathbf{z}(-T)\left(\sum_{h=j}^{m}\mathbf{e}(n_h)\right) &\ge z_{n_j}\mathbf{r}\left(\sum_{h=j}^{m}\mathbf{e}(n_h)\right)\\ &= z_{n_j}\left(\sum_{h=j}^{m} r_{n_h}\right)\\ &\ge z_{n_j}\left(\sum_{h=1}^{m-j+1} r_{[h]}\right).\end{aligned} \tag{6.34}$$

Combining (6.32) and (6.34), since $\sum_{h=1}^{m-j+1} r_{[h]} > 0$, we obtain

$$z_{n_j} \le \min\{k, m-j+1\}\left(\sum_{h=1}^{m-j+1} r_{[h]}\right)^{-1}. \tag{6.35}$$

Adding over $j = 1, 2, \ldots, m$, (6.31) follows. This completes the proof of Lemma 6.2.

Proof of Theorem 6.1. By Lemma 6.1, we have $-T^{-1}\mathbf{e} \prec^{w} -(T_{\downarrow}^{*})^{-1}\mathbf{e}$, or, equivalently, $(-T^{-1}\mathbf{e})_{\uparrow} \prec^{w} -(T_{\downarrow}^{*})^{-1}\mathbf{e}$. Since the elements in $\alpha_{\downarrow}$ are in descending order, we obtain $\alpha_{\downarrow}(-T^{-1}\mathbf{e})_{\uparrow} \ge -\alpha_{\downarrow}(T_{\downarrow}^{*})^{-1}\mathbf{e}$, which leads to

$$E[X] = -\alpha T^{-1}\mathbf{e} \ge \alpha_{\downarrow}(-T^{-1}\mathbf{e})_{\uparrow} \ge -\alpha_{\downarrow}(T_{\downarrow}^{*})^{-1}\mathbf{e}. \tag{6.36}$$

Since the elements of the vector $-(T_{\downarrow}^{*})^{-1}\mathbf{e}$ are in ascending order, $-\alpha_{\downarrow}(T_{\downarrow}^{*})^{-1}\mathbf{e} \ge (-(T_{\downarrow}^{*})^{-1}\mathbf{e})_1 = (r_1 + r_2 + \cdots + r_m)^{-1} = -1/(\mathbf{e}'T\mathbf{e})$. This proves the first part of the theorem.

By Lemma 6.2, we have $-T^{-1}\mathbf{e} \prec_{w} -(T_{\uparrow}^{*})^{-1}\mathbf{e}$, or, equivalently, $(-T^{-1}\mathbf{e})_{\uparrow} \prec_{w} -(T_{\uparrow}^{*})^{-1}\mathbf{e}$. Since the elements of $\alpha_{\uparrow}$ are in ascending order, we obtain $\alpha_{\uparrow}(-T^{-1}\mathbf{e})_{\uparrow} \le -\alpha_{\uparrow}(T_{\uparrow}^{*})^{-1}\mathbf{e}$, which leads to

$$E[X] = -\alpha T^{-1}\mathbf{e} \le \alpha_{\uparrow}(-T^{-1}\mathbf{e})_{\uparrow} \le -\alpha_{\uparrow}(T_{\uparrow}^{*})^{-1}\mathbf{e}. \tag{6.37}$$

Since the elements of the vector $-(T_{\uparrow}^{*})^{-1}\mathbf{e}$ are in ascending order, $-\alpha_{\uparrow}(T_{\uparrow}^{*})^{-1}\mathbf{e} \le (-(T_{\uparrow}^{*})^{-1}\mathbf{e})_m = \sum_{i=1}^{m}(\sum_{j=1}^{m-i+1} r_{[j]})^{-1}$. This proves the second part and concludes the proof of Theorem 6.1.

Proof of Theorem 6.2. Under the conditions given in Theorem 6.2, we have $-T^{-1}\mathbf{e} \prec -(T_{\downarrow}^{*})^{-1}\mathbf{e}$ and $-T^{-1}\mathbf{e} \prec -(T_{\uparrow}^{*})^{-1}\mathbf{e}$. The rest of the proof is similar to that of Theorem 6.1. This completes the proof of Theorem 6.2.

Proof of Corollary 6.2. By Theorem 6.1, part (i) of Corollary 6.1 holds for $k = 1$, i.e., $E[X] \geq 1/\lambda$. We prove the result for $k > 1$ by induction. Consider the stochastic vector $\gamma = \alpha(-T^{-1})^k/(\alpha(-T^{-1})^k\mathbf{e})$. Note that $E[X^k] = k!\alpha(-T^{-1})^k\mathbf{e}$ and $E[X_{\min}^k] = k!/\lambda^k$ for $k \geq 1$. Applying Theorem 6.1 to $(\gamma, T) \in \Omega_\lambda$ we obtain

$$\frac{\alpha(-T^{-1})^{k+1}\mathbf{e}}{\alpha(-T^{-1})^k\mathbf{e}} = \gamma(-T^{-1})\mathbf{e} \geq \frac{1}{(-\mathbf{e}'T\mathbf{e})} = \frac{1}{\lambda}, \tag{6.38}$$

which leads to $E[X^{k+1}] \geq (k+1)E[X^k]/\lambda$. By induction, we obtain $E[X^{k+1}] \geq (k+1)!/\lambda^{k+1} = E[X_{\min}^{k+1}]$. This proves part (i) of Corollary 6.1. To prove part (ii), we first note that, by definition,

$$E[\mathrm{e}^{-sX}] = \sum_{n=0}^{\infty} \frac{(-s)^n E[X^n]}{n!} \tag{6.39}$$

if the summation exists. Then $E[\mathrm{e}^{-sX_{\min}}] \leq E[\mathrm{e}^{-sX}]$ for $s_{\text{lower}} < s \leq 0$, is obtained from part 1), for some negative number s_{lower}. Since both functions $E[\mathrm{e}^{-sX_{\min}}]$ and $E[\mathrm{e}^{sX}]$ equal one at $s = 0$, by continuous extension at $s = 0$, we obtain $E[\mathrm{e}^{-sX_{\min}}] \geq E[\mathrm{e}^{-sX}]$, for $0 \leq s \leq s_{\text{upper}}$ and some positive number s_{upper}. This completes the proof of Corollary 6.2.

Proof of Corollary 6.3. We consider the *PH* generator $-sI + T$ for $s \geq 0$. Lemma 6.1 indicates that $(sI - T)^{-1}\mathbf{e}$ is weakly supermajorized by $(sJ - T_{\downarrow}^{*})^{-1}\mathbf{e}$, where

$$J = \begin{pmatrix} m & & & & \\ -(m-1) & m-1 & & & \\ & \ddots & \ddots & & \\ & & -2 & 2 & \\ & & & -1 & 1 \end{pmatrix}, \tag{6.40}$$

and $T_{\downarrow}^{*}$ was defined in (6.1). Applying Theorem 6.1 to $(\alpha, -sI + T)$ we obtain that $\alpha(sI - T)^{-1}\mathbf{e}$ is greater than or equal to the first element in the column vector $(sJ - T_{\downarrow}^{*})^{-1}\mathbf{e}$. Since $s \geq 0$, $s\alpha(sI - T)^{-1}\mathbf{e}$ is greater than or equal to the first element in the column vector $s(sJ - T_{\downarrow}^{*})^{-1}\mathbf{e}$, which is given by

$$\frac{1}{m} - \frac{\theta}{m(s+\theta)}. \tag{6.41}$$

Note that $1 - s\alpha(sI - T)^{-1}\mathbf{e} = \alpha(sI - T)^{-1}(-T)\mathbf{e} = E[\mathrm{e}^{-sX}]$. Then (6.18) is obtained from (6.41). This completes the proof of Corollary 6.3.

Proof of Corollary 6.4. The proof is similar to that of Corollary 6.2. Details are omitted.

Proof of Corollary 6.5. The proof is similar to that of Corollary 6.3. Details are omitted.

Conclusion and Discussion

For some subsets of *PH* distributions, in this chapter, it is found that the exponential distributions and Coxian distributions are extremal distributions with respect to all the moments and the LSTs of *PH* distributions. The results have potential applications in several areas.

- The results can be useful in parameter estimation of *PH* distributions. For instance, the relationship $E[X^k] \geq k!/(-\mathbf{e}'T\mathbf{e})^k$, for $k \geq 1$, provides constraints on the parameters in T if the sample moments of the *PH* distribution X can be found (through other methods). The constraints can be used in nonlinear programs (e.g., EM algorithm) for parameter estimation of *PH* distributions [2, 8]. The potential of the results in this area is yet to be explored.
- The results can be used in optimization. Consider the case $\mathbf{e}'T = -\mu\mathbf{e}$, where $\mu > 0$. Without loss of generality, we assume $\mu = 1$. Then we obtain $\mathbf{e}'(-T)^{-1} = \mathbf{e}'$. Denote by $\mathbf{a}_1, \mathbf{a}_2, \ldots,$ and $\mathbf{a}_m$ the column vectors of $-T^{-1}$, which is nonnegative. Then the vector $\mathbf{e}'/m$ is in the polytope generated by $\{\mathbf{a}_1, \mathbf{a}_2, \ldots, \mathbf{a}_m\}$. Then Corollary 6.1 gives the optimal solution(s) to the following optimization problem:

$$\max / \min_{\{\alpha_i, \mathbf{a}_i,\ 1 \leq i \leq m\}} \left(\sum_{i=1}^{m} \alpha_i \mathbf{a}_i \right) \mathbf{e}$$
$$s.t.\ \ \alpha_i \geq 0,\ \sum_{i=1}^{m} \alpha_i = 1;$$
$$(\mathbf{a}_1,\ \mathbf{a}_2,\ \ldots, \mathbf{a}_m)T = I;$$
$$\mathbf{e}'T = -\mathbf{e}';$$
$$T \text{ is a } PH \text{ generator.} \tag{6.42}$$

  The objective of optimization problem (6.42) is to find a point in the polytope generated by extreme points $\{\mathbf{a}_1, \mathbf{a}_2, \ldots, \mathbf{a}_m\}$ such that the objective function is either minimized or maximized.
- Because the bounds obtained in the sections “Bounds on Phase-Type Distributions” and “Extremal Phase-Type Distributions” are either partially or completely independent of the transition structure within T, they have the potential to be used in resource allocation if the transitions are affected by resources allocated to different phases.

Naturally, the preceding applications are interesting topics for future research. In addition, the issues on the distribution functions of *PH* distributions and extremal *PH* distributions raised at the end of the section “Extremal Phase-Type Distributions” are of theoretical interest for further investigation.

Acknowledgements The authors would like to thank reviewers for their valuable comments and suggestions. The authors would also like to thank Mr. Zurun Xu for proofreading the paper.

References

1. Aldous, D., Shepp, L.: The least variable phase type distribution is Erlang. Stoch. Models **3**, 467–473 (1987)
2. Asmussen, S., Nerman, O., Olsson, M.: Fitting phase-type distributions via the EM algorithm. Scand. J. Stat. **23**, 419–441 (1996)
3. Bodrog, L., Heindl, A., Horvath, A., Horvath, G., Telek, M.: Current results and open questions on *PH* and *MAP* characterization. In: Dagstuhl Seminar Proceedings 07461 (D. Bini, B. Meini, V. Ramaswami, M-A Remiche, and P. Taylor, eds.). Numerical Methods for Structured Markov Chains, http://drops.dagstuhl.de/opus/volltexte/2008/1401. (2008)
4. Commault, C., Mocanu, S.: Phase-type distributions and representations: some results and open problems for system theory. Int. J. Control **76**, 566–580 (2003)
5. He, Q.M., Zhang, H.Q.: A note on unicyclic representations of phase type distributions. Stoch. Models **21**, 465–483 (2005)
6. He, Q.M., Zhang, H.Q.: Spectral polynomial algorithms for computing bi-diagonal representations for phase type distributions and matrix exponential distributions. Stoch. Models **22**, 289–317 (2006)
7. He, Q.M., Zhang, H.Q., Vera, J.C.: On some properties of bivariate exponential distributions. Stoch. Models **28**, 187–206 (2012)
8. Horvath, A., Telek, M.: PhFit: A general purpose phase type fitting tool. In: Tools 2002, pp. 82–91. LNCS, vol. 2324. Springer, London (2002)
9. Latouche, G., Ramaswami, V.: Introduction to Matrix Analytic Methods in Stochastic Modeling. SIAM, Philadelphia (1999)
10. Maier, R.S.: The algebraic construction of phase-type distributions. Stoch. Models **7**, 573–602 (1991)
11. Marshall, A.W., Olkin, I.: Inequalities: Theory of Majorization and Its Applications. Academic, New York (1979)
12. Minc, H.: Non-Negative Matrices. Wiley, New York (1988)
13. Neuts, M.F.: Probability distrubutions of phase type. In: Liber Amicorum Professor Emeritus H. Florin, pp. 173–206. Department of Mathematics, University of Louvain, Belgium (1975)
14. Neuts, M.F.: Matrix-Geometric Solutions in Stochastic Models: An Algorithmic Approach. Johns Hopkins University Press, Baltimore (1981)
15. O'Cinneide, C.A.: Characterization of phase-type distributions. Stoch. Models **6**, 1–57 (1990)
16. O'Cinneide, C.A.: Phase-type distributions and majorization. Ann. Appl. Probab. **1**, 219–227 (1991)
17. O'Cinneide, C.A.: Phase-type distributions: open problems and a few properties. Stoch. Models **15**, 731–757 (1999)
18. Sangüesa, C.: On the minimal value in Maier's property concerning phase-type distributions. Stoch. Models **26**, 124–140 (2010)
19. Yao, R.H.: A proof of the steepest increase conjecture of a phase-type density. Stoch. Models **18**, 1–6 (2002)

References

Chapter 7
Acceptance-Rejection Methods for Generating Random Variates from Matrix Exponential Distributions and Rational Arrival Processes

Gábor Horváth and Miklós Telek

Introduction

Despite the widespread use of Markovian traffic models, phase-type (PH) distributions [14], and Markov arrival processes (MAPs) [9], in simulations there are surprisingly few results available on the efficient generation of random variates of these models. Furthermore, there are practically no results available on the efficient generation of random variates of matrix-exponential (ME) distributions [11] and rational arrival processes (RAPs) [1], apart from the trivial and computationally heavy method based on the numerical inversion of the cumulative distribution function [3]. The aim of this chapter is to propose efficient numerical methods for random-variate generation based on ME distributions and various versions of RAPs. The few works dealing with efficient generation of PH distributed random variates are based on the stochastic interpretation of PH distributions. These methods simulate the Markov chain that defines the PH distribution until it reaches the absorbing state and generates the required random variates in an efficient way [15]. In the sequel, this procedure of simulating the underlying Markov chain is referred to as the *play method*. Markovian traffic models are defined by a set of matrices (including vectors as special matrices) referred to as representation. The representation is not unique. Different sets of matrices can represent the same model. More recently, it has been recognized that the computational complexity of the play method depends on the particular representation of the PH distribution [16, 17].

ME distributions and RAPs do not have a straightforward stochastic interpretation. Consequently, the methods available for generating random variates of Markovian traffic models cannot be used for their simulation. To overcome

G. Horváth (✉) • M. Telek
Department of Telecommunications Technical University of Budapest, 1521 Budapest, Hungary
e-mail: ghorvath@hit.bme.hu; telek@hit.bme.hu

G. Latouche et al. (eds.), *Matrix-Analytic Methods in Stochastic Models*, Springer Proceedings in Mathematics & Statistics 27, DOI 10.1007/978-1-4614-4909-6_7,

this difficulty, we propose a version of the acceptance–rejection method. The acceptance–rejection method is a widely used method in simulation [18]. It consists of two main steps – drawing random samples from an easy-to-compute distribution and accepting the sample with a sample-dependent probability such that the overall probability density of the accepted samples is identical to the required one. The computational complexity of this method depends on the sample efficiency, which is the ratio of the number of accepted and the number of generated samples. Using a general distribution (e.g., exponential) whose shape is different from the required one results in a low sample efficiency. We propose specific methods with higher sample efficiency.

It turns out that, similar to the case of Markovian traffic models, the representation of ME distributions and RAPs affects the sample efficiency and the computational complexity of generating random variates of these models. We evaluate the behavior of two particular representations with nice structural properties.

As is demonstrated among the numerical experiments, there are cases where the proposed method that is developed for simulating ME distributions and RAPs is more efficient for the simulation of Markovian models (PH distributions and MAPs) than the existing methods based on their stochastic interpretations.

A procedure to generate pseudorandom numbers uniformly distributed on $(0, 1)$ is part of all common programming languages and simulation packages. In this work we investigate the computational effort to generate random variates of ME distributions and RAPs using these uniformly distributed pseudorandom numbers. The complexity of various computational steps might differ in various programming environments. We define the computational complexity of the proposed methods as a function of the more complex computational steps (number of pseudorandom samples, log operations, exp operations).

The main part of the chapter is devoted to ME distributed random variate generation because it is a main building block of RAP simulation. The section "Matrix Exponential Distributions and RAPs" introduces ME distributions and RAPs, and the section "Generating Random Variates of Markovian Traffic Models" summarizes the steps and the complexity of generating random variates of Markovian traffic models. Following these preliminaries, the section "Generating Random Variates from Matrix-Exponential Distributions Having a Markovian Generator" introduces the proposed acceptance–rejection method. The section "Generating Matrix-Exponentially Distributed Random Variates Using Feedback-Erlang Blocks" specializes the acceptance–rejection method to particular representations that are efficient for random variate generation. The use of ME distributed random number generation for simulating various RAPs is explained in the section "Generating Random Variates from Various RAPs." To demonstrate the efficiency of the proposed methods, examples and related numerical experiments are presented in the section "Numerical Experiments."

Matrix-Exponential Distributions and Rational Arrival Processes

We start the summary of the preliminaries with the definition of ME and PH distributions; later we introduce RAPs and MAPs and their variants.

Definition 7.1. A real-valued row vector square matrix pair of size N, $(\tau, \mathbf{T})$, defines an ME distribution iff

$$F(x) = Pr(X < x) = 1 - \tau e^{\mathbf{T}x}\mathbb{1}, \qquad x \geq 0, \tag{7.1}$$

is a valid cumulative distribution function (CDF), i.e., $F(0) \geq 0$, $\lim_{x\to\infty} F(x) = 1$ and $F(x)$ is monotone increasing.

In (7.1), the row vector τ is referred to as the initial vector, the square matrix $\mathbf{T}$ as the generator, and $\mathbb{1}$ as the closing vector. Without loss of generality [11], throughout this chapter we assume that the closing vector is a column vector of ones, i.e., $\mathbb{1} = [1, 1, \ldots, 1]^{\mathrm{T}}$. Furthermore, we restrict our attention to the case where there is no probability mass at 0, i.e., $F(0) = 0$, or, equivalently, $\tau\mathbb{1} = 1$.

The probability density function (PDF) of the ME distribution defined by $(\tau, \mathbf{T})$ is

$$f(x) = \tau e^{\mathbf{T}x}(-\mathbf{T})\mathbb{1}. \tag{7.2}$$

To ensure that $\lim_{x\to\infty} F(x) = 1$, $\mathbf{T}$ must fulfill the necessary condition that the real parts of its eigenvalues are negative (consequently $\mathbf{T}$ is nonsingular).

The remaining constraint is the monotonicity of $F(x)$ or, equivalently, the nonnegativity of $f(x)$. This constraint is the most difficult to check. The simulation methods proposed below implement control checks to indicate if this condition is violated during the simulation run.

Definition 7.2. If τ is nonnegative and $\mathbf{T}$ has negative diagonal and nonnegative off-diagonal elements, then $(\tau, \mathbf{T})$ is said to be Markovian and defines a PH distribution.

PH distributions can be interpreted as a time duration in which a Markov chain having N transient and an absorbing state arrives to the absorbing state. In the case of a non-Markovian representation, however, there is no such simple stochastic interpretation available.

If $N = 2$, then the class of ME distributions is identical with the class of PH distributions, but if $N > 2$, then the class of PH distributions is a proper subset of the class of ME distributions [10].

A RAP is a point process in which the interarrival times are ME distributed [1, 12].

Definition 7.3. A square matrix pair of size N, $(\mathbf{H_0}, \mathbf{H_1})$, satisfying $(\mathbf{H_0} + \mathbf{H_1})\mathbb{1} = \mathbf{0}$ defines a stationary RAP iff the joint density function of the interarrival times

$$f(x_1, \ldots, x_k) = \tau e^{\mathbf{H_0}x_1}\mathbf{H_1}e^{\mathbf{H_0}x_2}\mathbf{H_1}\ldots e^{\mathbf{H_0}x_k}\mathbf{H_1}\mathbb{1} \tag{7.3}$$

is nonnegative for all $k \geq 1$ and $x_1, x_2, \ldots, x_k \geq 0$ and τ is the unique solution of $\tau(-\mathbf{H_0})^{-1}\mathbf{H_1} = \tau$, $\tau\mathbb{1} = 1$.

If the solution $\tau(-\mathbf{H_0})^{-1}\mathbf{H_1} = \tau$, $\tau\mathbb{1} = 1$ is not unique, then $(\mathbf{H_0}, \mathbf{H_1})$ does not define the stationary behavior of the process.

RAPs inherit several properties from ME distributions. The real parts of the eigenvalues of matrix $\mathbf{H_0}$ are negative; consequently, the matrix is nonsingular. There is a real eigenvalue with maximal real part. Similar to the case of ME distributions, the nonnegativity of the joint density function is hard to check, and the proposed simulation methods contain run-time checks to indicate if the nonnegativity of the joint density is violated. The first interarrival time of the RAP is ME distributed with initial vector τ and square matrix $\mathbf{H_0}$. Vector τ and the off-diagonal blocks of matrix $\mathbf{H_0}$ may contain negative elements. If $\mathbf{H_1} = -\mathbf{H_0}\mathbb{1}\tau$, then the consecutive interarrivals are independent and identically distributed, that is, the RAP is a renewal process with ME distributed interarrivals.

Definition 7.4. If $\mathbf{H_1} \geq 0$ and all nondiagonal elements of $\mathbf{H_0}$ are nonnegative, then the matrix pair $(\mathbf{H_0}, \mathbf{H_1})$ is said to be Markovian and define a MAP.

The joint density function (7.3) of a MAP is always positive and $\tau \geq 0$. In the case of MAPs one can interpret the nondiagonal elements of matrix $\mathbf{H_0}$ and the elements of $\mathbf{H_1}$ as transition rates corresponding to hidden and visible events, respectively. Vector τ can be interpreted as the state of the MAP at time zero.

The extension of plain (single-arrival, single-event type) MAPs to MAPs with batch arrivals (BMAPs) [9] and with different types of arrivals (MMAPs) [7] can be applied to RAPs as well. This extension results in a batch rational arrival process (BRAP) and a marked rational arrival process (MRAP) [2], respectively. The stochastic behavior of MRAPs and BRAPs is practically the same. In what follows, we discuss MRAPs only.

Definition 7.5. A set of square matrices of size N, $(\mathbf{H_0}, \mathbf{H_1}, \ldots, \mathbf{H_K})$, satisfying $\sum_{k=0}^{K} \mathbf{H_k}\, \mathbb{1} = \mathbf{0}$, defines a stationary MRAP with K event types iff the joint density function of the arrival sequence (consecutive interarrival times and event types)

$$f(x_1, k_1, \ldots, x_j, k_j) = \tau e^{\mathbf{H_0} x_1} \mathbf{H_{k_1}} e^{\mathbf{H_0} x_2} \mathbf{H_{k_2}} \ldots e^{\mathbf{H_0} x_j} \mathbf{H_{k_j}} \mathbb{1} \tag{7.4}$$

is nonnegative for all $j \geq 1$ and $x_1, x_2, \ldots, x_j \geq 0$, $1 \leq k_1, k_2, \ldots, k_j \leq K$ and τ is the unique solution of $\tau(-\mathbf{H_0})^{-1} \sum_{k=1}^{K} \mathbf{H_k} = \tau$, $\tau\mathbb{1} = 1$.

If the solution $\tau(-\mathbf{H_0})^{-1} \sum_{k=1}^{K} \mathbf{H_k} = \tau$, $\tau\mathbb{1} = 1$ is not unique, then $(\mathbf{H_0}, \mathbf{H_1}, \ldots, \mathbf{H_K})$ does not define the stationary behavior of the process.

The class of MRAPs contains MMAPs since an MRAP is an MMAP if $\tau \geq 0$, $\mathbf{H_k} \geq 0$ for $k = 1, \ldots, K$ and all nondiagonal elements of $\mathbf{H_0}$ are nonnegative.

For later use we also define the initial vector after the first event. If a RAP with representation $(\mathbf{H_0}, \mathbf{H_1})$ starts with initial vector α and the first arrival happens at time x, then the initial vector characterizing the second arrival is $\alpha e^{\mathbf{H_0} x} \mathbf{H_1} / \alpha e^{\mathbf{H_0} x} \mathbf{H_1} \mathbb{1}$. If an MRAP with representation $(\mathbf{H_0}, \mathbf{H_1}, \ldots, \mathbf{H_K})$ starts with initial vector α and the first event happens at time x, then the probability that the event is of type k is $\alpha e^{\mathbf{H_0} x} \mathbf{H_k} \mathbb{1} / \sum_{j=1}^{K} \alpha e^{\mathbf{H_0} x} \mathbf{H_j} \mathbb{1}$. Furthermore, if an MRAP with representation $(\mathbf{H_0}, \mathbf{H_1}, \ldots, \mathbf{H_K})$ starts with initial vector α, then the first arrival happens at time x and is of type k; then the initial vector characterizing the second arrival is $\alpha e^{\mathbf{H_0}x}\mathbf{H_k} / \alpha e^{\mathbf{H_0}x}\mathbf{H_k}\mathbb{1}$.

The preceding matrix representations of the introduced processes are not unique. Various similarity transformations allow for generating different matrix representations of a given process. Similarity transformations exist for matrix representations of identical size [5] and different sizes [19]. We recall one of the possible similarity transformations for MRAPs from [19] without proof. Similar transformations for RAPs and ME distributions can be obtained as special cases [4].

Theorem 7.1. *If there is a matrix* $\mathbf{W} \in \mathbb{R}^{n,m}$, $m \geq n$ *such that* $\mathbb{1}_n = \mathbf{W}\mathbb{1}_m$ *(where* $\mathbb{1}_n$ *is a column vector of size n),* $\mathbf{W}\mathbf{H_k} = \mathbf{G_k}\mathbf{W}$ *for* $k = 0, \ldots, K$*, then* $(\mathbf{H_0}, \ldots, \mathbf{H_K})$ *and* $(\mathbf{G_0}, \ldots, \mathbf{G_K})$ *define the same MRAP.*

Generating Random Variates of Markovian Traffic Models

A trivial way to generate PH and ME distributed random numbers is based on the numerical inversion of the CDF. This computationally heavy method can be replaced by more efficient ones if the distribution allows a simple stochastic interpretation, e.g., in the case of PH distributions. Due to the simple stochastic interpretation of PH distributions through Markov chains the generation of PH distributed random variates can be done without the inversion of the numerical matrix exponential function in (7.1). Simulation approaches based on the underlying Markov-chain interpretation are presented in [15–17]. Below we list some of the related results of these papers and introduce some concepts that are also used in the current work for efficient random number generation.

- General PH distributions: General PH distributions can be interpreted as time to absorption of a Markov chain with N transient states and an absorbing state. The behavior of the Markov chain can be simulated by drawing a random sample for the initial state and by repeatedly drawing random samples for the state sojourn times and successor states until the absorbing state is reached. This method is referred to as the play method. Drawing samples of the state sojourn times requires drawing exponentially distributed random numbers $[R_{\mathrm{Exp}(\lambda)}]$ that are generated by transforming a random number U uniformly distributed on $(0,1)$ as

$$R_{\mathrm{Exp}(\lambda)} = -\frac{\log U}{\lambda}. \tag{7.5}$$

  Choosing the initial or a successor state requires drawing an additional random number U uniformly distributed on $(0,1)$ and comparing with the partial sums of elements of the probability vector. The play method is efficient if the mean number of state transitions before absorption is low. More efficient ways of generating random samples from PH distributions are proposed and analyzed

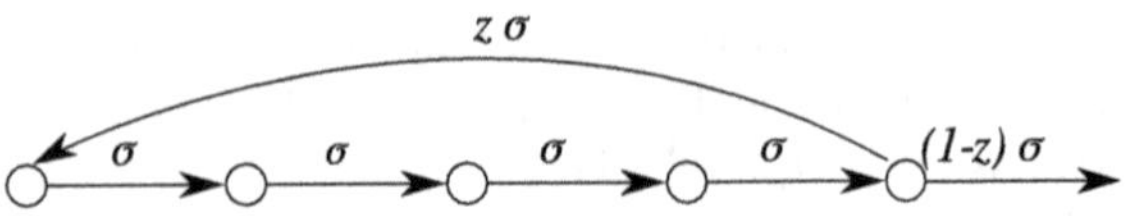

Fig. 7.1 A single feedback-Erlang block

in [15, 16]. Neuts and Pagano [15] recommends sampling the behavior of the discrete-time Markov chain embedded at state-transition instances, counting the number of visits to each state (each set of states with identical rate parameters) until absorption, and computing the PH distributed random sample as the sum of Erlang distributed random variables according to the number of visits and the rate of the associated state (set of states). Drawing Erlang distributed random variates requires only a single evaluation of the logarithm function, which is a considerable advantage:

$$R_{\mathrm{Erl}(\lambda,n)} = \sum_{i=1}^{n} -\frac{\log U_i}{\lambda} = -\frac{1}{\lambda}\log\prod_{i=1}^{n} U_i. \tag{7.6}$$

Reinecke et al. [16] recommend applying a similarity transformation of the original PH representation such that the transformed representation is cheaper to simulate. The complexity of these methods can further be improved by efficient discrete random variable sampling using the alias method [8].

- APH distributions: if the PH distribution has an acyclic representation, then even more efficient algorithms exist to generate random variates. Each APH can be transformed into one of the three canonical forms [6]. Assuming that an APH distribution is given in CF-1 form, a random variate is generated in two steps: first the initial state is drawn, then the time until absorption is sampled as the sum of exponentially distributed sojourn times of states between the initial and the absorbing state. Due to the structure of the CF-1 form, there is always exactly one successor state, so there is no need to draw a sample for choosing next states. Another important feature of the CF-1 form is the lack of cycles; thus the procedure terminates in at most as many steps as the phases of the APH.
- Hyper-Erlang (HEr) distribution: an HEr distribution is a convex combination of Erlang distributions. In the case of an HEr representation, first the Erlang branch must be chosen and then the Erlang distributed random number must be drawn.
- Hyperexponential (HE) distribution: an HE distribution is a convex combination of exponential distributions. An HE distribution is the most efficient representation of PH distributions with respect to random number generation. Only two operations are required: choose the branch and draw a sample for the selected exponential distribution.
- Feedback-Erlang block (FEB): an FEB is a series of independent, identical exponentially distributed phases with a single feedback from the last phase to the first one (Fig. 7.1). It is the main building block of the monocyclic representation introduced in [13]. An FEB has three parameters: the number of states n, the parameter of the exponential distribution σ, and the feedback probability z.

FEBs have the following advantages:

- They can represent complex eigenvalues in a Markovian way.
- They represent a real eigenvalue as a single exponential phase ($n = 1, z = 0$).
- Their eigenvalues are easy to obtain, which makes the construction of FEBs easy.
- It is efficient to draw random numbers from an FEB.

The generation of a sample from an FEB is similarly efficient to the generation of an Erlang distributed sample. First a geometrically distributed discrete random variate is sampled with parameter z, Δ, and after that

$$R_{\mathrm{FEB}(\sigma,n,z)} = -\frac{1}{\sigma}\log\prod_{i=1}^{n\Delta} U_i. \tag{7.7}$$

Generating Random Variates from Matrix-Exponential Distributions Having a Markovian Generator

In this section we present the main concept of the proposed acceptance–rejection method to generate random variates from an ME distribution. To apply this method, we assume that the representation of the ME distribution has a Markovian generator and a general initial vector (which might contain negative elements). The next section proposes such representations with Markovian generators. This section focuses only on the main idea of the proposed method. This acceptance–rejection approach is the basis of the subsequently introduced simulation of ME distributions and various RAPs.

Let $(\alpha, \mathbf{A})$ of size N be the representation of the ME distribution such that $\mathbf{A}$ is a Markovian generator matrix (nondiagonal elements are nonnegative, and the row sums are nonpositive). The PDF can be expressed as a nonconvex combination of PH distributions as follows:

$$f(x) = \alpha e^{\mathbf{A}x}(-\mathbf{A})\mathbb{1} = \sum_{i=1}^{N} \alpha_i \cdot \underbrace{\mathbb{e}_i e^{\mathbf{A}x}(-\mathbf{A})\mathbb{1}}_{g_i(x)}, \tag{7.8}$$

with $\mathbb{e}_i$ denoting a row vector of size N whose ith element is one and all other elements are zeros. Observe that $(\mathbb{e}_i, \mathbf{A})$ is a Markovian representation of the PH distribution with PDF $g_i(x)$; consequently $\int_0^\infty g_i(x)\mathrm{d}x = 1$.

To cope with the negative coefficients, we apply an acceptance–rejection method to generate a random variate as follows. The set of coefficients of the density function is divided into $\mathcal{A}_+$ and $\mathcal{A}_-$ such that $i \in \mathcal{A}_+$ if $\alpha_i \geq 0$ and $i \in \mathcal{A}_-$ otherwise. In this way $f(x)$ is separated into a positive part $[f_+(x)]$ and a negative part $[f_-(x)]$:

$$f(x) = \underbrace{\sum_{i\in\mathcal{A}_+} \alpha_i \cdot g_i(x)}_{f_+(x)} + \underbrace{\sum_{i\in\mathcal{A}_-} \alpha_i \cdot g_i(x)}_{f_-(x)}. \tag{7.9}$$

Note that $f_+(x) \geq 0,\ \forall x \geq 0$ and $f_-(x) \leq 0,\ \forall x \geq 0$ holds.

Algorithm 1 Algorithm for generating ME distributed random variates having a Markovian generator

---

1: Start: Draw a $\hat{f}_+(x)$ distributed random sample:
2:        $I =$ discrete random sample with distribution
            $p^* \sum_{i \in \mathcal{A}_+} \alpha_i \mathbb{e}_i$
3:        $R =$ random sample with pdf $g_I(x)$
4:          by any PH sampling method
5: **if** $\mathcal{A}_- = \emptyset$ **then**
6:     **return** R
7: **else**
8:     Calculate acceptance probability:
       $p_{accept}(R) = \frac{f_+(R)+f_-(R)}{f_+(R)}$
9:     **if** $p_{accept}(R) < 0$ **then**
10:        error **"INVALID DENSITY !!!"**
11:     **end if**
12:     Draw a uniform sample U
13:     **if** $U < p_{accept}(R)$ **then**
14:        **return** R
15:     **else**
16:        **goto** Start
17:     **end if**
18: **end if**

Multiplying by $p^* = 1/\sum_{j \in \mathcal{A}_+} \alpha_j$, the positive part gets normalized and we get

$$\hat{f}_+(x) = \sum_{i \in \mathcal{A}_+} \alpha_i \, p^* \cdot g_i(x), \tag{7.10}$$

which is a valid PH distribution with Markovian representation $(p^* \sum_{i \in \mathcal{A}_+} \alpha_i \mathbb{e}_i, \mathbf{A})$, where the initial vector is nonnegative and normalized. With these notations and definitions, the acceptance–rejection-based method to generate random numbers from $(\alpha, \mathbf{A})$ is formalized by Algorithm 1.

Theorem 7.2. *Algorithm 1 provides an $f(x)$ distributed random number and the mean number of required samples is geometrically distributed with parameter p^*, i.e., the probability that n samples are required is $(1 - p^*)^{n-1} p^*$.*

Proof. Let $f^*(x)$ be the probability density of the sample generated by Algorithm 1. In accordance with the standard proof of the acceptance–rejection method, we will show that $f^*(x) = f(x)$. The probability density that the first step of the algorithm results in sample R is $\hat{f}_+(R)$. The probability density that sample R is the accepted can be computed as

$$f^*(R) = \frac{\hat{f}_+(R)p_{accept}(R)}{\int_x \hat{f}_+(x)p_{accept}(x)dx}$$

$$= \frac{p^* f_+(R) \frac{f_+(R)+f_-(R)}{f_+(R)}}{\int_x p^* f_+(x) \frac{f_+(x)+f_-(x)}{f_+(x)}dx} = \frac{p^* f(R)}{p^* \int_x f(x)dx} = f(R). \tag{7.11}$$

The steps of the iterative procedure are independent. The probability of accepting a sample is $\int_x \hat{f}_+(x) p_{\text{accept}}(x)\mathrm{d}x = p^*$.

Generating Matrix-Exponentially Distributed Random Variates Using FEBs

As is shown in the previous section, "Generating Random Variates from Matrix-Exponential Distributions Having a Markovian Generator," there are several representations from which it is very efficient to draw random numbers. In this section we present two general representations with special structures that are composed by FEBs.

Hyper-Feedback-Erlang Representation

Definition 7.6. A Hyper-Feedback-Erlang (Hyper-FE) distribution is defined by an initial probability vector α and a transient generator having the following special structure (Fig. 7.2):

$$\mathbf{A} = \begin{bmatrix} \mathbf{M_1} & & & \\ & \mathbf{M_2} & & \\ & & \ddots & \\ & & & \mathbf{M_J} \end{bmatrix}, \tag{7.12}$$

where matrices $\mathbf{M_j}$ of size $n_j m_j \times n_j m_j$ are the subgenerators of several concatenated FEBs:

$$\mathbf{M_j}(\sigma_j, n_j, z_j, m_j) = \left[\begin{array}{cccc|c|cccc}
-\sigma_j & \sigma_j & & & & & & & \\
& \ddots & \ddots & & & & & & \\
& & -\sigma_j & \sigma_j & & & & & \\
z_j\sigma_j & & & -\sigma_j & (1-z_j)\sigma_j & & & & \\ \hline
& & & & \ddots & \ddots & & & \\ \hline
& & & & & -\sigma_j & \sigma_j & & \\
& & & & & & \ddots & \ddots & \\
& & & & & & & -\sigma_j & \sigma_j \\
& & & & & z_j\sigma_j & & & -\sigma_j
\end{array}\right]. \tag{7.13}$$

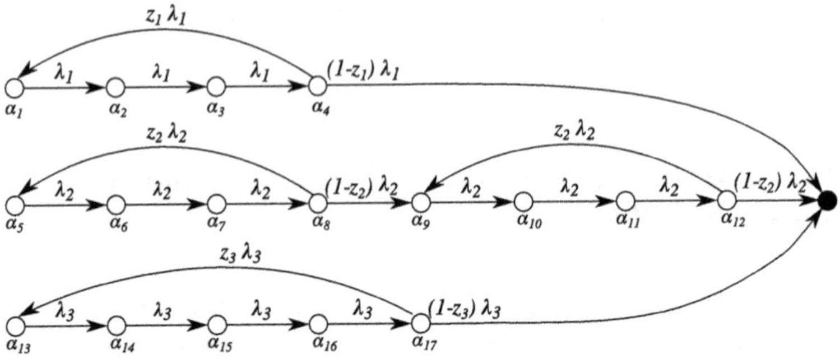

Fig. 7.2 Structure of Hyper-Feedback-Erlang distribution

Having a general non-Markovian representation of an ME distribution, $(\tau, \mathbf{T})$, we look for an equivalent representation $(\alpha, \mathbf{A})$, where $\mathbf{A}$ has a Hyper-FE structure. We denote the jth eigenvalue of $\mathbf{T}$ by λ_j (or, if it is a complex eigenvalue, the complex conjugate eigenvalue pair by $\lambda_j = a_j + b_j\mathrm{i}$ and $\overline{\lambda_j} = a_j - b_j\mathrm{i}$) and its multiplicity by ρ_j. The number of distinct real eigenvalues and complex conjugate eigenvalue pairs is J.

In the generator of the resulting Hyper-FE representation each matrix $\mathbf{M_j}$ in the block diagonal of $\mathbf{A}$ implements one real eigenvalue or a conjugate complex eigenvalue pair of $\mathbf{T}$. The construction of matrix $\mathbf{M_j}$ is performed as follows [13]:

- If λ_j is real, then the corresponding matrix degrades to an Erlang block; thus the parameters of $\mathbf{M_j}$ are

$$\sigma_j = \lambda_j, \quad n_j = 1, \quad z_j = 0, \quad m_j = \rho_j. \tag{7.14}$$

- If λ_j is complex, then the parameters of $\mathbf{M_j}$ are as follows:

$$n_j = \text{the smallest integer for which } a_j/b_j > \tan(\pi/n_j), \tag{7.15}$$

$$\sigma_j = \frac{1}{2}\left(2a_j - b_j \tan\frac{\pi}{n_j} + b_j \cot\frac{\pi}{n_j}\right), \tag{7.16}$$

$$z_j = \left(1 - \left(a_j - b_j \tan\frac{\pi}{n_j}\right)/(2\sigma_j)\right)^{n_j}, \tag{7.17}$$

$$m_j = \rho_j. \tag{7.18}$$

This construction ensures that $\mathbf{A}$ is a valid Markovian transient generator that has all the eigenvalues of $\mathbf{T}$ with the proper multiplicities. However, the FEBs, implementing the complex eigenvalues, introduce "extra" eigenvalues as well, but they do not cause problems because the initial vector α is set such that the "extra" eigenvalues have zero coefficients.

Initial vector α is obtained as follows [5]. Let n and m $(n \le m)$ be the size of $\mathbf{T}$ and $\mathbf{A}$, respectively. Compute matrix $\mathbf{W}$ of size $n \times m$ as the unique solution of

$$\mathbf{TW} = \mathbf{WA}, \quad \mathbf{W}\mathbb{1} = \mathbb{1}, \tag{7.19}$$

and, based on $\mathbf{W}$, the initial vector is

$$\alpha = \tau \cdot \mathbf{W}. \tag{7.20}$$

Vector α is decomposed into subvectors according to the block structure of $\mathbf{A}$, and the vector element associated with state i of block j is denoted by $\alpha_{j,i}$. Like (7.8), the probability density function can be then expressed as

$$f(x) = \alpha e^{\mathbf{A}x}(-\mathbf{A})\mathbb{1} = \sum_{j=1}^{J} \sum_{i=1}^{n_j m_j} \alpha_{j,i} \cdot \underbrace{\mathbb{e}_i e^{\mathbf{M_j}x}(-\mathbf{M_j})\mathbb{1}}_{g_{j,i}(x)}. \tag{7.21}$$

Observe that $(\mathbb{e}_i, \mathbf{M_j})$ is a Markovian representation of $g_{k,i}(x)$, from which it is very efficient to draw random numbers since it is composed by FEBs.

The method to obtain a random variate with density $g_{k,i}(x)$ denoted by $R_{g_{k,i}}$ is as follows:

$$L_{j,i} = n_j m_j - i + 1 + \sum_{\ell=\lceil i/n_j \rceil}^{m_j} n_j \cdot \left\lfloor \frac{\log U_\ell}{\log z_j} \right\rfloor,$$

$$R_{g_{j,i}} = -\frac{1}{\sigma_j} \log \prod_{\ell=1}^{L_{j,i}} U_\ell. \tag{7.22}$$

In this expression, $L_{j,i}$ corresponds to the number of steps (exponential distributions) taken before absorption. The first term, $n_j m_j - i + 1$, is the number of steps taken without feedback, while the sum represents the steps due to feedback: $\lfloor \log U_\ell / \log z_j \rfloor$ is the geometrically distributed random variate for the number of feedback loops, and n_j is the number of extra steps for a feedback loop.

In the case where $\alpha_{j,i} \ge 0, \forall i, j$, generating a random variate from $f(x)$ is simple: draw a discrete random sample with distribution α for the starting point of the Hyper-FE structure, and draw a $g_{j,i}(x)$ distributed random number according to (7.22).

However, if the initial vector has negative elements, then we apply the acceptance–rejection method to generate a random variate as described in the section "Generating Random Variates of Markovian Traffic Models." Utilizing the efficient Hyper-FE structure of $\mathbf{A}$ we generate efficiently the random variate in the third line of Algorithm 1.

In each iteration of the algorithm, before accepting a sample there is exactly one logarithm function computed to obtain a sample from an Erlang distribution of order $L_{j,i}$, and $(m_j - \lceil i/n_j \rceil + 1)$ logarithm functions are computed to draw the number of times a feedback loop is traversed in the FEBs. Note that it is not necessary to evaluate $\log z_j$ every time since it can be precalculated before starting the algorithm. The total number of logarithms evaluated is

$$\#\text{ilog} = \sum_{j=1}^{J} \sum_{i=1}^{n_j m_j} \alpha_{j,i} \cdot (2 + m_j - \lceil i/n_j \rceil). \tag{7.23}$$

As the average number of uniformly distributed random samples required in one iteration before accepting the sample we get

$$\#\text{iuni} = \sum_{j=1}^{J} \sum_{i=1}^{n_j m_j} \alpha_{j,i} \cdot \Bigg[\underbrace{\Big(1 + m_j - \lceil i/n_j \rceil\Big)}_{\text{to evaluate } L_{j,i}} + \underbrace{\Big(n_j m_j - i + 1 + (1 + m_j - \lceil i/n_j \rceil) n_j/(1 - z_j)\Big)}_{E(L_{j,i}) \text{ uniforms required by } R_{g_{j,i}}} \Bigg]. \tag{7.24}$$

Taking into consideration that the mean number of rejections until a sample is accepted is p^*, we have the following mean total number of basic operations:

$$\#\log = \frac{\#\text{ilog}}{p^*}, \quad \#\text{uni} = \frac{\#\text{iuni}}{p^*}. \tag{7.25}$$

Hypo-Feedback-Erlang Representation

Definition 7.7. A Hypo-Feedback-Erlang (Hypo-FE) distribution is defined by an initial probability vector α and a transient generator having the following special structure (Fig. 7.3):

$$\mathbf{A} = \begin{bmatrix} \mathbf{M_1} & \mathbf{M'_1} & & \\ & \mathbf{M_2} & \mathbf{M'_2} & \\ & & \ddots & \\ & & & \mathbf{M_J} \end{bmatrix}, \tag{7.26}$$

where matrices $\mathbf{M_j}$ are defined in (7.13) and

$$\mathbf{M'_j} = (-\mathbf{M_j})\mathbb{1} \cdot \mathbb{e}_1. \tag{7.27}$$

Matrices $\mathbf{M_j}$ are constructed in the same way as in the section "Hyper-FE Representation," and the initial vector is obtained by the same procedure.

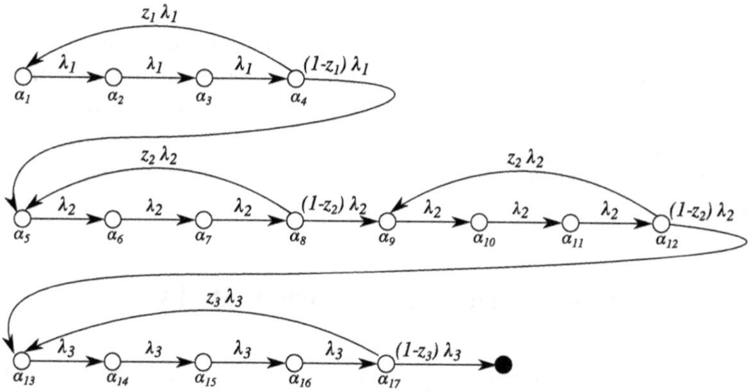

Fig. 7.3 Structure of Hypo-Feedback-Erlang distribution

As with the Hyper-FE structure, from the Hypo-FE structure it is also very efficient to draw random numbers:

$$L_{j,i} = n_j m_j - i + 1 + \sum_{\ell=\lceil i/n_j \rceil}^{m_j} n_j \cdot \left\lfloor \frac{\log U_\ell}{\log z_j} \right\rfloor, \tag{7.28}$$

$$R_{g_{j,i}} = -\frac{1}{\sigma_j} \log \prod_{\ell=1}^{L_{j,i}} U_\ell + \sum_{r=j+1}^{J} (-1) \frac{1}{\sigma_r} \log \prod_{\ell=1}^{L_{r,1}} U_\ell. \tag{7.29}$$

The only difference compared to the Hyper-FE structure is that after the initially selected block (j) is traversed, all consecutive blocks are traversed until the absorption.

The cost of generating a random sample from the Hypo-FE structure is calculated in a manner similar to the Hyper-FE case. The final expressions, including the cost of sample rejections, are

$$\#\text{log} = \frac{1}{p^*} \sum_{j=1}^{J} \sum_{i=1}^{n_j m_j} \alpha_{j,i} \cdot \left[(2 + m_j - \lceil i/n_j \rceil) + \sum_{r=j+1}^{J} (1 + m_r) \right], \tag{7.30}$$

$$\begin{aligned} \#\text{uni} = \frac{1}{p^*} \sum_{j=1}^{J} \sum_{i=1}^{n_j m_j} \alpha_{j,i} \cdot \Bigg[& \underbrace{(1 + m_j - \lceil i/n_j \rceil)}_{\text{to evaluate } L_{j,i}} \\ & + \underbrace{\left(n_j m_j - i + 1 + (1 + m_j - \lceil i/n_j \rceil) n_j / (1 - z_j) \right)}_{E(L_{j,i}) \text{ uniforms required by first term of } R_{g_{j,i}}} \\ & \underbrace{+ \sum_{r=j+1}^{J} \left(m_r + n_r m_r + n_r m_r / (1 - z_r) \right)}_{\text{uniforms required by the sum in } R_{g_{j,i}}} \Bigg]. \end{aligned} \tag{7.31}$$

It might appear that generating a Hypo-FE distributed sample is more expensive compared to a Hyper-FE distributed one due to the additional sum appearing in $R_{g_{j,i}}$ in (7.28). However, the initial vectors (α) of the two representations are different; consequently the mean number of rejections p^* is different as well. In some cases the Hyper-FE and in some other cases the Hypo-FE representations give the better performance, and the performance difference of the different representations can be significant.

Generating Random Variates from Various RAPs

The introduced random number generation method can be used to generate samples of various versions of RAPs. The simple case is where a RAP generates single arrivals of a single type. More complex cases, BRAPs or MRAPs, arise when batch arrivals or arrivals of different types are allowed.

Generating RAP Samples

When generating random variates from RAPs the state vector of the RAP must be stored between consecutive arrivals. Thus, the procedure consists of two steps: in the first step, the interarrival time is drawn (that is, ME distributed with parameters being the current state vector and $\mathbf{H_0}$), then the new state vector is calculated just after the arrival. Consider a RAP with representation (H_0, H_1); the following procedure generates a stationary series of random variates:

1: $\alpha = \tau$
2: **while** samples required **do**
3: $\quad R =$ a random sample from $\mathrm{ME}(\alpha, \mathbf{H_0})$
4: $\quad \alpha = \dfrac{\alpha e^{\mathbf{H_0} R} \mathbf{H_1}}{\alpha e^{\mathbf{H_0} R} \mathbf{H_1} \mathbb{1}}$
5: **end while**

The output of the algorithm is composed of the consecutive R values.

If, additionally, an initial vector of the RAP is known at time 0, then, instead of the stationary initial vector, this initial vector needs to be stored in α in the first step of the algorithm.

Generating MRAP Samples

Consider an MRAP (or, equivalently, a BMRAP) with representation $(H_0, H_1, \ldots, H_K)$; the following procedure generates stationary random samples of the process:

1: $\alpha = \tau$
2: **while** samples required **do**

3: $R =$ random sample from ME$(\alpha, \mathbf{H_0})$
4: **for** $k = 1$ to K **do**
5: $p_k = \alpha e^{\mathbf{H_0} R} \mathbf{H_k} \mathbb{1} / \sum_{j=1}^{K} \alpha e^{\mathbf{H_0} R} \mathbf{H_j} \mathbb{1}$
6: **if** $p_k < 0$ **then**
7: **error "INVALID PROCESS !!!"**
8: **end if**
9: **end for**
10: $B =$ random sample with distribution $\{p_1, \dots p_K\}$
11: **store** R, B
12: $\alpha = \dfrac{\alpha e^{\mathbf{H_0} R} \mathbf{H_B}}{\alpha e^{\mathbf{H_0} R} \mathbf{H_B} \mathbb{1}}$
13: **end while**

In this algorithm, each random sample is a pair representing the interarrival time R and the type of arrival (the batch size) B.

As with the previous RAP sample generation case, the first step of the algorithm needs to be modified if the process starts from an initial vector different from the stationary one.

If any of these algorithms is called with a set of vectors and matrices that do not represent a valid distribution or a valid process, then the procedure might throw out two kinds of error: either in line 10 of Algorithm 1 or in line 7 of the MRAP algorithm (in the case of an MRAP simulation). The first one is due to a negative density in the case of ME simulation or a sample path that results in a negative density in the case of RAP and MRAP simulation. The second one is due to a sample path by which the probability of a type k sample is negative. Indeed, simulation is one of the few available methods to check if a set of matrices defines a valid ME distribution or arrival process.

Numerical Experiments

The two methods presented in the paper were implemented in C++ using the Eigen3 linear algebra library. The implementation revealed that the most time-consuming step of the algorithm is the evaluation of $f_+(x)$ and $f_-(x)$ for every sample. This step is required only when the target distribution has a Hyper-FE or a Hypo-FE representation with some negative elements in the initial vector ($p^* < 1$). The computation of $f_+(x)$ and $f_-(x)$ requires the evaluation of an ME function. Our implementation uses a Jordan-decomposition-based solution for the matrix exponential. The decomposition step must be performed only once during the initialization of the computation. The repeated sampling of an ME distribution requires only the calculation of as many (scalar-) exponentials as the size of the representation of the distribution. The number of the computed scalar exponentials is #iexp. All the results in this section are obtained on an average PC with an Intel Core2 processor running at 3 GHz.

Table 7.1 Number of basic operations required in the case of random PH distributions

	Play method		Hyper-FE				Hypo-FE			
λ	#uni	#log	#iuni	#ilog	#iexp	p^*	#iuni	#ilog	#iexp	p^*
0.1	144.19	71.594	1.0074	1.0039	8	0.99724	16.017	7.6263	0	1
0.5	32.393	15.696	1.0377	1.0192	8	0.98685	13.791	6.4696	0	1
1	17.686	8.3432	1.0747	1.0378	8	0.97469	11.541	5.4732	0	1
2	10.703	4.8514	1.1331	1.0649	8	0.95899	8.8631	4.3851	0	1
4	7.0355	3.0178	1.1992	1.099	7.984	0.93797	6.1525	3.4279	0	1
8	5.1654	2.0827	1.1892	1.0945	7.808	0.94059	4.0318	2.7877	0	1

Generating PH Distributed Samples

In this section we examine how the efficiencies of the proposed procedures compare to the play method for PH distributions. For this reason we generated a large number of random PH distributions of order 8 and executed all the procedures. All the elements of the generator and the initial vector of the PH were uniformly distributed random numbers in $(0,1)$, except the transition rates to the absorbing state, which is considered to be a free parameter (denoted by λ). With this parameter we can control the number of steps before absorption in the play method.

The average number of basic operations is summarized by Table 7.1. In the case of the Hypo-FE- and Hyper-FE-based methods, the cost of computing the exponential function to calculate $f_+(x)$ and $f_-(x)$, if required, appears as well. The p^* parameter, indicating the mean number of rejected samples, is also given in the table. The basic operations #ilog, #iuni, and #iexp are meant for one iteration only. To obtain the total number of basic operations, they must be multiplied by the mean number of iterations, $1/p^*$. Interestingly, the 3,000 random PH distributions generated during the experiment had a valid Hypo-FE representation in all of the cases. In this way the Hypo-FE-based method did not calculate the acceptance probability, $f_+(x)$ and $f_-(x)$; thus no exponential functions were computed. The table shows that the per-iteration cost of the Hyper-FEs is the best among the compared procedures. However, most PH distributions do not have a Hyper-FE representation with $p^* = 1$. As λ increases, some PH distributions have a Hyper-FE representation with $p^* = 1$.

The results of the actual implementations are depicted in Fig. 7.4.

The figure indicates that the play method is very sensitive to the number of steps taken before absorption, while the Hypo-FE- and Hyper-FE-based methods provide an almost constant performance. Interestingly, in spite of the larger cost per iteration, the Hypo-FE-based method provides better performance than the Hyper-FE-based one in several cases because that representation gives better acceptance probability, and thus fewer rejections are required. We can conclude this numerical experiment of generating PH distributed random samples by stating that the Hypo-FE- and Hyper-FE-based methods provide a better performance than the play method if the PH takes several steps until absorption.

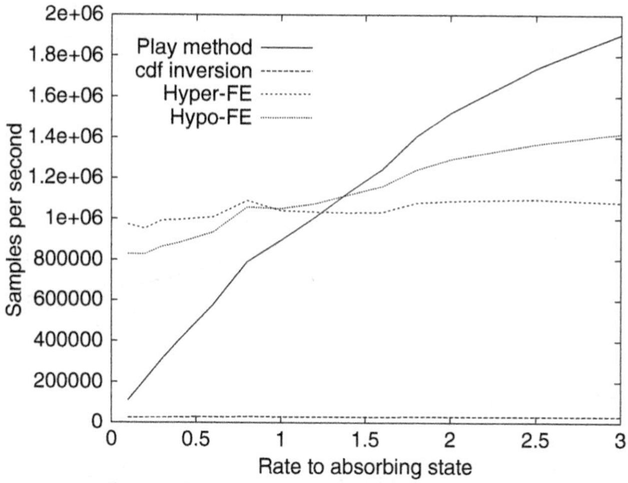

Fig. 7.4 Random samples per second in case of random PH distributions

Fig. 7.5 Probability density function of ME distribution with representation $(\tau, \mathbf{T})$

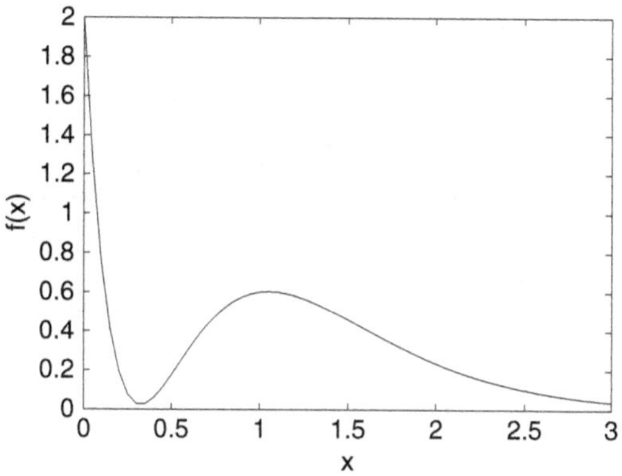

Generating ME Distributed Samples

Consider an ME distribution with representation $(\tau, \mathbf{T})$, where

$$\tau = \{7.69231, -6.69231, 0\}, \quad \mathbf{T} = \begin{pmatrix} -2 & 0 & 0 \\ 0 & -3 & 1 \\ 0 & -1 & -3 \end{pmatrix}.$$

Its PDF is depicted in Fig. 7.5. The eigenvalues of $\mathbf{T}$ are $\{-2, -3+1\mathrm{i}, -3-1\mathrm{i}\}$, and the corresponding FEBs (in both the Hyper-FE and the Hypo-FE representations) are

$$\mathbf{M_1} = -2, \qquad \mathbf{M_2} = \begin{bmatrix} -\sigma & \sigma & 0 \\ 0 & -\sigma & \sigma \\ z\sigma & 0 & \sigma \end{bmatrix}, \tag{7.32}$$

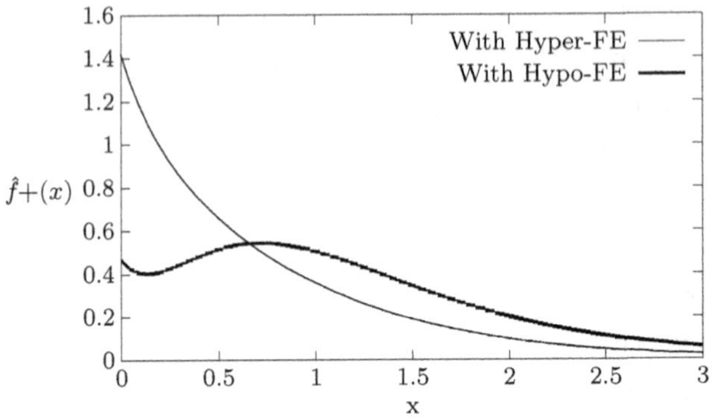

Fig. 7.6 Probability density function $\hat{f}_+(x)$

with $\sigma = 2.42265$ and $z = 0.108277$. The transformation matrix to the Hyper-FE and Hypo-FE representations are obtained based on (7.19).

$$W^{(\text{hyper})} = \begin{bmatrix} 1. & 0. & 0. & 0. \\ 0. & -0.46943 & 0.543647 & 0.925783 \\ 0. & -0.0281766 & -0.82339 & 1.85157 \end{bmatrix},$$

$$W^{(\text{hypo})} = \begin{bmatrix} -0.11547 & 0.0281766 & 0.16151 & 0.925783 \\ 0 & -0.46943 & 0.543647 & 0.925783 \\ 0 & -0.0281766 & -0.82339 & 1.85157 \end{bmatrix}.$$

Based on these transformation matrices the initial vectors of the Hyper-FE and Hypo-FE representations are

$$\alpha^{(\text{hyper})} = \begin{bmatrix} 7.69231 & 3.14157 & -3.63825 & -6.19563 \end{bmatrix},$$

$$\alpha^{(\text{hypo})} = \begin{bmatrix} -0.888231 & 3.35832 & -2.39587 & 0.925783 \end{bmatrix}.$$

Based on the initial vectors the mean number of required iterations can be obtained as

$$1/p^{*(\text{hyper})} = \sum_{i \in \mathcal{A}_+} \alpha^{(\text{hyper})} = 10.83388,$$

$$1/p^{*(\text{hypo})} = 4.284103;$$

thus, more than twice as many rejections occur when using the Hyper-FE structure.

To illustrate the behavior of Algorithm 1, Figs. 7.6 and 7.7 depict the density to draw samples from, $f_+(x)$, and the acceptance probability function, $p_{\text{accept}}(x)$, respectively. It can be observed that the $f_+(x)$ density corresponding to the Hypo-FE representation captures the behavior of the original PDF better; thus the acceptance probabilities are higher. It can also be observed that the original PDF approaches 0 at around $x = 0.32$. This behavior cannot be captured by the PH distribution of low

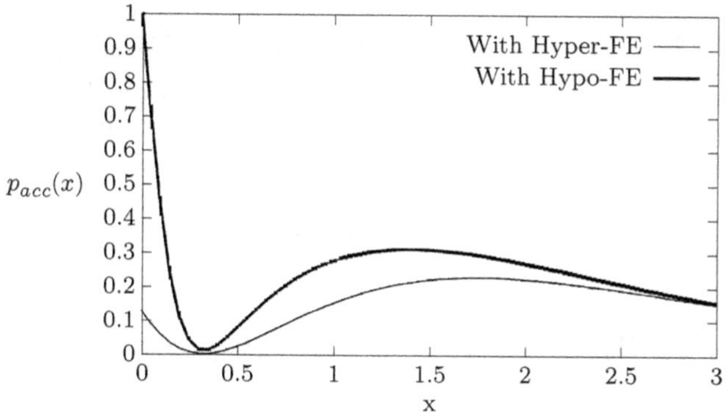

Fig. 7.7 Probability of accepting sample x

Table 7.2 Comparison of three methods for generating ME distributed samples

Method	Number of uniform random numbers generation	Number of log operations	Number of exp operations	p^*	Samples/s
CDF inversion	1	0	324.83	n/a	54,869
Hyper-FE	28.998	13.428	31.681	0.094693	179,560
Hypo-FE	27.51	8.022	12.033	0.24932	277,581

order, which is why the acceptance–rejection method is required. The acceptance probability function takes a very low value to ensure the low density of the samples around $x = 0.32$.

The number of basic operations per random sample (the number of uni, log and exp operations) and the overall performance of the methods (samples/s) are summarized in Table 7.2. The Hypo-FE-based method is five times faster than the CDF inversion-based method for this example.

Generating RAP Samples

From the section "Generating Random Variates from Various RAPs" it is obvious that random samples from a RAP can be generated efficiently once we have an efficient method to draw ME distributed random numbers. Through the previous two examples the behavior of the presented acceptance–rejection methods has been studied and compared in detail. Here we provide a simpler example to demonstrate the efficiency of our methods for generating samples from a RAP.

Table 7.3 Comparison of three methods for generating RAP samples

Method for ME	Samples/s
CDF inversion	55,872
Hyper-FE	336,247
Hypo-FE	329,224

The matrices of the RAP used in this example are as follows:

$$\mathbf{H_0} = \begin{bmatrix} -2 & 0 & 0 \\ 0 & -3 & 1 \\ 0 & -1 & -2 \end{bmatrix}, \quad \mathbf{H_1} = \begin{bmatrix} 1.8 & 0.2 & 0 \\ 0.2 & 1.8 & 0 \\ 0.2 & 1.8 & 1 \end{bmatrix}. \tag{7.33}$$

A significant performance hit over the ME distributed random number generators is that an ME function must be evaluated after a sample is draw to calculate the initial state vector for the next arrival. However, this time-consuming step is required in all methods for generating random variates. Consequently, we expect lower performance than in the case of ME distributed random sample generation, but according to Table 7.3, the Hyper-FE- and Hypo-FE-based methods are still six times faster than the CDF inversion-based one in this particular example. The number of basic operations is omitted since it varies with the initial vector in each step.

Conclusions

This chapter proposes acceptance–rejection-based numerical methods for generating ME, RAP, and MRAP samples. The key of the numerical efficiency of the acceptance–rejection-based methods is the high acceptance probability and the low computational cost of elementary random number generation. Numerical investigations show that both of these elements depend on the representation of the models. We investigated the efficiency of two FEB-based representations, which were relatively efficient in a wide range of cases, but optimal representations of these models, which make the simulation most efficient, are still open research problems.

References

1. Asmussen, S., Bladt, M.: Point processes with finite-dimensional conditional probabilities. Stoch. Process. Appl. **82**, 127–142 (1999)
2. Bean, N.G., Nielsen, B.F.: Quasi-birth-and-death processes with rational arrival process components. Stoch. Models **26**(3), 309–334 (2010)
3. Brown, E., Place, J., de Liefvoort, A.V.: Generating matrix exponential random variates. Simulation **70**, 224–230 (1998)

4. Buchholz, P., Telek, M.: Stochastic Petri nets with matrix exponentially distributed firing times. Perform. Eval. **67**(12), 1373–1385 (2010)
5. Buchholz, P., Telek, M.: On minimal representations of rational arrival processes. Ann. Oper. Res. (2011). doi:10.1007/s10479-011-1001-5 (to appear)
6. Cumani, A.: On the canonical representation of homogeneous Markov processes modelling failure-time distributions. Microelectron. Reliab. **22**, 583–602 (1982)
7. He, Q.M., Neuts, M.: Markov arrival processes with marked transitions. Stoch. Process. Appl. **74**, 37–52 (1998)
8. Kronmal, R., Peterson, A.: On the alias method for generating random variables from a discrete distribution. Am. Stat. **33**(4), 214–218 (1979)
9. Latouche, G., Ramaswami, V.: Introduction to Matrix-Analytic Methods in Stochastic Modeling. Society for Industrial and Applied Mathematics (1999)
10. van de Liefvoort, A.: The moment problem for continuous distributions. Technical report, WP-CM-1990-02, University of Missouri, Kansas City (1990)
11. Lipsky, L.: Queueing Theory: A Linear Algebraic Approach. MacMillan, New York (1992)
12. Mitchell, K.: Constructing a correlated sequence of matrix exponentials with invariant first order properties. Oper. Res. Lett. **28**, 27–34 (2001)
13. Mocanu, S., Commault, C.: Sparse representations of phase-type distributions. Comm. Stat. Stoch. Model **15**(4), 759–778 (1999)
14. Neuts, M.F.: Matrix-Geometric Solutions in Stochastic Models. An Algorithmic Approach. Dover, New York (1981)
15. Neuts, M.F., Pagano, M.E.: Generating random variates from a distribution of phase type. In: WSC '81: Proceedings of the 13th Conference on Winter Simulation, pp. 381–387. IEEE Press, Piscataway (1981)
16. Reinecke, P., Wolter, K., Bodrog, L., Telek, M.: On the cost of generating PH-distributed random numbers. In: International Workshop on Performability Modeling of Computer and Communication Systems (PMCCS), pp. 1–5. Eger, Hungary (2009)
17. Reinecke, P., Telek, M., Wolter, K.: Reducing the cost of generating APH-distributed random numbers. In: 15th International Conference on Measurement, Modelling and Evaluation of Computing Systems (MMB). Lecture Notes in Computer Science, vol. 5987, pp. 274–286. Springer, Essen (2010)
18. Robert, C., Casella, G.: Monte Carlo Statistical Methods. Springer, New York (2004)
19. Telek, M., Horváth, G.: A minimal representation of Markov arrival processes and a moments matching method. Perform. Eval. **64**(9–12), 1153–1168 (2007)

Chapter 8
Revisiting the Tail Asymptotics of the Double QBD Process: Refinement and Complete Solutions for the Coordinate and Diagonal Directions

Masahiro Kobayashi and Masakiyo Miyazawa

Introduction

We are concerned with a two-dimensional reflecting random walk on a nonnegative integer quadrant, which is the set of two-dimensional vectors (i, j) such that i, j are nonnegative integers. We assume that it is skip free in all directions, that is, its increments in each coordinate direction are at most one in absolute value. The boundary of the quadrant is partitioned into three faces: the origin and the two coordinate axes in the quadrant. We assume that the transition probabilities of this random walk are homogeneous on each boundary face, but they may change on different faces or the interior of the quadrant, that is, inside of the boundary.

This reflecting random walk is referred to as a double quasi-birth-and-death (QBD) process in [18]. This process can be used to describe a two-node queueing network under various settings such as server collaboration and simultaneous arrivals and departures, and its stationary distribution is important for the performance evaluation of such a network model. The existence of the stationary distribution, that is, stability, is nicely characterized, but the stationary distribution is hard to obtain analytically except for some special cases. Because of this and its own importance, research interest has been directed at its tail asymptotics.

Until now, the tail asymptotics for the double QBD have been obtained in terms of its modeling primitives under the most general setting by Miyazawa [18], while less explicit results have been obtained for more general two-dimensional reflecting random walks by Borovkov and Mogul'skii [2]. Foley and McDonald [10, 11] studied the double QBD under some limitations. Recently, Kobayashi and Miyazawa [13] modified the double QBD process in such a way that upward jumps may be unbounded; they also studied its tail asymptotics. This process, called a double $M/G/1$ type, includes the double QBD process as a special case. For

M. Kobayashi • M. Miyazawa (✉)
Tokyo University of Science, Noda, Chiba 278-8510, Japan
e-mail: masahilow@gmail.com; miyazawa@is.noda.tus.ac.jp

G. Latouche et al. (eds.), *Matrix-Analytic Methods in Stochastic Models*, Springer Proceedings in Mathematics & Statistics 27, DOI 10.1007/978-1-4614-4909-6_8,

special cases such as tandem and priority queues, the tail asymptotics were recently investigated in Guillemin and Leeuwaarden [12] and Li and Zhao [14, 15]. Li and Zhao [16] challenged the general double QBD process (see additional note at the end of this section).

Tail asymptotic problems have also been studied for a semimartingale reflecting Brownian motion (SRBM), which is a continuous-time-and-state counterpart of a reflecting random walk. For the two-dimensional SRBM, the rate function for large deviations was obtained under a certain extra assumption in Avram et al. [1]. Dai and Miyazawa [3] derived more complete answers but for stationary marginal distributions.

Thus, we now have many studies of the tail asymptotics for two-dimensional reflecting and related processes (see, e.g., Miyazawa [19] for a survey). Nevertheless, many problems remain unsolved even for the double QBD process. The exact tail asymptotics of the stationary marginal distributions in the coordinate directions are one such problem. Here, a sequence of nonnegative numbers $\{p(n); n = 0, 1, 2\}$ is said to have exact tail asymptotic $\{h(n); n = 0, 1, \ldots\}$ if its ratio $p(n)/h(n)$ converges to a positive constant as n goes to infinity. We also write this asymptotic as

$$p(n) \sim h(n).$$

We will find $h(n) = n^{\kappa} a^{-n}$ or $n^{\kappa}(1 + b(-1)^n) a^{-n}$ with constants $\kappa = -\frac{3}{2}, -\frac{1}{2}, 0, 1$, $a > 1$ and $|b| \leq 1$ for marginal distributions (and for stationary probabilities on the boundaries).

We aim to completely solve the exact tail asymptotics of stationary marginal distributions in the coordinate and diagonal directions, provided a stationary distribution exists. It is known that the tail asymptotics of the stationary probabilities on each coordinate axis are one of their key features (e.g., see Miyazawa [19]). These asymptotics are studied by Borovkov and Mogul'skii [2] and Miyazawa [18]. The researchers used Markov additive processes generated by removing one of the boundary faces that is not the origin and related their asymptotics. However, there are some limitations in that approach.

In this chapter, we revisit the double QBD process using a different approach that has been recently developed [3, 13, 20]. This approach is purely analytic and is called an analytic function method. It is closely related to the kernel method used in various studies [12, 14, 15]. Its details and related topics are reviewed by Miyazawa [19].

The analytic function method [3, 13, 20] only uses moment-generating functions because they have nice analytic properties including convexity. However, a generating function is more convenient for a distribution of integers because they are polynomials. Thus, generating functions have been used in the kernel method.

In this chapter, we use both generating functions and moment-generating functions. We first consider the convergence domain of the moment-generating function of a stationary distribution, which is two-dimensional. This part mainly refers to recent results from Kobayashi and Miyazawa [13]. Once the domain is obtained, we switch from a moment-generating function to a generating function and consider analytic behaviors around its dominant singular points. A key is the so-called kernel function. We derive inequalities for it (Lemma 8.8), adapting the idea presented by Dai and Mayazawa [3]. This is a crucial step in the present approach, which enables us to apply analytic extensions not using the Riemann surface that has been typically used in the kernel method. We then apply the inversion technique for generating functions and derive the exact tail asymptotics of the stationary tail probabilities on the coordinate axes.

The asymptotic results are exemplified by a two-node queueing network with simultaneous arrivals. This model is an extension of a two-parallel-queue model with simultaneous arrivals. For the latter, the tail asymptotics of its stationary distribution in the coordinate directions are obtained in from Flatto and Hahn [8] and Flatto and McKean [9]. We modify this model in such a way that a customer who has completed service may be routed to another queue with a given probability. Thus, our model is more like a Jackson network, but it does not have a product-form stationary distribution because of simultaneous arrivals. We will discuss how we can see the tail asymptotics from the modeling primitives.

This chapter is composed of seven sections. In the section "Double QBD Process and the Convergence Domain," we introduce the double QBD process and summarize existence results using moment-generating functions. The section "Analytic Function Method" considers generating functions for stationary probabilities on the coordinate axes. Analytic behaviors around their dominant singular points are studied. We then apply the inversion technique and derive exact asymptotics in the sections "Exact Tail Asymptotics for the Nonarithmetic Case" and "Exact Tail Asymptotics for the Arithmetic Case." An example for simultaneous arrivals is considered in the section "Application to a Network with Simultaneous Arrivals." We discuss some remaining problems in "Concluding Remarks."

(*Additional note*) After the first submission of this chapter, we have learned that Li and Zhao [16] studied the same exact tail asymptotic problem, including the case where the tail asymptotics is periodic. This periodic case was absent in our original submission and added in the present chapter. Thus, we benefited from Li and Zhao's work. However, our approach is different from theirs, although both use analytic functions and its asymptotic inversions. That is, the crucial step in Li and Zhao's case [16] is analytic extensions on a Riemann surface studied by Fayoelle et al. [6], whereas we use the convergence domain obtained by Kobayashi and Miyazawa [13] and the key lemma. Another difference is in the sorting of tail asymptotic results. Their presentation is purely analytic while we use the geometrical classifications of [13, 18] (see also Miyazawa [19]).

Double QBD Process and the Convergence Domain

The double QBD process was introduced and studied by Miyazawa [18]. Here we briefly introduce it and present results of the tail asymptotics of its stationary distribution. We use the following set of numbers:

$$\mathbb{Z} = \text{ the set of all integers,} \qquad \mathbb{Z}_+ = \{j \in \mathbb{Z}; j \geq 0\};$$
$$\mathbb{U} = \{(i,j) \in \mathbb{Z}^2; i,j = 0,1,-1\};$$
$$\mathbb{R} = \text{ the set of all real numbers,} \qquad \mathbb{R}_+ = \{x \in \mathbb{R}; x \geq 0\};$$
$$\mathbb{C} = \text{ the set of all complex numbers.}$$

Let $S = \mathbb{Z}_+^2$, which is a state space for the double QBD process. Define the boundary faces of S as

$$S_0 = \{(0,0)\}, \quad S_1 = \{(i,0) \in \mathbb{Z}_+^2; i \geq 1\}, \quad S_2 = \{(0,i) \in \mathbb{Z}_+^2; i \geq 1\}.$$

Let $\partial S = \cup_{i=0}^2 S_i$ and $S_+ = S \setminus \partial S$. We refer to ∂S and S_+ as the boundary and interior of S, respectively.

Let $\{\mathbf{Y}_\ell; \ell = 0,1,\ldots\}$ be a skip-free random walk on $\mathbb{Z}^2$. That is, its increments $\mathbf{X}_\ell^{(+)} \equiv \mathbf{Y}_\ell - \mathbf{Y}_{\ell-1}$ take values in $\mathbb{U}$ and are independent and identically distributed. By $\mathbf{X}^{(+)}$ we simply denote a random vector that has the same distribution as $\mathbf{X}_\ell^{(+)}$. Define a discrete-time Markov chain $\{\mathbf{L}_\ell\}$ with state space S by the following transition probabilities:

$$P(\mathbf{L}_{\ell+1} = \mathbf{j} | \mathbf{L}_\ell = \mathbf{i}) = \begin{cases} P(\mathbf{X}^{(+)} = \mathbf{j} - \mathbf{i}), \mathbf{j} \in S, \mathbf{i} \in S_+, \\ P(\mathbf{X}^{(k)} = \mathbf{j} - \mathbf{i}), \ \mathbf{j} \in S, \mathbf{i} \in S_k, k = 0,1,2, \end{cases}$$

where $\mathbf{X}^{(k)}$ is a random vector taking values in $\{(i_1,i_2) \in \mathbb{U}; i_{3-k} \geq 0\}$ for $k = 1,2$ and in $\{(i_1,i_2) \in \mathbb{U}; i_1, i_2 \geq 0\}$ for $k = 0$. Hence, we can write

$$\mathbf{L}_{\ell+1} = \mathbf{L}_\ell + \sum_{k=0,1,2,+} \mathbf{X}_\ell^{(k)} 1(\mathbf{L}_\ell \in S_k), \qquad \ell = 0,1,2,\ldots, \tag{8.1}$$

where $1(\cdot)$ is the indicator function of the statement "$\cdot$" and $\mathbf{X}_\ell^{(k)}$ has the same distribution as that of $\mathbf{X}^{(k)}$ for each $k = 0,1,2,+$, and is independent of everything else.

Thus, $\{\mathbf{L}_\ell\}$ is a skip-free reflecting random walk on the nonnegative integer quadrant S, which is called a double QBD process because its QBD transition structure is unchanged when the level and background states are exchanged.

We denote the moment-generating functions of $\mathbf{X}^{(k)}$ by γ_k, that is, for $\theta \equiv (\theta_1, \theta_2) \in \mathbb{R}^2$,

$$\gamma_k(\theta) = E(\mathrm{e}^{\langle \theta, \mathbf{X}^{(k)} \rangle}), \qquad k = 0,1,2,+,$$

where $\langle \mathbf{a}, \mathbf{b} \rangle = a_1 b_1 + a_2 b_2$ for $\mathbf{a} = (a_1, a_2)$ and $\mathbf{b} = (b_1, b_2)$. As usual, $\mathbb{R}^2$ is considered to be a metric space with Euclidean norm $\|\mathbf{a}\| \equiv \sqrt{\langle \mathbf{a}, \mathbf{a} \rangle}$. In particular, a vector $\mathbf{c}$ is called a directional vector if $\|\mathbf{c}\| = 1$. In this chapter, we assume that

(i) The random walk $\{\mathbf{Y}_\ell\}$ is irreducible,
(ii) The reflecting process $\{\mathbf{L}_\ell\}$ is irreducible and aperiodic, and
(iii) Either $E(X_1^{(+)}) \neq 0$ or $E(X_2^{(+)}) \neq 0$ for $\mathbf{X}^{(+)} = (X_1^{(+)}, X_2^{(+)})$.

Remark 8.1. If $E(X_1^{(+)}) = E(X_2^{(+)}) = 0$, then it is known that the stationary distribution of $\{\mathbf{L}_\ell\}$ cannot have a light tail, that is, it cannot geometrically (or exponentially) decay in all directions (see Fayoelle et al. [5] and Remark 3.1

of Kobayashi and Miyazawa [13]). Thus, assumption (iii) is not a restrictive assumption for considering the light tail.

Under these assumptions, tractable conditions are obtained for the existence of the stationary distribution in the book [5]. They recently have been corrected by Kobayashi and Miyazawa [13]. We refer to this corrected version below.

Lemma 8.1 (Lemma 2.1 of [13]). *Assume conditions (i)–(iii), and let*

$$\begin{aligned}\mathbf{m} &= (E(X_1^{(+)}), E(X_2^{(+)})),\\ \mathbf{m}_{\perp}^{(1)} &= (E(X_2^{(1)}), -E(X_1^{(1)})),\\ \mathbf{m}_{\perp}^{(2)} &= (-E(X_2^{(2)}), E(X_1^{(2)})).\end{aligned}$$

Then the reflecting random walk $\{\mathbf{L}_\ell\}$ has a stationary distribution if and only if any one of the following three conditions holds [13]).

$$m_1 < 0, m_2 < 0, \langle \mathbf{m}, \mathbf{m}_{\perp}^{(1)} \rangle < 0, \langle \mathbf{m}, \mathbf{m}_{\perp}^{(2)} \rangle < 0; \tag{8.2}$$

$$m_1 \geq 0, m_2 < 0, \langle \mathbf{m}, \mathbf{m}_{\perp}^{(1)} \rangle < 0; \textit{ in addition, } m_2^{(2)} < 0 \textit{ is needed if } m_1^{(2)} = 0; \tag{8.3}$$

$$m_1 < 0, m_2 \geq 0, \langle \mathbf{m}, \mathbf{m}_{\perp}^{(2)} \rangle < 0; \textit{in addition, } m_1^{(1)} < 0 \textit{ is needed if } m_2^{(1)} = 0. \tag{8.4}$$

Throughout the chapter, we also assume this stability condition. That is,

(iv) Any one of (8.2), (8.3), or (8.4) holds.

In addition to conditions (i)–(iv), we will use the following conditions to distinguish some periodical nature of the tail asymptotics:

(v-a) $P(\mathbf{X}^{(+)} \in \{(1,1),(-1,1),(0,0),(1,-1),(-1,-1)\}) < 1.$
(v-b) $P(\mathbf{X}^{(1)} \in \{(1,1),(0,0),(-1,1)\}) < 1.$
(v-c) $P(\mathbf{X}^{(2)} \in \{(1,1),(0,0),(1,-1)\}) < 1.$

These conditions are said to be nonarithmetic in the interior and boundary faces $1,2$, respectively, while the conditions under which they do not hold are called arithmetic. The remark below explains why they are so called.

Remark 8.2. To see the meaning of these conditions, let us consider random walk $\{\mathbf{Y}_\ell\}$ on $\mathbb{Z}^2$. We can view this random walk as a Markov additive process in the kth coordinate direction if we consider the kth entry of $\mathbf{Y}_\ell$ as an additive component and the other entry as a background state ($k = 1,2$). Then, condition (v-a) is exactly the nonarithmetic condition of this Markov additive process in each coordinate direction (see [21] for a definition of the period of a Markov additive process). For random walk $\{\mathbf{Y}_\ell\}$, if the Markov additive process in one direction is nonarithmetic, then the one in the other direction is also nonarithmetic.

We can give similar interpretations for (v-b) and (v-c). That is, for each $k = 1,2$ consider a random walk with increments subject to the same distribution as $\mathbf{X}^{(k)}$. This random walk is also viewed as a Markov additive process with an additive component in the kth coordinate direction. Then, (v-b) and (v-c) are the nonarithmetic conditions of this Markov additive process for $k = 1, 2$, respectively.

Remark 8.3. These conditions were recently studied by Li and Zhao in [16]. The authors of that study call a probability distribution on $\mathbb{U} \equiv \{(i, j); i, j = -1, 0, 1\}$ X-shaped if its support is included in

$$\{(1,1), (-1,1), (0,0), (1,-1), (-1,-1)\}.$$

Thus, conditions (v-a), (v-b), and (v-c) are for $\mathbf{X}^{(+)}$, $\mathbf{X}^{(1)}$, and $\mathbf{X}^{(2)}$, respectively, not X-shaped.

We denote the stationary distribution of $\{\mathbf{L}_\ell; \ell = 0, 1, \ldots\}$ by ν and let $\mathbf{L}$ be a random vector subject to ν. Then, it follows from (8.1) that

$$\mathbf{L} \simeq \mathbf{L} + \sum_{k=0,1,2,+} \mathbf{X}^{(k)} 1(\mathbf{L} \in S_k), \tag{8.5}$$

where "$\simeq$" stands for the equality in distribution. We introduce four moment-generating functions concerning ν. For $\theta \in \mathbb{R}^2$,

$$\begin{aligned} \varphi(\theta) &= E(e^{\langle \theta, \mathbf{L} \rangle}), \\ \varphi_+(\theta) &= E(e^{\langle \theta, \mathbf{L} \rangle} 1(\mathbf{L} \in S_+)), \\ \varphi_k(\theta_k) &= E(e^{\theta_k L_k} 1(\mathbf{L} \in S_k)), \qquad k = 1, 2. \end{aligned}$$

Then, from (8.5) and the fact that

$$\varphi(\theta) = \varphi_+(\theta) + \sum_{k=1}^{2} \varphi_k(\theta_k) + \nu(\mathbf{0})$$

we can easily derive the stationary equation

$$\begin{aligned} &(1 - \gamma_+(\theta))\varphi_+(\theta) + (1 - \gamma_1(\theta))\varphi_1(\theta_1) \\ &\quad + (1 - \gamma_2(\theta))\varphi_2(\theta_2) + (1 - \gamma_0(\theta))\nu(\mathbf{0}) = 0 \end{aligned} \tag{8.6}$$

as long as $\varphi(\theta)$ is finite. Clearly, this finiteness holds for $\theta \leq \mathbf{0}$.

To find the maximal region for (8.6) to be valid, we define the convergence domain of φ as

$$\mathcal{D} = \text{the interior of } \{\theta \in \mathbb{R}^2; \varphi(\theta) < \infty\}.$$

This domain is obtained by Kobayashi and Miyazawa [13]. To present this result, we introduce notations.

From (8.6) we can see that the curves $1 - \gamma_k(\theta) = 0$ for $k = +, 1, 2$ are keys for $\varphi(\theta)$ to be finite. Thus, we let

$$\begin{aligned} \Gamma_k &= \{\theta \in \mathbb{R}^2; \gamma_k(\theta) < 1\}, \\ \partial\Gamma_k &= \{\theta \in \mathbb{R}^2; \gamma_k(\theta) = 1\}, \qquad k = 1, 2, +. \end{aligned}$$

We denote the closure of Γ_k by $\overline{\Gamma}_k$. Since γ_k is a convex function, Γ_k and $\overline{\Gamma}_k$ are convex sets. Furthermore, condition (i) implies that Γ_+ is bounded, that is, it is included in a ball in $\mathbb{R}^2$. Let

$$\begin{aligned}
\theta^{(k,\mathrm{r})} &= \arg_{\theta\in\mathbb{R}^2} \sup\{\theta_k;\theta \in \Gamma_+ \cap \Gamma_k\}, \quad k = 1,2,\\
\theta^{(k,\min)} &= \arg_{\theta\in\mathbb{R}^2} \inf\{\theta_k;\theta \in \Gamma_+\},\\
\theta^{(k,\max)} &= \arg_{\theta\in\mathbb{R}^2} \sup\{\theta_k;\theta \in \Gamma_+\}.
\end{aligned}$$

These extreme points play key roles in obtaining the convergence domain. It is notable that $\theta^{(k,\mathrm{r})}$ is not the zero vector $\mathbf{0}$ because stability condition (iv) implies that, for each $k = 1,2$, $\Gamma_+ \cap \Gamma_k$ contains $\theta = (\theta_1, \theta_2)$ such that $\theta_k > 0$ (see Lemma 2.2 of [13]).

We further need the following points:

$$\theta^{(k,\Gamma)} = \begin{cases} \theta^{(k,\mathrm{r})}, & \gamma_k(\theta^{(k,\max)}) > 1,\\ \theta^{(k,\max)}, & \gamma_k(\theta^{(k,\max)}) \le 1, \end{cases} \qquad k = 1,2.$$

According to Miyazawa [18] (see also [3]), we classify the model into the following three categories:

$$\begin{array}{ll}
\text{Category I} & \theta_1^{(2,\Gamma)} < \theta_1^{(1,\Gamma)} \text{ and } \theta_2^{(1,\Gamma)} < \theta_2^{(2,\Gamma)},\\
\text{Category II} & \theta_1^{(2,\Gamma)} < \theta_1^{(1,\Gamma)} \text{ and } \theta_2^{(1,\Gamma)} \ge \theta_2^{(2,\Gamma)},\\
\text{Category III} & \theta_1^{(2,\Gamma)} \ge \theta_1^{(1,\Gamma)} \text{ and } \theta_2^{(1,\Gamma)} < \theta_2^{(2,\Gamma)}.
\end{array}$$

Note that it is impossible to have $\theta_1^{(2,\Gamma)} \ge \theta_1^{(1,\Gamma)}$ and $\theta_2^{(1,\Gamma)} \ge \theta_2^{(2,\Gamma)}$ at once because $\theta_1^{(2,\Gamma)} \ge \theta_1^{(1,\Gamma)}$ and the convexity of Γ_+ imply that $\theta_2^{(1,\Gamma)} \le \theta_2^{(2,\Gamma)}$ (see Sect. 4 of [18]). We further note that $\theta_2^{(1,\Gamma)} \ge \theta_2^{(2,\Gamma)}$ can be replaced by $\theta_2^{(1,\Gamma)} = \theta_2^{(2,\Gamma)}$ in category II. Similarly, $\theta_1^{(2,\Gamma)} \ge \theta_1^{(1,\Gamma)}$ can be replaced by $\theta_1^{(2,\Gamma)} = \theta_1^{(1,\Gamma)}$ in category III.

Define the vector τ as

$$\tau = \begin{cases} (\theta_1^{(1,\Gamma)}, \theta_2^{(2,\Gamma)}), & \text{for category I},\\ (\overline{\xi}_1(\theta_2^{(2,\mathrm{r})}), \theta_2^{(2,\mathrm{r})}), & \text{for category II},\\ (\theta_1^{(1,\mathrm{r})}, \overline{\xi}_2(\theta_1^{(1,\mathrm{r})})), & \text{for category III}, \end{cases}$$

where $\overline{\xi}_k(\theta_{3-k}) = \sup\{\theta_k; (\theta_1, \theta_2) \in \Gamma_+\}$. This definition of τ shows that categories I–III are convenient.

We are now ready to present results on the convergence domain $\mathcal{D}$ and the tail asymptotics obtained by Kobayashi and Miyazawa [13]. As was mentioned in the section "Introduction," they are obtained for the more general reflecting random walk. Thus, some of their conditions automatically hold for the double QBD process (Fig. 8.1).

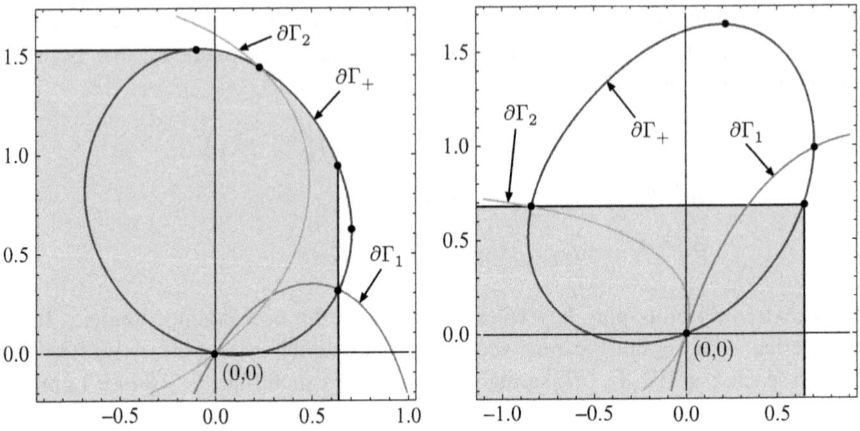

Fig. 8.1 The *light-green* areas are domains $\mathcal{D}$ for categories I and II

Lemma 8.2 (Theorem 3.1 of [13]).

$$\mathcal{D} = \{\theta \in \mathbb{R}^2; \theta < \tau \text{ and } \exists \theta' \in \Gamma_+ \text{ such that } \theta < \theta'\}. \tag{8.7}$$

Theorem 8.1 (Theorem 4.2 of [13]). *Under conditions (i)–(iv), we have, for* $k = 1, 2$,

$$\lim_{n\to\infty} \frac{1}{n} \log P(L_k \geq n, L_{3-k} = 0) = -\tau_k, \tag{8.8}$$

and, for any directional vector $\mathbf{c} \geq \mathbf{0}$,

$$\lim_{n\to\infty} \frac{1}{x} \log P(\langle \mathbf{c}, \mathbf{L} \rangle \geq x) = -\alpha_{\mathbf{c}}, \tag{8.9}$$

where we recall that $\alpha_{\mathbf{c}} = \sup\{x \geq 0; x\mathbf{c} \in \mathcal{D}\}$. *Furthermore, if* $\gamma(\alpha_{\mathbf{c}}\mathbf{c}) = 1$ *and if* $\gamma_k(\alpha_{\mathbf{c}}\mathbf{c}) \neq 1$ *and* $\alpha_{\mathbf{c}} c_k \neq \theta_{kk}^{(\infty)}$ *for* $k = 1, 2$, *then we have the following exact asymptotics:*

$$\lim_{x\to\infty} e^{\alpha_{\mathbf{c}} x} P(\langle \mathbf{c}, \mathbf{L} \rangle \geq x) = b_{\mathbf{c}}. \tag{8.10}$$

In this chapter, we aim to refine these asymptotics to be exact when $\mathbf{c}$ is $(1,0)$, $(0,1)$, or $(1,1)$. Recall that a sequence of nonnegative numbers $\{p(n); n \in \mathbb{Z}_+\}$ is said to have the exact asymptotic $(1 + b(-1)^n)n^{-\kappa}\alpha^{-n}$ for constants κ and $\alpha > 1$ if there exist real numbers $b \in [-1, 1]$ and a positive constant c such that

$$\lim_{n\to\infty} (1 + b(-1)^n) n^{\kappa} \alpha^n p(n) = c. \tag{8.11}$$

We note that, if $b = 0$, then this asymptotic is equivalent to

$$\lim_{n\to\infty}(1+b(-1)^n)n^{\kappa}\alpha^n\sum_{\ell=n}^{\infty}p(\ell)=c' \tag{8.12}$$

for some $c' > 0$. Thus, if $b = 0$, then there is no difference on the exact asymptotic between $P(L_k \geq n)$ and $P(L_k = n)$. In what follows, we are mainly concerned with the latter type of exact asymptotics.

Analytic Function Method

Our basic idea for deriving exact asymptotics is to adapt the method used in [3], which extends the moment-generating functions to complex variable analytic functions and obtains the exact tail asymptotics from analytic behavior around their singular points. A similar method is called a kernel method in some literature [12, 14–16]. Here we call it an analytic function method because our approach uses the convergence domain $\mathcal{D}$ heavily, which is not the case for the kernel method. See [19] for more details.

There is one problem in adapting the method of [3] because the moment-generating functions $\gamma_k(\theta)$ are not polynomials, while the corresponding functions of SRBM are polynomial. If they are not polynomials, the analytic function approach is hard to apply. This problem is resolved if we use generating functions instead of moment-generating functions. We here thanks for the skip-free assumption.

Convergence Domain of a Generating Function

Let us convert results on moment-generating functions to those on generating function using a mapping from $\mathbf{z} \equiv (z_1, z_2) \in \mathbb{C}$ to $\mathbf{g}(\mathbf{z}) \equiv (e^{z_1}, e^{z_2}) \in \mathbb{C}$. In particular, for $\theta \in \mathbb{R}^2$, $\mathbf{g}(\theta) \in (\mathbb{R}_+^{\mathrm{o}})^2$, where $\mathbb{R}_+^{\mathrm{o}} = (0,\infty)$. We use the following notations for $k = 1, 2$:

$$\left(u_1^{(k,\min)}, u_2^{(k,\min)}\right) = \mathbf{g}\left(\theta^{(k,\min)}\right), \qquad \left(u_1^{(k,\max)}, u_2^{(k,\max)}\right) = \mathbf{g}\left(\theta^{(k,\max)}\right);$$

$$\left(u_1^{(k,\mathrm{r})}, u_2^{(k,\mathrm{r})}\right) = \mathbf{g}\left(\theta^{(k,\mathrm{r})}\right), \qquad \left(u_1^{(k,\Gamma)}, u_2^{(k,\Gamma)}\right) = \mathbf{g}\left(\theta^{(k,\Gamma)}\right).$$

$$(\tilde{\tau}_1, \tilde{\tau}_2) = \mathbf{g}(\tau),$$

We now transfer the results on the moment-generating functions in the section "Double QBD Process and the Convergence Domain" to those on the generating functions. For this, we define

$$\tilde{\mathcal{D}} = \{\mathbf{g}(\theta) \in \mathbb{R}_+^2; \theta \in \mathcal{D}\},$$

$$\tilde{\Gamma}_k = \{\mathbf{g}(\theta) \in \mathbb{R}_+^2; \theta \in \Gamma_k\}, \quad k = 1, 2, +.$$

Define the following generating functions. For $k = 0,1,2,+$

$$\tilde{\gamma}_k(\mathbf{z}) = E\left(z_1^{X_1^{(k)}} z_2^{X_2^{(k)}}\right), \quad \mathbf{z} \equiv (z_1, z_2) \in \mathbb{C}^2,$$

which exists except for $z_1 = 0$ or $z_2 = 0$. Similarly,

$$\begin{aligned}
\tilde{\varphi}(\mathbf{z}) &= E(z_1^{L_1} z_2^{L_2}),\\
\tilde{\varphi}_+(\mathbf{z}) &= E(z_1^{L_1} z_2^{L_2} 1(\mathbf{L} \in S_+)),\\
\tilde{\varphi}_k(z_k) &= E(z_k^{L_k} 1(\mathbf{L} \in S_k)), \qquad k = 1,2.
\end{aligned}$$

as long as they exist.

Obviously, these generating functions are obtained from the corresponding moment-generating functions using the inverse mapping $\mathbf{g}^{-1}$:

$$\begin{aligned}
\tilde{\gamma}_k(\mathbf{z}) &= \gamma_k(\log z_1, \log z_2), \quad k = 0,1,2,+,\\
\tilde{\varphi}(\mathbf{z}) &= \varphi(\log z_1, \log z_2),\\
\tilde{\varphi}_+(\mathbf{z}) &= \varphi_+(\log z_1, \log z_2),\\
\tilde{\varphi}_k(z) &= \varphi_k(\log z), \qquad k = 1,2.
\end{aligned}$$

Then stationary Eq. (8.6) can be written as

$$\begin{aligned}
&(1 - \tilde{\gamma}_+(\mathbf{z}))\tilde{\varphi}_+(\mathbf{z}) + (1 - \tilde{\gamma}_1(\mathbf{z}))\tilde{\varphi}_1(z_1)\\
&\quad + (1 - \tilde{\gamma}_2(\mathbf{z}))\tilde{\varphi}_2(z_2) + (1 - \tilde{\gamma}_0(\mathbf{z}))\nu(\mathbf{0}) = 0.
\end{aligned} \tag{8.13}$$

It is easy to see that

$$\begin{aligned}
\tilde{\Gamma}_k &\equiv \{\mathbf{u} \in \mathbb{R}_+^2; \mathbf{u} > \mathbf{0}, \tilde{\gamma}_k(\mathbf{u}) < 1\}, \quad k = 1,2,+,\\
\tilde{\mathcal{D}} &\equiv \{\mathbf{u} \in \mathbb{R}_+^2; \mathbf{u} > \mathbf{0}, \tilde{\varphi}(\mathbf{u}) < \infty\}.
\end{aligned}$$

These sets may not be convex because two-dimensional generating functions may not be convex (Fig. 8.2).Nevertheless, they still have nice properties because the generating functions are polynomials with nonnegative coefficients. To make this specific, we introduce the following terminology.

Definition 8.1. A subset A of $\mathbb{R}^2$ is said to be nonnegative-directed (or coordinate-directed) convex if $\lambda\mathbf{x} + (1-\lambda)\mathbf{y} \in A$ for any number $\lambda \in [0,1]$ and any $\mathbf{x}, \mathbf{y} \in A$ such that $\mathbf{y} - \mathbf{x} \geq \mathbf{0}$ (or $\mathbf{y} - \mathbf{x}$ in either one of the coordinate axes, respectively).

We then immediately have the following facts.

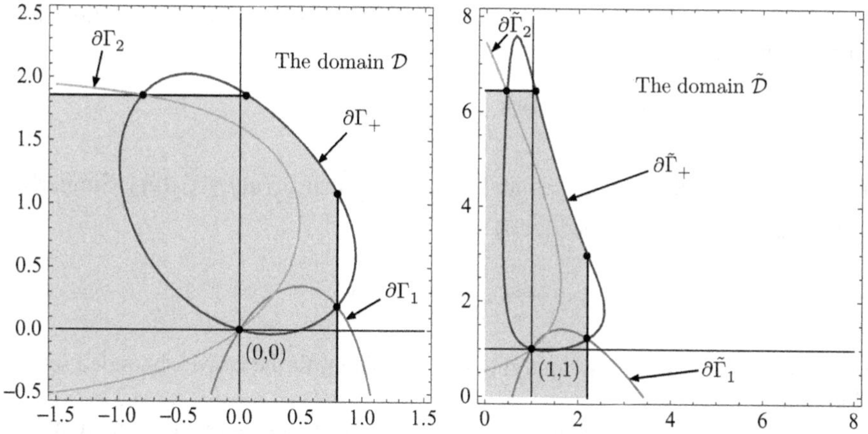

Fig. 8.2 Examples of $\mathcal{D}$ and the corresponding $\tilde{\mathcal{D}}$, which may not be convex, where $(p_{21}, p_{01}, p_{11}) = (0.1, 0.1, 0.7), (p_{20}, p_{00}, p_{10}) = (1.5, 0.5, 0.5), (p_{22}, p_{02}, p_{12}) = (2, 3, 1)$ for $p_{ij} = P(\mathbf{X}^{(+)} = (i, j))$

Lemma 8.3. *$\tilde{\mathcal{D}}$ is nonnegative-directed convex, and $\tilde{\Gamma}_k$ is coordinate-directed convex for $k = +, 0, 1, 2$.*

Note that (8.13) is valid for $\mathbf{z} \in \mathbb{C}^2$ satisfying $|\mathbf{z}| \in \tilde{\mathcal{D}}$ because $|\tilde{\varphi}(\mathbf{z})| \le \tilde{\varphi}(|\mathbf{z}|)$. Furthermore,

$$\begin{aligned} \{\mathbf{z} \in \mathbb{C}^2; |\mathbf{z}| \in \tilde{\mathcal{D}}\} &= \left\{\mathbf{z} \in \mathbb{C}^2; E(e^{L_1 \log|z_1| + L_2 \log|z_2|}) < \infty\right\} \\ &= \{\mathbf{g}(\log|z_1| + i \arg z_1, \log|z_2| + i \arg z_2); \mathbf{z} \in \mathbb{C}^2, (\log|z_1|, \log|z_2|) \in \mathcal{D}\} \\ &= \mathbf{g}(\{\mathbf{z} \in \mathbb{C}^2, (\Re z_1, \Re z_2) \in \mathcal{D}\}), \end{aligned}$$

where $|\mathbf{z}| = (|z_1|, |z_2|)$. Hence, the domain $\mathcal{D}$ is well transferred to $\tilde{\mathcal{D}}$. We will work on $\tilde{\mathcal{D}}$ to find the analytic behaviors of $\tilde{\varphi}_1(z)$ and $\tilde{\varphi}_2(z)$ around their dominant singular points. This is different from the kernel method, which works directly on the set of complex vectors $\mathbf{z}$ satisfying $\tilde{\gamma}_+(\mathbf{z}) = 1$ and applies deeper complex analysis such as analytic extension on a Riemann surface (e.g., see [6]). We avoid it using the domain $\tilde{\mathcal{D}}$.

A Key Function for Analytic Extension

Once the domain $\tilde{\mathcal{D}}$ is obtained, the next step is to study the analytic behaviors of the generating function $\tilde{\varphi}_k$ for $k = 1, 2$. For this, we use a relation between them by letting $\tilde{\gamma}_+(\mathbf{z}) - 1 = 0$ in stationary Eq. (8.13), which removes $\tilde{\varphi}_+(\mathbf{z})$. For this, let us consider the solution $u_2 > 0$ of $\tilde{\gamma}_+(u_1, u_2) = 1$ for each fixed $u_1 > 0$. Since this equation is quadratic concerning u_2 and $\tilde{\mathcal{D}} \subset (\mathbb{R}^o)^2_+$, it has two positive solutions for each u_1 satisfying

$$u_1^{(1,\min)} \le u_1 \le u_1^{(1,\max)}.$$

Denote these solutions by $\underline{\zeta}_2(u_1)$ and $\overline{\zeta}_2(u_1)$ such that $\underline{\zeta}_2(u_1) \le \overline{\zeta}_2(u_1)$. Similarly, $\underline{\zeta}_1(u_2)$ and $\overline{\zeta}_1(u_2)$ are defined for u_2 satisfying

$$u_2^{(2,\min)} \le u_2 \le u_2^{(2,\max)}.$$

One can see these facts also applying the mapping $\mathbf{g}$ to the convex bounded set $\mathcal{D}$ (Lemma 8.3).

We now adapt the arguments in [3]. For this, we first examine the function $\underline{\zeta}_2$. Let

$$p_{*k}(u) = E(u^{X_1^{(+)}} 1(X_2^{(+)} = k)),$$
$$p_{k*}(u) = E(u^{X_2^{(+)}} 1(X_1^{(+)} = k)), \quad k = 0, 1, -1.$$

Then $\tilde{\gamma}_+(u_1, u_2) = 1$ can be written as

$$u_2^2 p_{*1}(u_1) - u_2(1 - p_{*0}(u_1)) + p_{*-1}(u_1) = 0. \tag{8.14}$$

Hence, we have, for $u \in [u_1^{(1,\min)}, u_1^{(1,\max)}]$,

$$\underline{\zeta}_2(u) = \frac{1 - p_{*0}(u) - \sqrt{D_2(u)}}{2p_{*1}(u)}, \tag{8.15}$$

where

$$D_2(u) = (1 - p_{*0}(u))^2 - 4p_{*1}(u)p_{*-1}(u) \ge 0.$$

Since $D_2(u_1^{(1,\min)}) = D_2(u_1^{(1,\max)}) = 0$ and $u^2 D_2(u)$ is a polynomial of order 4 at most and order 2 at least by condition (i), $u^2 D_2(u)$ can be factorized as

$$u^2 D_2(u) = (u - u_1^{(1,\min)})(u_1^{(1,\max)} - u)h_2(u),$$

where $h_2(u) \ne 0$ for $u \in (u_1^{(1,\min)}, u_1^{(1,\max)})$. This fact can be verified by the mapping $\mathbf{g}$ from Γ_+ to $\tilde{\Gamma}_+$.

To obtain tail asymptotics, we will use analytic functions. So far, we would like to analytically extend the function $\underline{\zeta}_2$ from the real interval to a sufficiently large region in the complex plane $\mathbb{C}$. For this, we prepare a series of lemmas. We first note the following fact.

Lemma 8.4 (Lemma 2.3.8 of [6]). *All the solutions of $z^2D_2(z) = 0$ for $z \in \mathbb{C}$ are real numbers.*

In the light of the preceding arguments, this lemma immediately leads to the following fact.

Lemma 8.5. *$z^2D_2(z) = 0$ for $z \in \mathbb{C}$ has no solution in the region such that $|z| \in (u_1^{(1,\min)}, u_1^{(1,\max)})$.*

We will also use the following two lemmas, which show how the periodic nature of the random walk $\{\mathbf{Y}_\ell\}$ is related to the branch points (see Remark 8.2 on the periodic nature). They are proved in the appendices "Proof of Lemma 8.6" and "Proof of Lemma 8.7," respectively.

Lemma 8.6. *The equation*

$$D_2(z) = 0, \qquad |z| = u_1^{(1,\max)}, \, z \in \mathbb{C}, \tag{8.16}$$

has only one solution $z = u_1^{(1,\max)}$ if and only if (v-a) holds. Otherwise, it has two solutions $z = \pm u_1^{(1,\max)}$, and $u^2D_2(u)$ is an even function.

Lemma 8.7. *There are $x, y > 0$ such that*

$$\tilde{\gamma}_+(x,y) = 1, \qquad \tilde{\gamma}_+(-x,-y) = 1 \tag{8.17}$$

if and only if (v-a) does not hold.

Remark 8.4. Lemma 8.6 is essentially the same as Remark 3.1 of [16], which is obtained as a corollary of their Lemma 3.1, which is immediate from Lemmas 2.3.8 of [6].

By Lemmas 8.4 and 8.5, $\underline{\zeta}_2(u)$ on $(u_1^{(1,\min)}, u_1^{(1,\max)})$ is extendable as an analytic function of a complex variable to the region $\tilde{\mathcal{G}}_0(u_1^{(1,\min)}, u_1^{(1,\max)})$, where

$$\tilde{\mathcal{G}}_0(a,b) = \{z \in \mathbb{C}; z \notin (-\infty, a] \cup [b, \infty)\}, \qquad a, b \in \mathbb{R},$$

and has a single branch point $u_1^{(1,\max)}$ on $|z| = u_1^{(1,\max)}$ if (v-a) holds and two branch points $\pm u_1^{(1,\max)}$ there otherwise by Lemmas 8.6 and 8.5. Both branch points have order 2. We denote this extended analytic function by $\underline{\zeta}_2(z)$. That is, we use the same notation for an analytically extended function. We identify it by its argument. The following lemma is a key for our arguments. The idea of this lemma is similar to Lemma 6.3 of [3], but its proof is entirely different from that lemma.

Lemma 8.8. *(a) $\underline{\zeta}_2$ of (8.15) is analytically extended on $\tilde{\mathcal{G}}_0(u_1^{(1,\min)}, u_1^{(1,\max)})$.*

(b) For $z \in \mathbb{C}$ satisfying $|z| \in (u_1^{(1,\min)}, u_1^{(1,\max)}]$,

$$|\underline{\zeta}_2(z)| \le \underline{\zeta}_2(|z|) \le u_2^{(1,\max)}, \tag{8.18}$$

where the second inequality is strict if $|z| < u_1^{(1,\max)}$.

(c) *If either* $m_2^{(1)} = 0$ *or (v-b) holds, then*

$$\tilde{\gamma}_1(z, \underline{\zeta}_2(z)) = 1, \qquad |z| = u_1^{(1,r)}, \tag{8.19}$$

has no solution other than $z = u_1^{(1,r)}$.

(d) *Equation (8.19) has two solutions* $z = \pm u_1^{(1,r)}$ *if and only if* $m_2^{(1)} = 0$, *(v-a), and (v-b) do not hold.*

Proof. We have already proved (a). Thus, we only need to prove (b)–(d). We first prove (b). For this, it is sufficient to prove (8.18) for $|z| < u_1^{(1,\max)}$ by the continuity of $\underline{\zeta}_2(z)$ for $|z| \le u_1^{(1,\max)}$ at $z = u_1^{(1,\max)}$. Substituting complex numbers z_1 and z_2 into u_1 and u_2 of (8.14), we have

$$z_2^2 p_{*1}(z_1) + z_2 p_{*0}(z_1) + p_{*-1}(z_1) = z_2. \tag{8.20}$$

Obviously, this equation has the following solutions for each fixed z_1 such that $|z_1| \in (u_1^{(1,\min)}, u_1^{(1,\max)})$:

$$z_2 = \underline{\zeta}_2(z_1), \quad \overline{\zeta}_2(z_1). \tag{8.21}$$

We next take the absolute values of both sides of (8.20); then

$$|z_2|^2 p_{*1}(|z_1|) + |z_2| p_{*0}(|z_1|) + p_{*-1}(|z_1|) \ge |z_2|.$$

Thus, we get

$$|z_2|(\tilde{\gamma}_+(|z_1|, |z_2|) - 1) \ge 0.$$

By the definitions of $\underline{\zeta}_2(|z_1|)$ and $\overline{\zeta}_2(|z_1|)$, this inequality can be written as

$$(|z_2| - \underline{\zeta}_2(|z_1|))(|z_2| - \overline{\zeta}_2(|z_1|)) = |z_2|(\tilde{\gamma}_+(|z_1|, |z_2|) - 1) \ge 0.$$

Hence, $\underline{\zeta}_2(|z_1|) \le \overline{\zeta}_2(|z_1|)$ implies

$$|z_2| \le \underline{\zeta}_2(|z_1|) \quad or \quad \overline{\zeta}_2(|z_1|) \le |z_2|. \tag{8.22}$$

By (8.21), we can substitute $z_2 = \underline{\zeta}_2(z_1)$ into (8.22) and get

$$|\underline{\zeta}_2(z_1)| \le \underline{\zeta}_2(|z_1|) \quad or \quad \overline{\zeta}_2(|z_1|) \le |\underline{\zeta}_2(z_1)|, \qquad |z_1| \in (u_1^{(1,\min)}, u_1^{(1,\max)}). \tag{8.23}$$

Thus, (b) is obtained if we show that $\overline{\zeta}_2(|z_1|) \le |\underline{\zeta}_2(z_1)|$ *is impossible. Suppose the contrary of this, then there is a* $z_1^{(0)}$ *such that*

$$\underline{\zeta}_2(|z_1^{(0)}|) < |\underline{\zeta}_2(z_1^{(0)})|, \qquad |z_1^{(0)}| \in (u_1^{(1,\min)}, u_1^{(1,\max)}). \tag{8.24}$$

Since $|\underline{\zeta}_2(z)|$ *is continuous and converges to* $\underline{\zeta}_2(|z_1^{(0)}|)$ *as* z *goes to* $|z_1^{(0)}|$ *on the path where* $|z| = |z_1^{(0)}|$*, there must be a* $z_1^{(1)}$ *such that* $|z_1^{(1)}| = |z_1^{(0)}|$ *and*

$$\underline{\zeta}_2(|z_1^{(1)}|) < |\underline{\zeta}_2(z_1^{(1)})| < \overline{\zeta}_2(|z_1^{(1)}|).$$

Since $|z_1^{(1)}| = |z_1^{(0)}| \in (u_1^{(1,\min)}, u_1^{(1,\max)})$*, this contradicts (8.23), which proves (b).*

We next prove (c). Let

$$p_{*k}^{(1)}(z) = E(z^{X_1^{(1)}} 1(X_2^{(1)} = k)), \qquad k = 0, 1.$$

First, assume that $m_2^{(1)} = 0$*. This implies* $p_{*1}^{(1)}(z) = 0$*, and therefore (8.19) is reduced to* $p_{*0}^{(1)}(z) = 1$*. Hence, its solution is* $z = 1$ *or* $z = p_{-10}^{(1)}/p_{10}^{(1)} \ge 0$ *if* $p_{10}^{(1)} \ne 0$ *(otherwise,* $z = 1$ *is the only solution). Both are nonnegative numbers, and therefore (8.19) has no solution* z *such that*

$$|z| = u_1^{(1,r)}, \qquad z \ne u_1^{(1,r)}. \tag{8.25}$$

We next assume that $m_2^{(1)} \ne 0$*, which implies* $p_{*1}^{(1)}(z) \ne 0$*. Since (8.19) can be written as*

$$\underline{\zeta}_2(z) p_{*1}^{(1)}(z) + p_{*0}^{(1)}(z) = 1 \tag{8.26}$$

and $1 \le |w| + |1 - w|$ *for any* $w \in \mathbb{C}$*, we have*

$$|\underline{\zeta}_2(z)| = \left| \frac{1 - p_{*0}^{(1)}(z)}{p_{*1}^{(1)}(z)} \right| \ge \frac{1 - |p_{*0}^{(1)}(z)|}{|p_{*1}^{(1)}(z)|} \ge \frac{1 - p_{*0}^{(1)}(|z|)}{p_{*1}^{(1)}(|z|)} = \underline{\zeta}_2(|z|). \tag{8.27}$$

If (8.25) holds, then both sides of this inequality are identical for $z = \gamma pmu_1^{(1,r)}$ *if and only if (v-b) does not hold. Hence, if (v-b) holds, then* $|\underline{\zeta}_2(z)| > \underline{\zeta}_2(|z|)$*, and therefore (8.19) has no solution satisfying (8.25) because of (8.18).*

We finally prove (d). For this we assume that neither $m_2^{(1)} = 0$ *nor (v-b) holds. In this case,* $p_{01}^{(1)} = p_{(-1)0}^{(1)} = p_{10}^{(1)} = 0$*, so it follows from (8.26) that*

$$\underline{\zeta}_2(z) = \frac{(1 - p_{00}^{(1)})z}{p_{-11}^{(1)} + p_{11}^{(1)} z^2}.$$

Hence, if (8.25) holds, then we must have $z = -u_1^{(1,r)}$ *because of (8.18) and (8.27). By the preceding equation, we also have* $\underline{\zeta}_2(-u_1^{(1,r)}) = -\underline{\zeta}_2(u_1^{(1,r)})$*. Hence, we need to check whether* $(-u_1^{(1,r)}, -\underline{\zeta}_2(u_1^{(1,r)}))$ *is the solution of* $\gamma_+(x,y) = 1$*. By Lemma 8.7,*

Table 8.1 The solutions of (8.16) and (8.19), where ○, ×, and — indicate "yes," "no," and "'irrelevant"

Nonarithmetic: (v-a)	○	×	×	×	×
Nonarithmetic: (v-b)	–	○	○	×	×
$m_2^{(1)} = 0$	–	○	×	○	×
The solutions of (8.16)	$u_1^{(1,\max)}$		$\pm u_1^{(1,\max)}$		
The solutions of (8.19)	$u_1^{(1,r)}$		$u_1^{(1,r)}$		$\pm u_1^{(1,r)}$

$z = -u_1^{(1,r)}$ is the solution of (8.19) if and only if (v-a) does not hold. Combining this with (b) and (c) completes the proof of (d). □

For the convenience of later reference, we summarize the results in (c) and (d) of Lemma 8.8 in Table 8.1. Similar results can be obtained in the direction of the second axes using (v-b) and $m_1^{(2)} = 0$ instead of (v-a) and $m_2^{(1)} = 0$. Since the results are symmetric, we omit them. We remark that Li and Zhao [16] have not considered the cases $m_1^{(2)} = 0$ and $m_2^{(1)} = 0$, which seems to be overlooked.

Nature of the Dominant Singularity

We consider the complex variable functions $\tilde{\varphi}_1(z_1)$ and $\tilde{\varphi}_2(z_2)$. Recall that

$$\tilde{\varphi}(\mathbf{z}) = \tilde{\varphi}_+(\mathbf{z}) + \tilde{\varphi}_1(z_1) + \tilde{\varphi}_2(z_2) + \nu(\mathbf{0}). \tag{8.28}$$

Obviously, $\tilde{\varphi}(\mathbf{z})$ is analytic for $\mathbf{z} \in \mathbb{C}^2$, such that $(|z_1|, |z_2|) \in \tilde{\mathcal{D}}$, and singular on the boundary of $\tilde{\mathcal{D}}$. This implies that $\tilde{\varphi}_i(z_i)$ is analytic for $|z_i| < \tilde{\tau}_i$ and has a point on the circle $|z| = \tilde{\tau}_i$. This is easily seen from (8.28) with $z_j = 0$ for $j = 3 - i$. Furthermore, $z_i = \tilde{\tau}_i$ must be a singular point for $i = 1, 2$ by Pringsheim's theorem (see, e.g., Theorem 17.13 in volume 1 of [17]). In addition to this point, we need to find all singular points on $|z| = \tilde{\tau}_i$ to get the tail asymptotics, as we will see. As expected from Lemma 8.6, $z = -\tilde{\tau}_i$ may be another singular point, which occurs only when (v-a) does not hold.

We focus on these singular points instead of searching for singular points on $|z| = \tilde{\tau}_i$ and show that there is no other singular point on the circle because of the analytic behavior of $\tilde{\varphi}_i(z)$. Since the results are symmetric for $\tilde{\varphi}_1(z)$ and $\tilde{\varphi}_2(z)$, we only consider $\tilde{\varphi}_1(z)$ in this section.

For this, we use stationary Eq. (8.13), which is valid on $\tilde{\mathcal{D}}$. Plugging $(z_1, z_2) = (z, \underline{\zeta}_2(z))$ into (8.13) yields, for $|z| \in (u_1^{(1,\min)}, \tilde{\tau}_1)$,

$$\tilde{\varphi}_1(z) = \frac{(\tilde{\gamma}_2(z, \underline{\zeta}_2(z)) - 1)\tilde{\varphi}_2(\underline{\zeta}_2(z))}{1 - \tilde{\gamma}_1(z, \underline{\zeta}_2(z))} + \frac{(\tilde{\gamma}_0(z, \underline{\zeta}_2(z)) - 1)\nu(\mathbf{0})}{1 - \tilde{\gamma}_1(z, \underline{\zeta}_2(z))}. \tag{8.29}$$

In light of this equation, the dominant singularity of $\tilde{\varphi}_1(z)$ is caused by $\underline{\zeta}_2(z)$, $\tilde{\varphi}_2(\underline{\zeta}_2(z))$, or

$$\tilde{\gamma}_1(z, \underline{\zeta}_2(z)) = 1. \tag{8.30}$$

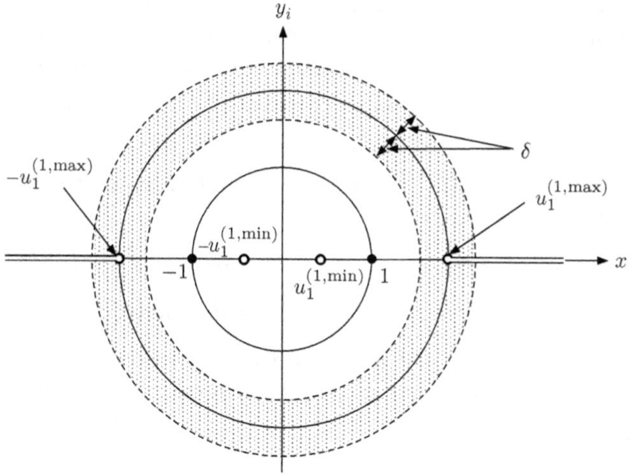

Fig. 8.3 *Shaded area*: $\tilde{\mathcal{G}}_\delta^-(-u_1^{(1,\max)}) \cap \tilde{\mathcal{G}}_\delta^+(u_1^{(1,\max)})$

In addition to $\tilde{\mathcal{G}}_0(a,b)$, we will use the following sets to consider analytic regions (Fig. 8.3):

$$\tilde{C}_\delta(u) = \{z \in \mathbb{C}; u - \delta < |z| < u + \delta, z \neq u\}, \qquad u, \delta > 0,$$

$$\tilde{\mathcal{G}}_\delta^+(u) = \tilde{\mathcal{G}}_0(u_1^{(1,\min)}, u) \cap \tilde{C}_\delta(u), \qquad u_1^{(1,\min)} < u,$$

$$\tilde{\mathcal{G}}_\delta^-(u) = \tilde{\mathcal{G}}_0(u, -u_1^{(1,\min)}) \cap \tilde{C}_\delta(u), \qquad u < -u_1^{(1,\min)}.$$

Remark 8.5. One may wonder whether (8.18) in Lemma 8.8 is sufficient for verifying the analyticity of $\tilde{\varphi}_1(z)$ in $\tilde{\mathcal{G}}_\delta^+(u_1^{(1,\max)})$ when $\tilde{\tau}_1 = u_1^{(1,\max)}$. This will turn out to be no problem because of (8.29).

In what follows, we first consider the case where (v-a) holds, then we consider the other case.

Singularity for the Nonarithmetic Case

Assume the nonarithmetic condition (v-a). We consider the analytic behavior of $\tilde{\varphi}_1(z)$ around the singular point $z = \tilde{\tau}_1$. This behavior will show that there is no other singular point on $|z| = \tilde{\tau}_1$. We separately consider the three causes discussed above.

(8Ia) The solution of (8.30): This equation has six solutions at most because it can be written as a polynomial equation of order 6. $z = 1, u_1^{(1,\mathrm{r})}$ are clearly the solutions. Because $\tilde{\varphi}_1(z)$ of (8.29) is analytic for $|z| < \tilde{\tau}_1$, (8.30) cannot have a solution such that $|z| < \tilde{\tau}_1$, except for the points where the numerator of the right-hand side of (8.29) vanishes. This must be finitely many because the numerator vanishes otherwise by the uniqueness of analytic extension. On the other hand, (8.30) has no solution on the circle $|z| = u_1^{(1,r)}$, except for $z = u_1^{(1,r)}$, by Lemma 8.8.

Thus, the compactness of the circle implies that, if $\tilde{\tau}_1 = u_1^{(1,r)} < u_1^{(1,\max)}$, then (8.30) has no solution on $\tilde{C}_\delta(u_1^{(1,r)})$ for some $\delta > 0$. Hence, we have the following fact from (8.29).

Lemma 8.9. *Assume that $\tilde{\tau}_1 = u_1^{(1,r)} < u_1^{(1,\max)}$ and $\tilde{\varphi}_2(\underline{\zeta}_2(z))$ is analytic at $|z| = u_1^{(1,r)}$. Then, $\tilde{\varphi}_1(z)$ has a simple pole at $z = u_1^{(1,r)}$ and is analytic on $\tilde{C}_\delta(u_1^{(1,r)})$.*

Remark 8.6. For categories I and III, the analytic condition on $\tilde{\varphi}_2(\underline{\zeta}_2(z))$ in this lemma is always satisfied because Lemma 8.8 and the category condition, $\underline{\zeta}_2(\tilde{\tau}_1) < \tilde{\tau}_2$, imply, for $|z| = u_1^{(1,r)}$,

$$|\tilde{\varphi}_2(\underline{\zeta}_2(z))| \le \tilde{\varphi}_2(|\underline{\zeta}_2(z)|) \le \tilde{\varphi}_2(\underline{\zeta}_2(|z|)) = \tilde{\varphi}_2(u_2^{(1,r)}) < \infty.$$

If $\tilde{\tau}_1 = u_1^{(1,r)} = u_1^{(1,\max)}$, then the analytic behavior of $\tilde{\varphi}_1(z)$ around $z = u_1^{(1,r)}$ is a bit complicated because $\underline{\zeta}_2(z)$ is also singular there. We will consider this case in the section "Exact Tail Asymptotics for the Nonarithmetic Case."

(8Ib) The singularity of $\underline{\zeta}_2(z)$: By Lemma 8.8, this function is analytic on $\tilde{\mathscr{G}}_0(u_1^{(1,\min)}, u_1^{(1,\max)})$ and singular at $z = u_1^{(1,\max)}$, which is a branch point.

(8Ic) The singularity of $\tilde{\varphi}_2(\underline{\zeta}_2(z))$: This function is singular at $z = \tilde{\tau}_1$ if $\underline{\zeta}_2(\tilde{\tau}_1) = \tilde{\tau}_2$. Otherwise, it is singular at $z = u_1^{(1,\max)}$ because $\underline{\zeta}_2(z)$ is singular there. Furthermore, we may simultaneously have $\underline{\zeta}_2(\tilde{\tau}_1) = \tilde{\tau}_2$ and $\tilde{\tau}_1 = u_1^{(1,\max)}$. Thus, we need to consider these three cases: $\tilde{\tau}_1 = u_1^{(1,\max)}$ for categories I and III, and $\tilde{\tau}_1 < u_1^{(1,\max)}$ or $\tilde{\tau}_1 = u_1^{(1,\max)}$ for category II. For this, we will use the following fact, which is essentially the same as Lemma 4.2 of [20].

Lemma 8.10. *$\overline{\zeta}_1(e^\theta)$ is a concave function of $\theta \in [\theta_2^{(2,\min)}, \theta_2^{(2,\max)}]$, $\overline{\zeta}_1'(u_2^{(1,\max)}) = 0$, $\overline{\zeta}_1''(u_2^{(1,\max)}) < 0$, and*

$$\lim_{\substack{z \to u_1^{(1,\max)} \\ z \in \tilde{\mathscr{G}}_0(u_1^{(1,\min)}, u_1^{(1,\max)})}} \frac{u_2^{(1,\max)} - \underline{\zeta}_2(z)}{(u_1^{(1,\max)} - z)^{\frac{1}{2}}} = \frac{\sqrt{2}}{\sqrt{-\overline{\zeta}_1''(u_2^{(1,\max)})}}. \tag{8.31}$$

Proof. *The first part is immediate from the facts that Γ_+ is a convex set and $u_1^{(1,\max)} = e^{\theta_1^{(1,\max)}}$. By Taylor expansion of $\overline{\zeta}_1(z_2)$ at $z_2 = u_2^{(1,\max)} < u_2^{(2,\max)}$,*

$$\overline{\zeta}_1(z_2) = u_1^{(1,\max)} + \frac{1}{2}\overline{\zeta}_1''(u_2^{(1,\max)})(z_2 - u_2^{(1,\max)})^2 + o(|z_2 - u_2^{(1,\max)}|^2).$$

Letting $z_2 = \underline{\zeta}_2(z)$ *in this equation yields (8.31) since* $\overline{\zeta}_1(\underline{\zeta}_2(z)) = z$ *for* z *to be sufficiently close to* $u_1^{(1,\max)}$. □

Another useful asymptotic is as follows.

Lemma 8.11. *If* $u_1^{(1,\max)} = u_1^{(1,\mathrm{r})}$, *then for any* $\delta > 0$,

$$\lim_{\substack{z \to u_1^{(1,\max)} \\ z \in \tilde{\mathcal{G}}_\delta^+(u_1^{(1,\max)})}} \frac{(u_1^{(1,\max)} - z)^{\frac{1}{2}}}{1 - \tilde{\gamma}_1(z, \underline{\zeta}_2(z))} = \frac{\sqrt{-\overline{\zeta}_1''(u_2^{(1,\max)})}}{\sqrt{2} p_{*1}^{(1)}(u_1^{(1,\mathrm{r})})}. \tag{8.32}$$

Proof. By the condition $u_1^{(1,\max)} = u_1^{(1,\mathrm{r})}$, we have

$$\begin{aligned}1 - \tilde{\gamma}_1(z, \underline{\zeta}_2(z)) &= \tilde{\gamma}_1(\mathbf{u}^{(1,\max)}) - \tilde{\gamma}_1(z, \underline{\zeta}_2(z)) \\ &= (u_2^{(1,\max)} - \underline{\zeta}_2(z)) p_{*1}^{(1)}(u_1^{(1,\mathrm{r})}) \\ &\quad + \underline{\zeta}_2(z)(p_{*1}^{(1)}(u_1^{(1,\mathrm{r})}) - p_{*1}^{(1)}(z)) + p_{*0}^{(1)}(u_1^{(1,\mathrm{r})}) - p_{*0}^{(1)}(z).\end{aligned}$$

Hence, if we divide both sides by $(u_1^{(1,\max)} - z)^{\frac{1}{2}}$, then Lemma 8.10 yields (8.31) because $p_{*1}^{(1)}(z)$ and $p_{*0}^{(1)}(z)$ are analytic except for $z = 0$. □

We now consider the three cases separately.

(8Ic-1) $\underline{\zeta}_2(\tilde{\tau}_1) < \tilde{\tau}_2$, equivalently, categories I or III, and $\tilde{\tau}_1 = u_1^{(1,\max)}$: In this case, $\tilde{\varphi}_2(z)$ is analytic for $z \in \tilde{C}_\delta(u_2^{(1,\max)})$ for some $\delta > 0$ because $u_2^{(1,\max)} = \underline{\zeta}_2(u_1^{(1,\max)}) = \underline{\zeta}_2(\tilde{\tau}_1) < \tilde{\tau}_2$. Hence, by Taylor expansion, we have, for $|z| < \tilde{\tau}_2$,

$$\tilde{\varphi}_2(z) = \tilde{\varphi}_2(u_2^{(1,\max)}) + \tilde{\varphi}_2'(u_2^{(1,\max)})(z - u_2^{(1,\max)}) + o(|z - u_2^{(1,\max)}|). \tag{8.33}$$

Thus, the analytic behavior of $\tilde{\varphi}_2(\underline{\zeta}_2(z))$ around $z = u_1^{(1,\max)}$ is determined by that of $\underline{\zeta}_2(z) - u_2^{(1,\max)}$. Since $u_2^{(1,\max)} = \underline{\zeta}_2(u_1^{(1,\max)}) < \tilde{\tau}_2$ by the conditions of (8Ic-1), Lemma 8.10 yields

$$\begin{aligned}\tilde{\varphi}_2(\underline{\zeta}_2(z)) = \tilde{\varphi}_2(u_2^{(1,\max)}) - \frac{\sqrt{2}\tilde{\varphi}_2'(u_2^{(1,\max)})}{\sqrt{-\overline{\zeta}_1''(u_2^{(1,\max)})}}(u_1^{(1,\max)} - z)^{\frac{1}{2}} \\ + o(|z - u_2^{(1,\max)}|^{\frac{1}{2}}).\end{aligned} \tag{8.34}$$

Thus, $\tilde{\varphi}_2(\underline{\zeta}_2(z))$ has a branch point of order 2 at $z = \tilde{\tau}_1 = u_1^{(1,\max)}$ and is analytic on $\tilde{\mathcal{G}}_\delta^+(u_1^{(1,\max)})$ for some $\delta > 0$.

(8Ic-2) $\underline{\zeta}_2(\tilde{\tau}_1) = \tilde{\tau}_2$ and $\tilde{\tau}_1 < u_1^{(1,\max)}$: This is only for category II. Hence, $\tilde{\tau}_2 = u_2^{(2,\mathrm{r})} < u_2^{(2,\max)}$, and therefore the $\tilde{\varphi}_2$-version of Lemma 8.9 is available. Thus, $\tilde{\varphi}_2(z)$ has a simple pole at $z = u_2^{(2,\mathrm{r})}$. Here, that $u_2^{(2,\mathrm{r})}$ is the solution of the equation

$$\tilde{\gamma}_2(\underline{\zeta}_1(z), z) = 1 \tag{8.35}$$

is crucial. Furthermore, $\underline{\zeta}_2(z)$ is analytic at $z = \tilde{\tau}_1$. Hence, $\tilde{\varphi}_2(\underline{\zeta}_2(z))$ has a simple pole at $z = \tilde{\tau}_1$ and is analytic on $\tilde{C}_\delta(u_1^{(1,\max)})$ for some $\delta > 0$.
(8Ic-3) $\underline{\zeta}_2(\tilde{\tau}_1) = \tilde{\tau}_2$ and $\tilde{\tau}_1 = u_1^{(1,\max)}$: This is also only for category II. This case is similar to (8Ic-2) except that $\underline{\zeta}_2(z)$ has a branch point at $z = \tilde{\tau}_1 = u_1^{(1,\max)}$. Since $\tilde{\varphi}_2(z)$ has a simple pole at $z = \tilde{\tau}_2$, we have, by Lemma 8.10,

$$\tilde{\varphi}_2(\underline{\zeta}_2(z)) \sim (u_1^{(1,\max)} - z)^{-\frac{1}{2}},$$

and $\tilde{\varphi}_2(\underline{\zeta}_2(z))$ is analytic on $\tilde{\mathcal{G}}_\delta^+(u_1^{(1,\max)})$ for some $\delta > 0$.

Singularity for the Arithmetic Case

We next consider the case where (v-a) does not hold. That is, the Markov additive process for the interior is arithmetic. In this case, the singularity of $\tilde{\varphi}_1(z)$ at $z = \tilde{\tau}_1$ occurs similarly to its occurrence in the section "Singularity for the Nonarithmetic Case." In addition to this singular point, we may have another singular point $-\tilde{\tau}_1$, as can be seen in Table 8.1. For this, we separately consider two subcases:

(B1) Either (v-b) or $m_2^{(1)} = 0$ holds. (B2) Neither (v-b) nor $m_2^{(1)} = 0$ holds.

In some cases, we need further classification:

(C1) Either (v-c) or $m_1^{(2)} = 0$ holds. (C2) Neither (v-b) nor $m_1^{(2)} = 0$ holds.

Consider (B1). From Table 8.1, the solutions of (8.16) are $z = \pm u_1^{(1,\max)}$, and the solution of (8.19) is $z = u_1^{(1,\mathrm{r})}$. There is no other solution. We consider cases similar to (8Ia), (8Ib), (8Ic-2), (8Ic-1), and (8Ic-3) of the section "Singularity for the Nonarithmetic Case."

(8Ia') The solution of (8.30): This case is exactly the same as in the section "Singularity for the Nonarithmetic Case" because $z = -u_1^{(1,\mathrm{r})}$ is not the solution of (8.19). Hence, Lemma 8.9 also holds true.
(8Ib') The singularity of $\underline{\zeta}_2(z)$ at $|z| = u_1^{(1,\max)}$: It is singular at $z = \pm u_1^{(1,\max)}$.
(8Ic') The singularity of $\tilde{\varphi}_2(\underline{\zeta}_2(z))$ at $|z| = \tilde{\tau}_1$: For $z = \tilde{\tau}_1$, the story is the same as in the section "Singularity for the Nonarithmetic Case."Hence, we only

consider the case where $z = -\tilde{\tau}_1$. From (8.15) and the condition that (v-a) does not hold, we have

$$\underline{\zeta}_2(-\tilde{\tau}_1) = -\frac{1-p_{00}}{2(p_{-11}+p_{11}\tilde{\tau}_1^2)}\tilde{\tau}_1 = -\underline{\zeta}_2(\tilde{\tau}_1). \tag{8.36}$$

Hence, $|\underline{\zeta}_2(-\tilde{\tau}_1)| = \underline{\zeta}_2(\tilde{\tau}_1) > 0$, and

$$|\underline{\zeta}_1(\underline{\zeta}_2(-\tilde{\tau}_1))| = |\underline{\zeta}_1(-\underline{\zeta}_2(\tilde{\tau}_1))| = \underline{\zeta}_1(\underline{\zeta}_2(\tilde{\tau}_1)).$$

Since $\underline{\zeta}_1(\underline{\zeta}_2(\tilde{\tau}_1)) < \tilde{\tau}_1$, $\tilde{\varphi}_1(\underline{\zeta}_1(\underline{\zeta}_2(z)))$ is analytic around $z = -\tilde{\tau}_1$. Furthermore, Lemma 8.10 and (8.34) are still valid if we replace $u_i^{(1,\max)}$ by $-u_i^{(1,\max)}$ for $i = 1,2$. However, this $z = -\tilde{\tau}_1$ cannot be the solution of (8.30) because of (B1). Thus, we must partially change the arguments in the section "Singularity for the Nonarithmetic Case."

(8Ic'-1) $\underline{\zeta}_2(\tilde{\tau}_1) < \tilde{\tau}_2$ and $\tilde{\tau}_1 = u_1^{(1,\max)}$: This is only for categories I and III, and $\tilde{\varphi}_2(\underline{\zeta}_2(z))$ has a branch point of order 2 at $z = -u_1^{(1,\max)}$ and is analytic on $\tilde{\mathcal{G}}_\delta^-(-u_1^{(1,\max)}) \cap \tilde{\mathcal{G}}_\delta^+(u_1^{(1,\max)})$ for some $\delta > 0$ because it also has a branch point at $z = u_1^{(1,\max)}$.

(8Ic'-2) $\underline{\zeta}_2(\tilde{\tau}_1) = \tilde{\tau}_2$ and $\tilde{\tau}_1 < u_1^{(1,\max)}$: This is only for category II. Since $\underline{\zeta}_2(z)$ is analytic at $z = \tilde{\tau}_1$, $\tilde{\varphi}_2(\underline{\zeta}_2(z))$ is analytic at $z = -\tilde{\tau}_1$ if (C1) holds. Otherwise, if (C2) holds, it has a simple pole at $z = -\tilde{\tau}_1$ because $\underline{\zeta}_2(-\tilde{\tau}_1) = -\underline{\zeta}_2(\tilde{\tau}_1)$ is the solution of (8.35).

(8Ic'-3) $\underline{\zeta}_2(\tilde{\tau}_1) = \tilde{\tau}_2$ and $\tilde{\tau}_1 = u_1^{(1,\max)}$: This is only for category II, and the situation is similar to (8c'-2), except that the singularity is caused by $\underline{\zeta}_2(z)$ at $z = -\tilde{\tau}_1$. To verify this fact, we rework $\tilde{\varphi}_2(\underline{\zeta}_2(z))$. Similarly to (8.29), we have, for $|z| \in (u_2^{2,\min)}, \tilde{\tau}_2)$,

$$\tilde{\varphi}_2(z) = \frac{(\tilde{\gamma}_1(\underline{\zeta}_1(z),z)-1)\tilde{\varphi}_1(\underline{\zeta}_1(z))}{1-\tilde{\gamma}_2(\underline{\zeta}_1(z),z)} + \frac{(\tilde{\gamma}_0(\underline{\zeta}_2(z),z)-1)\nu(\mathbf{0})}{1-\tilde{\gamma}_2(\underline{\zeta}_1(z),z)}.$$

Substituting $\underline{\zeta}_2(z)$ into z of this equation, we have

$$\begin{aligned}\tilde{\varphi}_2(\underline{\zeta}_2(z)) = &\frac{(\tilde{\gamma}_1(\underline{\zeta}_1(\underline{\zeta}_2(z)),\underline{\zeta}_2(z))-1)\tilde{\varphi}_1(\underline{\zeta}_1(\underline{\zeta}_2(z)))}{1-\tilde{\gamma}_2(\underline{\zeta}_1(\underline{\zeta}_2(z)),\underline{\zeta}_2(z))}\\ &+\frac{(\tilde{\gamma}_0(\underline{\zeta}_2(\underline{\zeta}_2(z)),\underline{\zeta}_2(z))-1)\nu(\mathbf{0})}{1-\tilde{\gamma}_2(\underline{\zeta}_1(\underline{\zeta}_2(z)),\underline{\zeta}_2(z))}.\end{aligned} \tag{8.37}$$

By the assumptions of (8c-3), if (C2) holds, then $\tilde{\varphi}_2(z)$ has a simple pole at $z = -\tilde{\tau}_2$, and therefore $\tilde{\varphi}_2(\underline{\zeta}_2(z)) \sim (-u_1^{(1,\max)} - z)^{-\frac{1}{2}}$ around $z = -u_1^{(1,\max)}$ by Lemma 8.10. Otherwise, if (C1) holds, then we need to consider $\tilde{\varphi}_1(\underline{\zeta}_1(\underline{\zeta}_2(z)))$ in (8.37) due to the singularity of $\underline{\zeta}_2(z)$ at $z = -\tilde{\tau}_1 = -u_1^{(1,\max)}$, where $\tilde{\varphi}_1(\underline{\zeta}_1(z))$ is analytic at $z = -u_2^{(1,\max)} = -\underline{\zeta}_2(u_1^{(1,\max)})$ because

$$|\underline{\zeta}_1(\underline{\zeta}_2(-\tilde{\tau}_1))| = |\underline{\zeta}_1(\underline{\zeta}_2(\tilde{\tau}_1))| < \tilde{\tau}_1.$$

Hence, $\tilde{\varphi}_1(\underline{\zeta}_1(\underline{\zeta}_2(z))) - \tilde{\varphi}_1(-\underline{\zeta}_1(u_2^{(1,\max)})) \sim (-u_1^{(1,\max)} - z)^{\frac{1}{2}}$. On the other hand, $\tilde{\gamma}_1(\underline{\zeta}_1(\underline{\zeta}_2(z)), \underline{\zeta}_2(z)) - 1 \sim (-u_1^{(1,\max)} - z)^{\frac{1}{2}}$ because (v-a) does not hold. Combining these asymptotics in (8.37), we have $\tilde{\varphi}_2(\underline{\zeta}_2(z)) - \tilde{\varphi}_2(-\tilde{\tau}_2) \sim (-u_1^{(1,\max)} - z)^{\frac{1}{2}}$ around $z = -u_1^{(1,\max)}$ by Lemma 8.10.

We next consider (B2). From Table 8.1, the solutions of (8.16) are $z = \pm u_1^{(1,\max)}$, and the solutions of (8.19) are $z = \pm u_1^{(1,\mathrm{r})}$. In this case, the arguments for $z = -\tilde{\tau}_1$ are completely parallel to those for $z = \tilde{\tau}_1$ except for the cases (8Ic'-2) and (8Ic'-3). The latter two cases are also parallel if (C2) holds. Otherwise, $\tilde{\varphi}_2(z)$ is analytic at $z = -\tilde{\tau}_2$.

Asymptotic Inversion Formula

From these singularities, we derive exact tail asymptotics of the stationary distribution. For this, we use a Tauberian-type theorem for generating functions.

Lemma 8.12 (Theorem VI.5 of [7]). *Let f be a generating function of a sequence of real numbers $\{p(n); n = 0, 1, \ldots\}$. If $f(z)$ is singular at finitely many points $a_1, a_2, \ldots, a_m$ on the circle $|z| = \rho$ for some $\rho > 0$ and positive integer m and analytic on the set*

$$\Delta_i \equiv \{z \in \mathbb{C}; |z| < r_i, z \neq a_i, |\arg(z - a_i)| > \omega_i\}, \quad i = 1, 2, \ldots, m,$$

for some ω_i and r_i such that $\rho < r_i$ and $0 \leq \omega_i < \frac{\pi}{2}$, and if

$$\lim_{\Delta_i \ni z \to a_i} (a_i - z)^{\kappa_i} f(z) = b_i, \qquad i = 1, 2, \ldots, m, \tag{8.38}$$

for $\kappa_i \notin \{0, -1, -2, \ldots\}$ and some constant $b_i \in \mathbb{R}$, then

$$\lim_{n \to \infty} \left(\sum_{i=1}^{m} \frac{n^{\kappa_i - 1}}{\Gamma(\kappa_i)} a_i^{-n} \right)^{-1} p(n) = b \tag{8.39}$$

for some real number b, where $\Gamma(z)$ is the gamma function for complex number z (see Sect. 52 of volume II of [17]).

Recall the asymptotic notation "$\sim$" introduced in the introduction. With this notation, (8.39) can be written as

$$p(n) \sim \sum_{i=1}^{m} \frac{n^{\kappa_i - 1}}{\Gamma(\kappa_i)} a_i^{-n},$$

where $\Gamma(\frac{1}{2}) = \sqrt{\pi}$ and $\Gamma(-\frac{1}{2}) = -2\sqrt{\pi}$.

We will apply Lemma 8.12 in the following cases: For $m = 1$, $a_1 = u_1^{(1,\mathrm{r})}$, and $\kappa_1 = 1, 2$, $a_1 = u_1^{(1,\max)}$ and $\kappa_1 = \pm\frac{1}{2}$. For $m = 2$, $a_1 = \pm u_1^{(1,\mathrm{r})}$, and $\kappa_1 = 1, 2$, $a_1 = \pm u_1^{(1,\max)}$ and $\kappa_1 = -\frac{1}{2}$.

Exact Tail Asymptotics for the Nonarithmetic Case

Throughout this section, we assume the nonarithmetic condition (v-a). We first derive exact asymptotics for the stationary probabilities $\nu(n,0)$ and $\nu(0,n)$ on the boundary faces. Because of symmetry, we are only concerned with $\nu(n,0)$.

Boundary Probabilities for Nonarithmetic Case

We separately consider the two cases where $u_2^{(1,\Gamma)} < u_2^{(2,\Gamma)}$ and $u_2^{(1,\Gamma)} \geq u_2^{(2,\Gamma)}$, which correspond to categories I (or III) and II, respectively. In this subsection, we prove the following two theorems.

Theorem 8.2. *Under conditions (i)–(iv) and (v-a), for categories I and III, $\tilde{\tau}_1 = u_1^{(1,\Gamma)}$, and $P(L_1 = n, L_2 = 0)$ has the following exact asymptotic $h_1(n)$:*

$$h_1(n) = \begin{cases} \tilde{\tau}_1^{-n}, & u_1^{(1,\Gamma)} \neq u_1^{(1,\max)}, \\ n^{-\frac{1}{2}} \tilde{\tau}_1^{-n}, & u_1^{(1,\Gamma)} = u_1^{(1,\max)} = u_1^{(1,\mathrm{r})}, \\ n^{-\frac{3}{2}} \tilde{\tau}_1^{-n}, & u_1^{(1,\Gamma)} = u_1^{(1,\max)} \neq u_1^{(1,\mathrm{r})}. \end{cases} \tag{8.40}$$

By symmetry, the corresponding results are also obtained for $P(L_1 = 0, L_2 = n)$ for categories I and II.

Theorem 8.3. *Under conditions (i)–(iv) and (v-a), for category II, $\tilde{\tau}_2 = u_2^{(2,\mathrm{r})}$, and $P(L_1 = n, L_2 = 0)$ has the following exact asymptotic $h_1(n)$:*

$$h_1(n) = \begin{cases} \tilde{\tau}_1^{-n}, & \tilde{\tau}_1 < u_1^{(1,\Gamma)}, \text{ or} \\ & \tilde{\tau}_1 = u_1^{(1,\Gamma)} = u_1^{(1,\max)} = u_1^{(1,\mathrm{r})}, \\ n\tilde{\tau}_1^{-n}, & \tilde{\tau}_1 = u_1^{(1,\Gamma)} \neq u_1^{(1,\max)}, \\ n^{-\frac{1}{2}} \tilde{\tau}_1^{-n}, & \tilde{\tau}_1 = u_1^{(1,\Gamma)} = u_1^{(1,\max)} \neq u_1^{(1,\mathrm{r})}. \end{cases} \tag{8.41}$$

By symmetry, the corresponding results are also obtained for $P(L_1 = 0, L_2 = n)$ for categories III.

Remark 8.7. Theorems 8.2 and 8.3 exactly correspond with Theorem 6.1 of [4] (see also Theorems 2.1 and 2.3 of [3]). This is not surprising because of the similarity of the stationary equations, although moment-generating functions are used in [3,4].

Remark 8.8. These theorems fill missing cases for the exact asymptotics of Theorem 4.2 of [18]. Furthermore, they correct two errors there. Both of them are for category II. The exact asymptotic is geometric for $\tilde{\tau}_1 = u_1^{(1,\Gamma)} = u_1^{(1,\max)} = u_1^{(1,\mathrm{r})}$ and not geometric for $\tilde{\tau}_1 = u_1^{(1,\Gamma)} \neq u_1^{(1,\max)}$ (Theorem 8.3). However, in Theorem 4.2 of [18], they are not geometric [see (43d3) there] and geometric [see (4c) there], respectively. Thus, these should be corrected.

Proof of Theorem 8.2. We assume category I or III. This is equivalent to $u_2^{(1,\Gamma)} < u_2^{(2,\Gamma)}$ and $\tilde{\tau}_1 = u_1^{(1,\Gamma)}$. Furthermore, we always have $\underline{\zeta}_2(u_1^{(1,\Gamma)}) = u_2^{(1,\Gamma)} < \tilde{\tau}_2$, and therefore $\tilde{\varphi}_2(\underline{\zeta}_2(z))$ is analytic at $z = u_1^{(1,\Gamma)}$. We consider three cases separately.
(8IIa) $u_1^{(1,\Gamma)} < u_1^{(1,\max)}$: This case implies that $u_1^{(1,\mathrm{r})} < u_1^{(1,\max)}$ and $\tilde{\gamma}_1(\mathbf{u}^{(1,\max)}) > 1$, and therefore $\mathbf{u}^{(1,\mathrm{r})} = \mathbf{u}^{(1,\Gamma)}$. Hence, by Lemma 8.9, $\tilde{\varphi}_1$ of (8.29) satisfies the conditions of Lemma 8.12 under the setting (8.38) with $a_1 = u_1^{(1,\mathrm{r})}$, $\kappa = 1$. Thus, letting

$$b = \frac{(\tilde{\gamma}_2(\mathbf{u}^{(1,\mathrm{r})}) - 1)\tilde{\varphi}_2(u_2^{(1,\mathrm{r})}) + (\tilde{\gamma}_0(\mathbf{u}^{(1,\mathrm{r})}) - 1)\nu(\mathbf{0})}{\frac{d}{du}\tilde{\gamma}_1(u, \underline{\zeta}_2(u))|_{u=u_1^{(1,\mathrm{r})}}},$$

which must be positive by (8.39) and the fact that $\tilde{\varphi}_1(z)$ is singular at $z = u_1^{(1,\mathrm{r})}$, we have

$$\lim_{n\to\infty} \tilde{\tau}_1^n P(L_1 = n, L_2 = 0) = b.$$

(8IIb) $u_1^{(1,\Gamma)} = u_1^{(1,\max)}$, $u_1^{(1,\mathrm{r})} = u_1^{(1,\max)}$: In this case, category III is impossible, and $\tilde{\gamma}_1(\mathbf{u}^{(1,\max)}) = 1$. On the other hand, $\tilde{\varphi}_2(z)$ is analytic at $z = \underline{\zeta}_2(u_1^{(1,\max)}) < \tilde{\tau}_2$ because of category I. Hence, we can use the Taylor expansion (8.33), and therefore (8.29), (8.34), and Lemma 8.11 yield, for some $\delta > 0$,

$$\lim_{\mathcal{G}_\delta^+(u_1^{(1,\max)}) \ni z \to u_1^{(1,\max)}} (u_1^{(1,\max)} - z)^{\frac{1}{2}} \tilde{\varphi}_1(z) = b, \tag{8.42}$$

where

$$b = \left((\tilde{\gamma}_2(\mathbf{u}^{(1,\max)}) - 1)\tilde{\varphi}_2(u_2^{(1,\max)}) + (\tilde{\gamma}_0(\mathbf{u}^{(1,\max)}) - 1)\nu(\mathbf{0})\right) \frac{\sqrt{-\overline{\zeta}_1''(u_2^{(1,\max)})}}{\sqrt{2} p_{*1}^{(1)}(u_1^{(1,\mathrm{r})})}.$$

Hence, $\tilde{\varphi}_1$ satisfies the conditions of Lemma 8.12 under the setting (8.38) with $a_1 = u_1^{(1,\max)}$ and $\kappa_1 = \frac{1}{2}$, and therefore we have

$$\lim_{n\to\infty} n^{\frac{1}{2}} \tilde{\tau}_1^n P(L_1 = n, L_2 = 0) = \frac{b}{\sqrt{\pi}},$$

where the positivity of b is checked, similarly to case (8a) [see also case (8c) below].
(8IIc) $u_1^{(1,\Gamma)} = u_1^{(1,\max)}$, $u_1^{(1,\mathrm{r})} \neq u_1^{(1,\max)}$: In this case, category III is also impossible, and $\tilde{\gamma}_1(\mathbf{u}^{(1,\max)}) \neq 1$. Thus, we consider the setting (8.38) with $\kappa_1 = -\frac{1}{2}$. From (8.29) we have

$$\begin{aligned}
&\tilde{\varphi}_1(z) - \tilde{\varphi}_1(u_1^{(1,\max)}) \\
&= \frac{(\tilde{\gamma}_2(z,\underline{\zeta}_2(z)) - 1)\tilde{\varphi}_2(\underline{\zeta}_2(z)) + (\tilde{\gamma}_0(z,\underline{\zeta}_2(z)) - 1)\nu(\mathbf{0})}{1 - \tilde{\gamma}_1(z,\underline{\zeta}_2(z))} - \tilde{\varphi}_1(u_1^{(1,\max)}) \\
&= \frac{(\tilde{\gamma}_2(z,\underline{\zeta}_2(z)) - 1)(\tilde{\varphi}_2(\underline{\zeta}_2(z)) - \tilde{\varphi}_2(u_2^{(1,\max)}))}{1 - \tilde{\gamma}_1(z,\underline{\zeta}_2(z))} \\
&\quad + \frac{(\tilde{\gamma}_2(z,\underline{\zeta}_2(z)) - \tilde{\gamma}_2(\mathbf{u}^{(1,\max)}))\tilde{\varphi}_2(u_2^{(1,\max)})}{1 - \tilde{\gamma}_1(z,\underline{\zeta}_2(z))} + \frac{(\tilde{\gamma}_0(z,\underline{\zeta}_2(z)) - \tilde{\gamma}_0(\mathbf{u}^{(1,\max)}))\nu(\mathbf{0})}{1 - \tilde{\gamma}_1(z,\underline{\zeta}_2(z))} \\
&\quad + \left[\frac{(\tilde{\gamma}_2(\mathbf{u}^{(1,\max)}) - 1)\tilde{\varphi}_2(u_2^{(1,\max)})}{(1 - \tilde{\gamma}_1(z,\underline{\zeta}_2(z)))(1 - \tilde{\gamma}_1(\mathbf{u}^{(1,\max)}))} + \frac{(\tilde{\gamma}_0(\mathbf{u}^{(1,\max)}) - 1)\nu(\mathbf{0})}{(1 - \tilde{\gamma}_1(z,\underline{\zeta}_2(z)))(1 - \tilde{\gamma}_1(\mathbf{u}^{(1,\max)}))}\right] \\
&\quad \times \left(\tilde{\gamma}_1(z,\underline{\zeta}_2(z)) - \tilde{\gamma}_1(\mathbf{u}^{(1,\max)})\right).
\end{aligned} \tag{8.43}$$

We recall (8.34) that

$$\tilde{\varphi}_2(\underline{\zeta}_2(z)) - \tilde{\varphi}_2(u_2^{(1,\max)}) = -(u_1^{(1,\max)} - z)^{\frac{1}{2}} \frac{\sqrt{2}\tilde{\varphi}_2'(u_2^{(1,\max)})}{\sqrt{-\overline{\zeta}_1''(u_2^{(1,\max)})}} + o(|u_1^{(1,\max)} - z|^{\frac{1}{2}}).$$

From (8.31) we have

$$\begin{aligned}
\tilde{\gamma}_0(z,\underline{\zeta}_2(z)) - \tilde{\gamma}_0(\mathbf{u}^{(1,\max)}) &= (\underline{\zeta}_2(z) - \underline{\zeta}_2(u_1^{(1,\max)}))p_{*1}^{(0)}(z) \\
&\quad + \underline{\zeta}_2(u_1^{(1,\max)})(p_{*1}^{(0)}(z) - p_{*1}^{(0)}(u_1^{(1,\max)})) + p_{*0}^{(0)}(z) - p_{*0}^{(0)}(u_1^{(1,\max)}) \\
&= -\frac{\sqrt{2}p_{*1}^{(0)}(u_1^{(1,\max)})}{\sqrt{-\overline{\zeta}_1''(u_2^{(1,\max)})}}(u_1^{(1,\max)} - z)^{\frac{1}{2}} + o(|u_1^{(1,\max)} - z|^{\frac{1}{2}}).
\end{aligned}$$

Similarly,

$$\tilde{\gamma}_1(z, \underline{\zeta}_2(z)) - \tilde{\gamma}_1(\mathbf{u}^{(1,\max)})$$
$$= -\frac{\sqrt{2} p_{*1}^{(1)}(u_1^{(1,\max)})}{\sqrt{-\overline{\zeta}_1''(u_2^{(1,\max)})}}(u_1^{(1,\max)} - z)^{\frac{1}{2}} + o(|u_1^{(1,\max)} - z|^{\frac{1}{2}}),$$
$$\tilde{\gamma}_2(z, \underline{\zeta}_2(z)) - \tilde{\gamma}_2(\mathbf{u}^{(1,\max)})$$
$$= -\frac{\sqrt{2}\left(p_{*1}^{(2)}(u_1^{(1,\max)}) - \frac{p_{*-1}^{(2)}(u_1^{(1,\max)})}{\left(u_2^{(1,\max)}\right)^2}\right)}{\sqrt{-\overline{\zeta}_1''(u_2^{(1,\max)})}}(u_1^{(1,\max)} - z)^{\frac{1}{2}} + o(|u_1^{(1,\max)} - z|^{\frac{1}{2}}).$$

With the notation

$$c_1 = \frac{\sqrt{2}}{(1 - \tilde{\gamma}_1(\mathbf{u}^{(1,\max)}))\sqrt{-\overline{\zeta}_1''(u_2^{(1,\max)})}},$$
$$d_k = \frac{\partial}{\partial v}\tilde{\gamma}_k(u_1^{(1,\max)}, v)\Big|_{v=\underline{\zeta}_2(u_1^{(1,\max)})},$$

(8.43) yields, as $z \to u_1^{(1,\max)}$ satisfying that $z \in \tilde{\mathcal{G}}_\delta^+(u_1^{(1,\max)})$ for some $\delta > 0$,

$$\tilde{\varphi}_1(z) - \tilde{\varphi}_1(u_1^{(1,\max)}) = -c_1(u_1^{(1,\max)} - z)^{\frac{1}{2}}\Big(\big(\tilde{\gamma}_2(\mathbf{u}^{(1,\max)}) - 1\big)\tilde{\varphi}_2'(u_2^{(1,\max)})$$
$$+ d_2\tilde{\varphi}_2(u_2^{(1,\max)}) + d_0\nu(\mathbf{0}) + d_1\tilde{\varphi}_1(u_1^{(1,\max)})\Big) + o(|u_1^{(1,\max)} - z|^{\frac{1}{2}}). \quad (8.44)$$

Let

$$b = -\Big(\big(\tilde{\gamma}_2(\mathbf{u}^{(1,\max)}) - 1\big)\tilde{\varphi}_2'(u_2^{(1,\max)}) + d_2\tilde{\varphi}_2(u_2^{(1,\max)}) + d_0\nu(\mathbf{0}) + d_1\tilde{\varphi}_1(u_1^{(1,\max)})\Big).$$

Then, taking u_1 which is sufficiently close to $u_1^{(1,\max)}$ from below in (8.44), we can see that this b must be negative because $\tilde{\varphi}_1(u_1)$ is strictly increasing in $u_1 \in [0, u_1^{(1,\max)})$. Thus, (8.38) holds for the setting of (8.38) with $\kappa_1 = -\frac{1}{2}$, and therefore (8.39) leads to

$$\lim_{n\to\infty} n^{\frac{3}{2}}\tilde{\tau}_1^n P(L_1 = n, L_2 = 0) = -\frac{b}{2\sqrt{\pi}} > 0.$$

Thus, we have obtained all the cases of (8.40), and the proof is completed. □

Proof of Theorem 8.3. Assume category II. In this case, $\tilde{\tau}_2 = \underline{\zeta}_2(\tilde{\tau}_1)$, and $\tilde{\varphi}_2(z)$ has a simple pole at $z = \tilde{\tau}_2$ because of category II [see (8Ic-2)]. We need to consider the following cases.

(8IIa'): $\tilde{\tau}_1 < u_1^{(1,\Gamma)}$: In this case, $\tilde{\varphi}_2(\underline{\zeta}_2(z))$ has a simple pole at $z = \tilde{\tau}_1$. Since $\tilde{\varphi}_1(z)$ has no other singularity on $|z| = \tilde{\tau}_1$, it has a simple pole at $z = \tilde{\tau}_1$.

(8IIb'): $\tilde{\tau}_1 = u_1^{(1,\Gamma)}$: This case is further partitioned into the following subcases:

(8IIb'-1) $u_1^{(1,\Gamma)} \neq u_1^{(1,\max)}$: In this case, $\tilde{\tau}_1 = u_1^{(1,r)} < u_1^{(1,\max)}$, and therefore it is easy to see from (8.29) that $\tilde{\varphi}_1(z)$ has a double pole at $z = \tilde{\tau}_1$. Hence, we can apply the setting (8.38) with $a_1 = u_1^{(1,r)}$ and $\kappa = 2$.

(8IIb'-2) $u_1^{(1,\Gamma)} = u_1^{(1,\max)} \neq u_1^{(1,r)}$: (8.30) does not hold, and therefore (8.31) and the fact that $\tilde{\varphi}_2(z)$ has a simple pole at $z = \tilde{\tau}_2$ yield the same asymptotic as (8.42) but with a different b. Hence, we apply (8.38) with $\kappa_1 = \frac{1}{2}$.

(8IIb'-3) $u_1^{(1,\Gamma)} = u_1^{(1,\max)} = u_1^{(1,r)}$: In this case, we note the following facts.

(8IIb'-3-1) $\tilde{\varphi}_2(z)$ has a simple pole at $z = \tilde{\tau}_2$, and therefore Lemma 8.10 yields

$$\tilde{\varphi}_2(\underline{\zeta}_2(z)) \sim (u_1^{(1,\max)} - z)^{-\frac{1}{2}}.$$

(8IIb'-3-2) By Lemma 8.11, $1 - \tilde{\gamma}_1(z, \underline{\zeta}_2(z)) \sim (u_1^{(1,\max)} - z)^{\frac{1}{2}}$.

Hence, (8.29) yields $\tilde{\varphi}_1(z) \sim (u_1^{(1,\max)} - z)^{-1}$, and therefore we apply (8.38) with $a_1 = u_1^{(1,r)}$ and $\kappa = 1$.

Thus, similar to Theorem 8.2, we can obtain (8.41), which completes the proof. □

Marginal Distributions for the Nonarithmetic Case

We consider the asymptotics of $P(\langle \mathbf{c}, \mathbf{L} \rangle = x)$ as $x \to \infty$ for $\mathbf{c} = (1,0), (0,1), (1,1)$. For them, we use the generating functions $\tilde{\varphi}_+(z,1)$, $\tilde{\varphi}_+(1,z)$, and $\tilde{\varphi}_+(z,z)$. For simplicity, we denote them by $\psi_{10}(z)$, $\psi_{01}(z)$, $\psi_{11}(z)$, respectively. We note that generating functions are not useful for the other direction $\mathbf{c}$ because we cannot appropriately invert them. For general $\mathbf{c} > 0$, we should use moment-generating functions instead of generating functions. However, in this case, we need finer analytic properties to apply asymptotic inversion (e.g., see Appendix C of [3]). Thus, we leave it for future study.

From (8.13) and (8.28) we have, for $\mathbf{z} \in \mathbb{C}^2$ satisfying $(|z_1|, |z_2|) \in \tilde{\mathcal{D}}$,

$$\tilde{\varphi}(\mathbf{z}) = \left(1 + \frac{\tilde{\gamma}_1(\mathbf{z}) - 1}{1 - \tilde{\gamma}_+(\mathbf{z})}\right) \tilde{\varphi}_1(z_1) + \frac{\tilde{\gamma}_2(\mathbf{z}) - \tilde{\gamma}_+(\mathbf{z})}{1 - \tilde{\gamma}_+(\mathbf{z})} \tilde{\varphi}_2(z_2) + \frac{\tilde{\gamma}_0(\mathbf{z}) - \tilde{\gamma}_+(\mathbf{z})}{1 - \tilde{\gamma}_+(\mathbf{z})} \nu(\mathbf{0}). \quad (8.45)$$

Hence, the asymptotics of $P(\langle \mathbf{c}, \mathbf{L}\rangle = x)$ can be obtained for $\mathbf{c} = (1,0),(0,1),(1,1)$ by the analytic behavior of $\psi_{10}(z)$, $\psi_{01}(z)$, $\psi_{11}(z)$, respectively, around the singular points on the circles with radii $\rho_{\mathbf{c}}$, where

$$\rho_{(1,0)} = \sup\{u \geq 0; (u,1) \in \tilde{\mathcal{D}}\}, \quad \rho_{(0,1)} = \sup\{u \geq 0; (1,u) \in \tilde{\mathcal{D}}\},$$
$$\rho_{(1,1)} = \sup\{u \geq 0; (u,u) \in \tilde{\mathcal{D}}\}.$$

Since $\psi_{10}(z)$ and $\psi_{01}(z)$ are symmetric, we only consider $\psi_{10}(z)$ and $\psi_{11}(z)$. From (8.45) we have

$$\psi_{10}(z) = \left(1 + \frac{\tilde{\gamma}_1(z,1) - 1}{1 - \tilde{\gamma}_+(z,1)}\right) \tilde{\varphi}_1(z) + \frac{\tilde{\gamma}_2(z,1) - \tilde{\gamma}_+(z,1)}{1 - \tilde{\gamma}_+(z,1)} \tilde{\varphi}_2(1)$$
$$+ \frac{\tilde{\gamma}_0(z,1) - \tilde{\gamma}_+(z,1)}{1 - \tilde{\gamma}_+(z,1)} \nu(\mathbf{0}), \tag{8.46}$$
$$\psi_{11}(z) = \left(1 + \frac{\tilde{\gamma}_1(z,z) - 1}{1 - \tilde{\gamma}_+(z,z)}\right) \tilde{\varphi}_1(z) + \frac{\tilde{\gamma}_2(z,z) - \tilde{\gamma}_+(z,z)}{1 - \tilde{\gamma}_+(z,z)} \tilde{\varphi}_2(z)$$
$$+ \frac{\tilde{\gamma}_0(z,z) - \tilde{\gamma}_+(z,z)}{1 - \tilde{\gamma}_+(z,z)} \nu(\mathbf{0}). \tag{8.47}$$

We first consider the tail asymptotics for $\mathbf{c} = (1,0)$ under nonarithmetic condition (v-a). From (8.46) the singularity of $\psi_{11}(z)$ on the circle $|z| = \rho_{(1,0)}$ occurs by either that of $\tilde{\varphi}_1(z)$ or the solution of the following equation:

$$\tilde{\gamma}_+(z,1) = 1. \tag{8.48}$$

Since this equation is quadratic and the domain $\tilde{\mathcal{D}}$ contains vectors $\mathbf{x} > \mathbf{1} \equiv (1,1)$, Eq. (8.48) has a unique real solution greater than 1. We denote it by σ_+. We then have the following asymptotics (see also Fig. 8.4).

Theorem 8.4. *Under conditions (i)–(iv) and (v-a), let $h_1(n)$ be the exact asymptotic function given in Theorems 8.2 and 8.3; then $P(L_1 = n)$ has the following exact asymptotic $g_1(n)$ as $n \to \infty$:*

(a) If $\overline{\zeta}_2(u_1^{(1,\Gamma)}) < 1$, then $g_1(n) = \sigma_+^{-n}$.
(b) If $\overline{\zeta}_2(u_1^{(1,\Gamma)}) > 1$ and $\underline{\zeta}_2(u_1^{(1,\Gamma)}) \neq 1$, then $g_1(n) = h_1(n)$.
(c) If $\overline{\zeta}_2(u_1^{(1,\Gamma)}) > 1 = \underline{\zeta}_2(u_1^{(1,\Gamma)})$, then $g_1(n) = \tilde{\tau}_1^{-n}$.
(d) If $\overline{\zeta}_2(u_1^{(1,\Gamma)}) = 1 = \underline{\zeta}_2(u_1^{(1,\Gamma)})$, then $g_1(n) = \tilde{\tau}_1^{-n}$.
(e) If $\overline{\zeta}_2(u_1^{(1,\Gamma)}) = 1 > \underline{\zeta}_2(u_1^{(1,\Gamma)})$, then $g_1(n) = n\tilde{\tau}_1^{-n}$.

Remark 8.9. The corresponding but less complete results are obtained using moment-generating functions in Corollary 4.3 of [18].

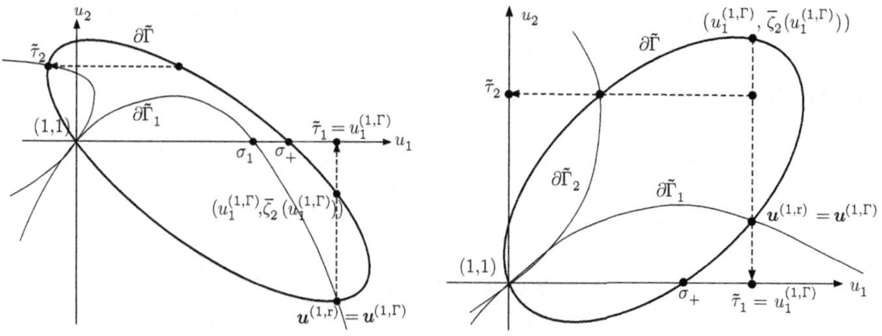

Fig. 8.4 *Left*: $\overline{\zeta}_2(u_1^{(1,\Gamma)}) < 1$; *right*: $\overline{\zeta}_2(u_1^{(1,\Gamma)}) > 1$ and $\underline{\zeta}_2(u_1^{(1,\Gamma)}) \neq 1$

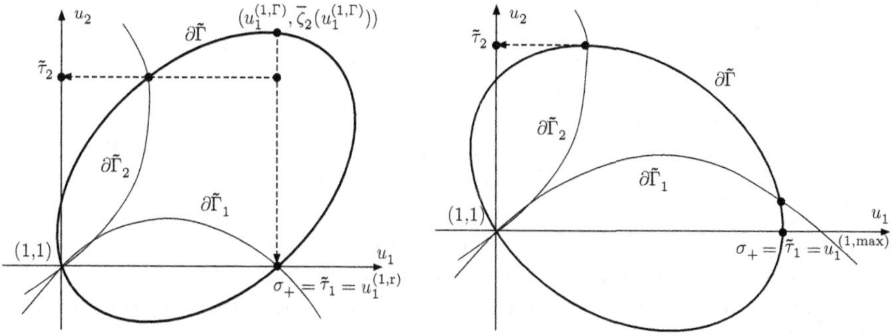

Fig. 8.5 *Left*: $\overline{\zeta}_2(u_1^{(1,\Gamma)}) > 1$ and $\underline{\zeta}_2(u_1^{(1,\Gamma)}) = 1$; *right*: $\overline{\zeta}_2(u_1^{(1,\Gamma)}) = \underline{\zeta}_2(u_1^{(1,\Gamma)}) = 1$

Before proving this theorem, we present asymptotics for the marginal distribution in the diagonal direction. Let σ_{d} be the real solution of

$$\tilde{\gamma}_+(u,u) = 1, \qquad u > 1,$$

which can be shown to be unique (Fig. 8.6). Because of symmetry, we assume without loss of generality that $\tilde{\tau}_1 \le \tilde{\tau}_2$. See Fig. 8.5 for the location of this point.

Theorem 8.5. *Under conditions (i)–(iv), (v-a), and $\tilde{\tau}_1 \le \tilde{\tau}_2$, let $h_1(n)$ be the exact asymptotic function given in Theorems 8.2 and 8.3; then $P(L_1 + L_2 = n)$ has the following exact asymptotic $g_+(n)$ as $n \to \infty$:*

(a) If $\sigma_{\mathrm{d}} < \tilde{\tau}_1$, then $g_+(n) = \sigma_{\mathrm{d}}^{-n}$.
(b) If $\sigma_{\mathrm{d}} > \tilde{\tau}_1$, then $g_+(n) = h_1(n)$.
(c) If $\sigma_{\mathrm{d}} = \tilde{\tau}_1 \neq u_1^{(1,\max)}$, then $g_+(n) = n\sigma_{\mathrm{d}}^{-n}$.
(d) If $\sigma_{\mathrm{d}} = \tilde{\tau}_1 = u_1^{(1,\max)} = \tilde{\tau}_2$, then $g_+(n) = n\sigma_{\mathrm{d}}^{-n}$
(e) If $\sigma_{\mathrm{d}} = \tilde{\tau}_1 = u_1^{(1,\max)} \neq \tilde{\tau}_2$, then $g_+(n) = \sigma_{\mathrm{d}}^{-n}$.

In what follows, we prove Theorem 8.4. The proof of Theorem 8.5 is similar, so we only outline it briefly.

Proof of Theorem 8.4. Let

$$\xi(z) = (\tilde{\gamma}_2(z,1) - \tilde{\gamma}_+(z,1))\tilde{\varphi}_2(1) + (\tilde{\gamma}_0(z,1) - \tilde{\gamma}_+(z,1))\nu(\mathbf{0});$$

then (8.46) can be written as

$$\psi_{10}(z) = \left(1 + \frac{\tilde{\gamma}_1(z,1) - 1}{1 - \tilde{\gamma}_+(z,1)}\right)\tilde{\varphi}_1(z) + \frac{\xi(z)}{1 - \tilde{\gamma}_+(z,1)}. \tag{8.49}$$

Since $\tilde{\gamma}_2(u,1) > 1, \tilde{\gamma}_0(u,1) > 1$ for $u > 0$ and

$$\frac{\partial}{\partial u}\tilde{\gamma}_1(u,1)\bigg|_{u=\sigma_1} < 0, \qquad \frac{\partial}{\partial u}\tilde{\gamma}_+(u,1)\bigg|_{u=\sigma_+} > 0 \text{ if } \underline{\zeta}_2(\sigma_+) = 0,$$

where σ_1 is a positive number satisfying that $\tilde{\gamma}_1(\sigma_1,1) = 1$, $\xi(\sigma_+) > 0$, and $\sigma_+ = \sigma_1$ implies that the prefactor of $\tilde{\varphi}_1(z)$ is positive at $z = \sigma_+$ if $\underline{\zeta}_2(\sigma_+) = 0$. With these observations in mind, we prove each case.

(a) Assume that $\overline{\zeta}_2(u_1^{(1,\Gamma)}) < 1$. This occurs if and only if $\sigma_+ = \rho_{10} < \tilde{\tau}_1$ (see the left-hand picture of Fig. 8.4). In this case, $\psi_{10}(z)$ must be singular at $z = \sigma_+$ because it is on the boundary of the convergence domain $\tilde{\mathcal{D}}$. Hence, it has a simple pole at $z = \sigma_+$, and therefore we have the exact geometric asymptotic.
(b) Assume that $\overline{\zeta}_2(u_1^{(1,\Gamma)}) > 1$ and $\underline{\zeta}_2(u_1^{(1,\Gamma)}) \neq 1$. This case occurs if and only if $\sigma_+ \neq \rho_{10} = \tilde{\tau}_1$ (see the right-hand picture of Fig. 8.4). In this case, $\tilde{\gamma}_1(\tilde{\tau}_1,1) \neq 1$, $\tilde{\gamma}_+(\tilde{\tau}_1,1) \neq 1$, and $\tilde{\gamma}_1(\tilde{\tau}_1,1) - 1$ has the same sign as $1 - \tilde{\gamma}_+(\tilde{\tau}_1,1)$. Hence, the prefactor of $\tilde{\varphi}_1(z)$ is analytic at $z = \tilde{\tau}_1$, and the singularity of $\psi_{10}(z)$ is determined by $\tilde{\varphi}_1(z)$. Thus, we have the same asymptotics as in Theorems 8.2 and 8.3.
(c) Assume that $\overline{\zeta}_2(u_1^{(1,\Gamma)}) > 1$ and $\underline{\zeta}_2(u_1^{(1,\Gamma)}) = 1$ (see the left-hand figure of Fig. 8.5). In this case, $\tilde{\gamma}_+(\tilde{\tau}_1,1) = \tilde{\gamma}_1(\tilde{\tau}_1,1) = 1$, and category II is impossible, and therefore, from (8.49) and Theorem 8.2, we have the exact geometric asymptotic.
(d) Assume that $\overline{\zeta}_2(u_1^{(1,\Gamma)}) = 1 = \underline{\zeta}_2(u_1^{(1,\Gamma)})$ (see the right-hand figure of Fig. 8.5). In this case, $\tilde{\tau}_1 = \sigma_+ = u_1^{(1,\max)}$, and therefore $\tilde{\gamma}_+(\tilde{\tau}_1,1) = 1$. We need to consider two subcases, $u_1^{(1,r)} = u_1^{(1,\max)}$ and $u_1^{(1,r)} \neq u_1^{(1,\max)}$. If $u_1^{(1,r)} = u_1^{(1,\max)}$, then $\tilde{\gamma}_1(\tilde{\tau}_1,1) = 1$ and $\tilde{\varphi}_1(z) \sim (\tilde{\tau}_1 - z)^{-\frac{1}{2}}$ by Theorem 8.2. Thus, we have $\psi_{10}(z) \sim (\tilde{\tau}_1 - z)^{-1}$ due to the second term of (8.49). Otherwise, if $u_1^{(1,r)} \neq u_1^{(1,\max)}$, then $\tilde{\gamma}_1(\tilde{\tau}_1,1) \neq 1$ implies that the prefactor of $\tilde{\varphi}_1(z)$ in (8.46) has a single pole at $z = \tilde{\tau}_1$ and that $\tilde{\varphi}_1(z) - \tilde{\varphi}_1(u_1^{(1,\max)}) \sim (\tilde{\tau}_1 - z)^{\frac{1}{2}}$. Again, from (8.49), we have $\psi_{10}(z) \sim (\tilde{\tau}_1 - z)^{-1}$. Thus, we have the exact geometric asymptotic in both cases.

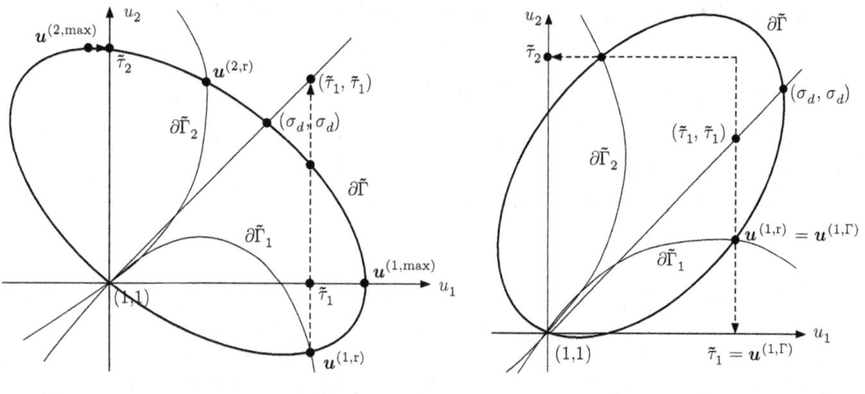

Fig. 8.6 *Left*: $\sigma_d < \tilde{\tau}_1$, *Right*: $\sigma_d > \tilde{\tau}_1$

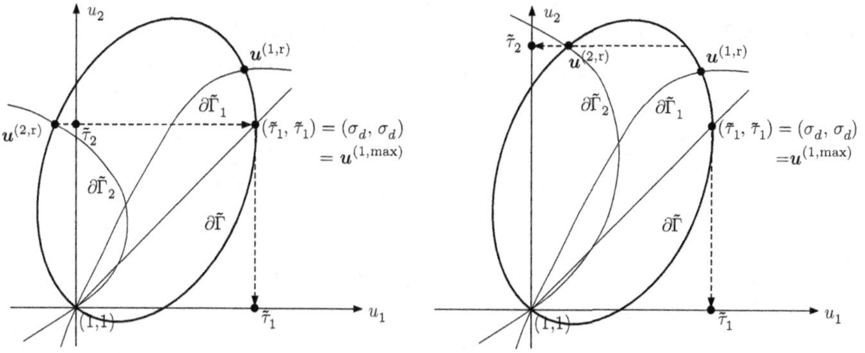

Fig. 8.7 *Left*: $\sigma_d = \tilde{\tau}_1 = u_1^{(1,\max)} = \tilde{\tau}_2$; *right*: $\sigma_d = \tilde{\tau}_1 = u_1^{(1,\max)} \neq \tilde{\tau}_2$

(e) Assume that $\overline{\zeta}_2(u_1^{(1,\Gamma)}) = 1 > \underline{\zeta}_2(u_1^{(1,\Gamma)})$. In this case, $\tilde{\tau}_1 = \sigma_+ = u_1^{(1,r)} < u_1^{(1,\max)}$, and we must have category I or III. Since $\tilde{\gamma}_+(\tilde{\tau}_1, 1) = 1$, $\tilde{\gamma}_1(\tilde{\tau}_1, 1) > 1$, and $\tilde{\varphi}_1(z)$ has a single pole at $z = \tilde{\tau}_1$, $\psi_{10}(z)$ in (8.46) has a double pole at $z = \tilde{\tau}_1$. This yields the desired asymptotic. □

The proof of Theorem 8.5 is more or less similar to that of $P(L_1 \geq n)$. From Figs. 8.6 and 8.7, we can see how the dominant singular point is located. Since its derivation is routine, we omit the detailed proof.

Exact Tail Asymptotics for the Arithmetic Case

Throughout this section, we assume that (v-a) does not hold. As in the section "Singularity for the Arithmetic Case," we separately consider two cases: (B1) either (v-b) or $m_2^{(1)} = 0$ holds, and (B2) neither (v-b) nor $m_2^{(1)} = 0$ holds, according to Table 8.1. In some cases, we need: (C1) either (v-c) or $m_1^{(2)} = 0$ holds, and (C2) neither (v-c) nor $m_1^{(2)} = 0$ holds.

Boundary Probabilities for Arithmetic Case with (B1)

In this case, we have the following asymptotics.

Theorem 8.6. *Under conditions (i)–(iv) and (B1), if (v-a) does not hold, then for categories I and III* $\tilde{\tau}_1 = u_1^{(1,\Gamma)}$, *and* $P(L_1 = n, L_2 = 0)$ *has the following exact asymptotic* $h_2(n)$. *For some constant* $b \in [-1,1]$

$$h_2(n) = \begin{cases} \tilde{\tau}_1^{-n}, & u_1^{(1,\Gamma)} \neq u_1^{(1,\max)}, \\ n^{-\frac{1}{2}}\,\tilde{\tau}_1^{-n}, & u_1^{(1,\Gamma)} = u_1^{(1,\max)} = u_1^{(1,\mathrm{r})}, \\ n^{-\frac{3}{2}}\,(1+b(-1)^n)\,\tilde{\tau}_1^{-n}, & u_1^{(1,\Gamma)} = u_1^{(1,\max)} \neq u_1^{(1,\mathrm{r})}. \end{cases} \tag{8.50}$$

By symmetry, the corresponding results are also obtained for $P(L_1 = 0, L_2 = n)$ *for categories I and II.*

Theorem 8.7. *Under conditions (i)–(iv) and (B1), if (v-a) does not hold, then, for category II,* $\tilde{\tau}_1 = \underline{\zeta}_2(\tilde{\tau}_2)$, $\tilde{\tau}_2 = u_2^{(2,\mathrm{r})}$, *and* $P(L_1 = n, L_2 = 0)$ *has the following exact asymptotic* $h_2(n)$. *For some constant* $b \in [-1,1]$

$$h_2(n) = \begin{cases} \tilde{\tau}_1^{-n}, & \tilde{\tau}_1 < u_1^{(1,\Gamma)} \text{ or} \\ & \tilde{\tau}_1 = u_1^{(1,\Gamma)} = u_1^{(1,\max)} = u_1^{(1,\mathrm{r})}, \\ n\tilde{\tau}_1^{-n}, & \tilde{\tau}_1 = u_1^{(1,\Gamma)} \neq u_1^{(1,\max)}, \\ n^{-\frac{1}{2}}\,\tilde{\tau}_1^{-n}, & \tilde{\tau}_1 = u_1^{(1,\Gamma)} = u_1^{(1,\max)} \neq u_1^{(1,\mathrm{r})}, \\ & \text{and (C1) holds.} \\ n^{-\frac{1}{2}}(1+b(-1)^n)\tilde{\tau}_1^{-n}, & \tilde{\tau}_1 = u_1^{(1,\Gamma)} = u_1^{(1,\max)} \neq u_1^{(1,\mathrm{r})}, \\ & \text{and (C2) holds.} \end{cases} \tag{8.51}$$

By symmetry, the corresponding results are also obtained for $P(L_1 = 0, L_2 = n)$ *for category III.*

Remark 8.10. As we will see in the proofs of these theorems, the asymptotics can be refined for those with the same geometric decay term $\tilde{\tau}_1^{-n}$. There is no difficulty in finding them, but they are cumbersome because we need additional cases. Thus, we omit their details.

Remark 8.11. One may wonder whether $b = \pm 1$ can occur in Theorems 8.6 and 8.7. If this is the case, then the tail asymptotics are purely periodic. Closely look at the coefficients of the asymptotic expansion of the terms in (8.37); it is unlikely to occur because $|\tilde{\varphi}_2(-\underline{\zeta}_2(\tilde{\tau}_1))| < \tilde{\varphi}_2(\underline{\zeta}_2(\tilde{\tau}_1))$. Thus, we conjecture that $|b| < 1$ is always the case.

By Table 8.1, $\tilde{\varphi}_1(z)$ may be singular at $z = -\tilde{\tau}_1$ on $|z| = \tilde{\tau}_1$. On the other hand, $\tilde{\varphi}_1(z)$ has the same singularity at $z = \tilde{\tau}_1$ as in the nonarithmetic case, so we can only focus on the singularity at $z = -\tilde{\tau}_1$. We note that $z = -u_1^{(1,\mathrm{r})}$ cannot be the solution of (8.19) under the assumptions of Theorems 8.6 and 8.7. With this fact in mind, we give proofs.

Proof of Theorem 8.6. We consider the singularity of $\tilde{\varphi}_1(z)$ at $z = -\tilde{\tau}_1$ by (8.29) using the arguments in the sections "Singularity for the Arithmetic Case" and "Exact Tail Asymptotics for the Nonarithmetic Case." Note that $\tilde{\tau}_1 = u_1^{(1,\Gamma)}$ because the category is either I or III. We need to consider the following three cases.

(8IIIa) $u_1^{(1,\Gamma)} \neq u_1^{(1,\max)}$: This case is equivalent to $u_1^{(1,\Gamma)} < u_1^{(1,\max)}$, and it follows from (8.29) that $\tilde{\varphi}_1(z)$ is analytic at $z = -u_1^{(1,r)}$. Hence, there is no singularity contribution by $z = -u_1^{(1,r)}$.

(8IIIb) $u_1^{(1,\Gamma)} = u_1^{(1,\max)}$, $u_1^{(1,r)} = u_1^{(1,\max)}$: In this case, as $z \to -u_1^{(1,\max)}$ in such a way that $z \in \tilde{\mathcal{G}}_\delta^+(-u_1^{(1,\max)})$ for some $\delta > 0$,

$$\tilde{\varphi}_2\left(\underline{\zeta}_2(z)\right) - \tilde{\varphi}_2\left(\underline{\zeta}_2(-u_1^{(1,\max)})\right) \sim \left(-u_1^{(1,\max)} - z\right)^{\frac{1}{2}},$$

but $1 - \tilde{\gamma}_1(z, \underline{\zeta}_2(z))$ does not vanish at $z = -u_1^{(1,\max)}$, and therefore

$$\tilde{\varphi}_1(z) - \tilde{\varphi}_1\left(-u_1^{(1,\max)}\right) \sim \left(-u_1^{(1,\max)} - z\right)^{\frac{1}{2}}.$$

This yields the asymptotic function $n^{-\frac{3}{2}} \tilde{\tau}_1^{-n}$, but this function is dominated by the slower asymptotic function $n^{-\frac{1}{2}} \tilde{\tau}_1^{-n}$ due to the singularity at $z = u_1^{(1,\max)}$.

(8IIIc) $u_1^{(1,\Gamma)} = u_1^{(1,\max)}$, $u_1^{(1,r)} \neq u_1^{(1,\max)}$: In this case, the solution of (8.19) has no essential role, so $\tilde{\varphi}_1(z)$ has the same analytic behavior at $z = -u_1^{(1,\max)}$ as at $z = u_1^{(1,\max)}$ in (8IIc) in the proof of Theorem 8.2.

Thus, combining with the asymptotics in Theorem 8.2, we complete the proof. □

Proof of Theorem 8.7. Because of category II, $\tau_2 = \underline{\zeta}_2(\tilde{\tau}_1)$, and therefore $\underline{\zeta}_2(-\tilde{\tau}_1) = -\underline{\zeta}_2(\tilde{\tau}_1) = -\tilde{\tau}_2$ by the assumption that (v-a) does not hold. We consider the singularity at $z = -\tilde{\tau}_1$ for the following cases with this in mind.

(8IIIa') $\tilde{\tau}_1 < u_1^{(1,\Gamma)}$: This case is included in (8Ic'-2). Hence, if (C1) holds, then $\tilde{\varphi}_2(\underline{\zeta}_2(z))$, and therefore $\tilde{\varphi}_1(z)$ are analytic at $z = -u_1^{(1,r)}$. Otherwise, if (C2) holds, then $\tilde{\varphi}_2(\underline{\zeta}_2(z))$ has a simple pole at $z = -u_1^{(1,r)}$. However, in (8.29), $\tilde{\varphi}_2(\underline{\zeta}_2(z))$ has the prefactor $\tilde{\gamma}_2(z, \underline{\zeta}_2(z)) - 1$, which vanishes at $z = -u_1^{(1,r)}$ because of (C2). Hence, the pole of $\tilde{\varphi}_2(\underline{\zeta}_2(z))$ is cancelled, and therefore $\tilde{\varphi}_1(z)$ is analytic at $z = -u_1^{(1,r)}$. Thus, neither case has a contribution by $z = -u_1^{(1,r)}$.

(8IIIb') $\tilde{\tau}_1 = u_1^{(1,\Gamma)} \neq u_1^{(1,\max)}$: In this case, $\tilde{\tau}_1 = u_1^{(1,r)}$. If (C2) holds, then $\tilde{\varphi}_2(z)$ has a simple pole at $z = -\tilde{\tau}_2$, and therefore, as in (8IIIb'-3-1),

$$\tilde{\varphi}_2\left(\underline{\zeta}_2(z)\right) \sim \left(-u_1^{(1,\max)} - z\right)^{-\frac{1}{2}},$$

but (8IIb'-3-2) is not the case, and therefore this yields the asymptotic function $n^{-\frac{1}{2}}\tilde{\tau}_1^{-n}$. However, this asymptotic term is again dominated by $\tilde{\tau}_1^{-n}$ due to the singularity at $z = u_1^{(1,\max)}$. On the other hand, if (C1) holds, then there is no singularity contribution by $z = -\tilde{\tau}_1$. Hence, we have the same asymptotics as in the corresponding case of Theorem 8.3.

(8IIIc') $\tilde{\tau}_1 = u_1^{(1,\Gamma)} = u_1^{(1,\max)}$: This is the case of (8c'-3). As we discussed there, if (C2) holds, then $\tilde{\varphi}_2(\underline{\zeta}_2(z)) \sim (-u_1^{(1,\max)} - z)^{-\frac{1}{2}}$ around $z = -u_1^{(1,\max)}$. Because of (B1), there is no other singularity contribution in (8.29), and therefore we also have $\tilde{\varphi}_1(z) \sim (-u_1^{(1,\max)} - z)^{-\frac{1}{2}}$ around $z = -u_1^{(1,\max)}$. This results in the asymptotic $n^{-\frac{1}{2}}\tilde{\tau}_1^{-n}$. On the other hand, if (C1) holds, we similarly have $\tilde{\varphi}_1(z) - \tilde{\varphi}_1(-u_1^{(1,\max)}) \sim (-u_1^{(1,\max)} - z)^{\frac{1}{2}}$. This implies the asymptotic $n^{-\frac{3}{2}}\tilde{\tau}_1^{-n}$. To combine this with the corresponding asymptotics obtained in Theorem 8.3, we consider two subcases.

- (8IIIc'-1) $u_1^{(1,\max)} = u_1^{(1,r)}$: In this case, the asymptotics caused by $z = \tilde{\tau}_1$ is $n\tilde{\tau}_1^{-n}$, and therefore the asymptotic due to $z = -u_1^{(1,\max)}$ is ignorable.
- (8IIIc'-2) $u_1^{(1,\max)} \neq u_1^{(1,r)}$: In this case, the asymptotic caused by $z = \tilde{\tau}_1$ is $n^{-\frac{1}{2}}\tilde{\tau}_1^{-n}$. Hence, we have two different cases. If (C1) holds, then the contribution by $z = -\tilde{\tau}_1$ is ignorable. Otherwise, if (C2) holds, then we have an additional asymptotic term: $(-1)^n n^{-\frac{1}{2}}\tilde{\tau}_1^{-n}$.

Thus, the proof is completed. □

Boundary Probabilities for Arithmetic Case with (B2)

We next consider case (B2). As noted in the section "Singularity for the Arithmetic Case," in this case, $\tilde{\varphi}_1(z)$ is singular at $z = \pm u_1^{(1,r)}$, and both singular points have essentially the same properties. Thus, we have the following theorems.

Theorem 8.8. *Under conditions (i)–(iv) and (B2), if (v-a) does not hold, then for categories I and III, $\tilde{\tau}_1 = u_1^{(1,\Gamma)}$, and $P(L_1 = n, L_2 = 0)$ has the following exact asymptotic $h_3(n)$. For some constants $b_i \in [-1,1]$ for $i = 1,2,3$*

$$h_3(n) = \begin{cases} (1 + b_1(-1)^n)\tilde{\tau}_1^{-n}, & u_1^{(1,\Gamma)} \neq u_1^{(1,\max)}, \\ n^{-\frac{1}{2}}(1 + b_2(-1)^n)\tilde{\tau}_1^{-n}, & u_1^{(1,\Gamma)} = u_1^{(1,\max)} = u_1^{(1,r)}, \\ n^{-\frac{3}{2}}(1 + b_3(-1)^n)\tilde{\tau}_1^{-n}, & u_1^{(1,\Gamma)} = u_1^{(1,\max)} \neq u_1^{(1,r)}. \end{cases} \tag{8.52}$$

By symmetry, the corresponding results are also obtained for $P(L_1 = 0, L_2 = n)$ for categories I and II.

Theorem 8.9. *Under conditions (i)–(iv) and (B2), if (v-a) does not hold, then, for category II, $\tilde{\tau}_2 = u_2^{(2,\mathrm{r})}$, and $P(L_1 = n, L_2 = 0)$ has the following exact asymptotic $h_3(n)$. For some constants $b_i \in [-1,1]$ for $i = 1,2,3$*

$$h_3(n) = \begin{cases} (1+b_1(-1)^n)\tilde{\tau}_1^{-n}, & \tilde{\tau}_1 < u_1^{(1,\Gamma)} \text{ or} \\ & \tilde{\tau}_1 = u_1^{(1,\Gamma)} = u_1^{(1,\max)} = u_1^{(1,\mathrm{r})}, \\ n(1+b_2(-1)^n)\tilde{\tau}_1^{-n}, & \tilde{\tau}_1 = u_1^{(1,\Gamma)} \neq u_1^{(1,\max)}, \\ n^{-\frac{1}{2}}(1+b_3(-1)^n)\tilde{\tau}_1^{-n}, & \tilde{\tau}_1 = u_1^{(1,\Gamma)} = u_1^{(1,\max)} \neq u_1^{(1,\mathrm{r})}. \end{cases} \tag{8.53}$$

By symmetry, the corresponding results are also obtained for $P(L_1 = 0, L_2 = n)$ for categories III.

Marginal Distributions for Arithmetic Case

Under the arithmetic condition that (v-a) does not hold, we consider the tail asymptotics of the marginal distributions. Basically, the results are the same as in Theorems 8.4 and 8.5, in which Theorems 8.2 and 8.3 should be replaced by Theorems 8.6 and 8.7 for case (B1) and Theorems 8.8 and 8.9 for case (B2). Thus, we omit their details.

Application to a Network with Simultaneous Arrivals

In this section, we apply the asymptotic results to a queueing network with two nodes numbered 1 and 2. Assume that customers simultaneously arrive at both nodes from the outside subject to the Poisson process at the rate λ. For $i = 1,2$, service times at node i are independent and identically distributed with the exponential distribution with mean μ_i^{-1}. Customers who have finished their services at node 1 go to node 2 with probability p. Similarly, customers departing from queue 2 go to queue 1 with probability q. This routing is independent of everything else. Customers what are not routed to the other queue leave the network. We refer to this queueing model as a two-node Jackson network with simultaneous arrival.

Obviously, this network is stable, that is, it has a stationary distribution, if and only if

$$\frac{\lambda(1+q)}{1-pq} < \mu_1, \qquad \frac{\lambda(1+p)}{1-pq} < \mu_2. \tag{8.54}$$

This fact can also be checked by stability condition (iv).

We are interested in how the tail asymptotics of the stationary distribution of this network are changed. If $p = q = 0$, this model is studied in [8, 9]. As we will see subsequently, this model can be described by a double QBD process, and therefore we know the solutions to the tail asymptotic problem. However, this does not mean that the solutions are analytically tractable. Thus, we will consider what kind of difficulty arises in applications of our tail asymptotic results.

Let $L_i(t)$ be the number of customers at node i at time t. It is easy to see that $\{(L_1(t), L_2(t)); t \in \mathbb{R}_+\}$ is a continuous-time Markov chain. Because the transition rates of this Markov chain are uniformly bounded, we can construct a discrete-time Markov chain given by uniformization, which has the same stationary distribution. We denote this discrete-time Markov chain by $\{\mathbf{L}_n = (L_{1\ell}, L_{2\ell}); \ell \in \mathbb{Z}_+\}$, where it is assumed without loss of generality that

$$\lambda + \mu_1 + \mu_2 = 1.$$

Obviously, $\{\mathbf{L}_n; \ell \in \mathbb{Z}_+\}$ is a double QBD process. We denote a random vector subject to the stationary distribution of this process by $\mathbf{L} \equiv (L_1, L_2)$, as we did in the section "Double QBD Process and the Convergence Domain."

To apply our asymptotic results, we first compute generating functions. For $\mathbf{u} = (u_1, u_2) \in \mathbb{R}^2$

$$\tilde{\gamma}_+(\mathbf{u}) = \lambda u_1 u_2 + \mu_1 p u_1^{-1} u_2 + \mu_2 q u_1 u_2^{-1} + \mu_1(1-p)u_1^{-1} + \mu_2(1-q)u_2^{-1}, \quad (8.55)$$

$$\tilde{\gamma}_1(\mathbf{u}) = \lambda u_1 u_2 + \mu_1 p u_1^{-1} u_2 + \mu_1(1-p)u_1^{-1} + \mu_2, \quad (8.56)$$

$$\tilde{\gamma}_2(\mathbf{u}) = \lambda u_1 u_2 + \mu_2 q u_1 u_2^{-1} + \mu_2(1-q)u_2^{-1} + \mu_1. \quad (8.57)$$

We next find the extreme point $\mathbf{u}^{(1,\mathrm{r})} = (u_1^{(1,\mathrm{r})}, u_2^{(1,\mathrm{r})})$. This is obtained as the solution to the equations

$$\tilde{\gamma}_+(\mathbf{u}) = \tilde{\gamma}_1(\mathbf{u}) = 1.$$

Applying (8.55) and (8.56) to the first equation we have

$$u_2 = u_1 q + (1-q). \quad (8.58)$$

Substituting (8.58) into $\tilde{\gamma}_1(\mathbf{u}) = 1$ we have

$$\lambda u_1^2(u_1 q + 1 - q) + \mu_1 p(u_1 q + 1 - q) + \mu_1(1-p) + \mu_2 u_1 = u_1.$$

Assume that $q > 0$. Then u_1 has the following solutions:

$$u_1 = 1, \frac{-\lambda \pm \sqrt{\lambda^2 + 4\lambda q \mu_1(1-pq)}}{2\lambda q}.$$

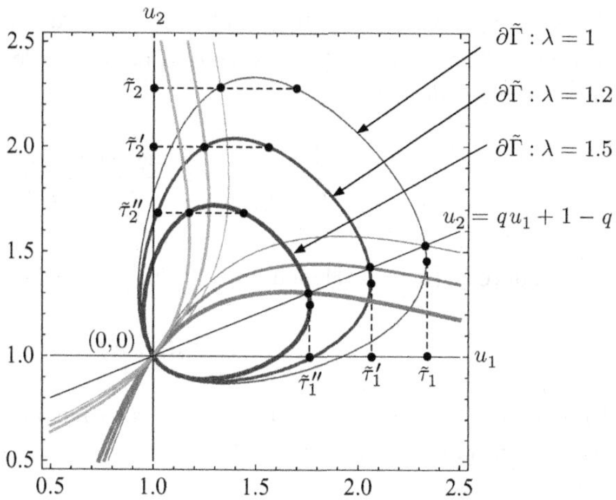

Fig. 8.8 Effect of the arrival rate: λ is changed from 1 to 1.2 and 1.5 (*thicker curves*), while $\mu_1 = 5$, $\mu_2 = 4$, $p = 0.25$, $q = 0.4$ are unchanged

We are only interested in the solution $u_1 > 1$, which must be $u_1^{(1,\mathrm{r})}$, that is,

$$u_1^{(1,\mathrm{r})} = \frac{-\lambda + \sqrt{\lambda^2 + 4\lambda q \mu_1 (1 - pq)}}{2\lambda q}. \tag{8.59}$$

We next consider the maximal point $\mathbf{u}^{(1,\max)}$ of $\tilde{\gamma}(\mathbf{u}) = 1$. This can be obtained to solve the equations

$$\tilde{\gamma}_+(\mathbf{u}) = 1, \quad \frac{du_1}{du_2} = 0.$$

These equations are equivalent to

$$\lambda u_1 + \mu_1 p u_1^{-1} - \mu_2 q u_1 u_2^{-2} - \mu_2 (1-q) u_2^{-2} = 0, \tag{8.60}$$

$$\lambda u_1 u_2 + \mu_1 p u_1^{-1} u_2 + \mu_2 q u_1 u_2^{-1} + \mu_1 (1-p) u_1^{-1} + \mu_2 (1-q) u_2^{-1} = 1. \tag{8.61}$$

Theoretically we know that these equations have two solutions such that $\mathbf{u} > \mathbf{0}$, which must be $\mathbf{u}^{(1,\min)}$ and $\mathbf{u}^{(1,\max)}$. We can numerically obtain them, but their analytic expressions are not easy to obtain. Furthermore, even if they are obtained, they would be analytically intractable.

To circumvent this difficulty, we propose to draw figures. Today we have at our disposal excellent software such as *Mathematica* to draw two-dimensional figures. Then we can manipulate figures and could discover how modeling parameters change the tail asymptotics. This is essentially the same as numerical computations. However, figures are more informative to see how changes occur (see, e.g., Fig. 8.8).

We finally consider a simpler case to find analytically tractable results. Assume that $q = 0$ but $p > 0$. $q = 0$ implies that

$$\mathbf{u}^{(1,\mathrm{r})} = (\rho_1^{-1}, 1),$$

where $\rho_1 = \frac{\lambda_1}{\mu_1}$. Obviously, ρ_1 must be the decay rate of $P(L_1 = n)$. This can be also verified by Theorem 8.4. However, it may not be the decay rate of $P(L_1 = n, L_2 = 0)$. In fact, we can derive

$$\left.\frac{du_1}{du_2}\right|_{\mathbf{u}=\mathbf{u}^{(1,\mathrm{r})}} = \frac{\mu_2 - (\mu_1 + \lambda p)}{\lambda(1 - \rho_1^{-1})}$$

on the curve $\tilde{\gamma}_+(\mathbf{u}) = 1$. Hence, $u_1^{(1,\Gamma)} = u_1^{(1,\mathrm{r})}$ if and only if

$$\mu_2 \geq \mu_1 + \lambda p_1. \tag{8.62}$$

Thus, if (8.62) holds, then $P(L_1 = n, L_2 = 0)$ has an exact geometric asymptotic. Otherwise, we have, by Theorem 8.2,

$$\lim_{n\to\infty} n^{-\frac{3}{2}} (u_1^{(1,\max)})^{-n} P(L_1 = n, L_2 = 0) = b. \tag{8.63}$$

We can see that $\rho_1^{-1} < u_1^{(1,\max)}$, but $u_1^{(1,\max)}$ is only numerically obtained by solving (8.60) and (8.61).

Concluding Remarks

We derived the exact asymptotics for a stationary distribution applying the analytic function method based on the convergence domain. We here discuss which problems can be studied by this method and what is needed to develop it further.

Technical issue. In the analytic function method, a key ingredient is that the function $\underline{\zeta}_2(z)$ is analytic and suitably bounded for an appropriate region, as we showed in Lemma 8.8. For this, we use the fact that $\underline{\zeta}_2(z)$ is the solution of a quadratic equation, which is equivalent for the random walk to be skip free in the interior of the quadrant. The quadratic equation (or polynomial equation in general) is also a key for the alternative approach based on an analytic extension on a Riemann surface. If the random walk is not skip free, then it would be harder to get a right analytic function. However, the non-skip-free case is also interesting. Thus, it is challenging to overcome this difficulty. A completely different approach might be needed here.

Probabilistic interpretation. We employed a purely analytic method and gave no stochastic interpretations except a few, although the asymptotic results are stochastic. However, probabilistic interpretations may be helpful. For example, one might wonder what the probabilistic meanings of the function $\underline{\zeta}_2$ and (8.29) are. We believe there should be something here. If sound meanings are provided, then we may better explain Lemma 8.8 and may resolve the technical issues discussed previously.

Modeling extensions. We think the present approach is applicable to a higher-dimensional model as well as a generalized reflecting random walk proposed in [19] as long as the skip-free assumption is satisfied. One might also consider relaxing the irreducibility condition on the random walk in the interior of the quadrant. However, this is essentially equivalent to reducing the dimension, so there should be no difficulty in considering it. Another extension is to modulate the double QBD process or multidimensional reflecting random walk in general by a background Markov chain. The tail asymptotic problem becomes harder, but there should be a way to use the present analytic function approach at least for the two-dimensional case with finitely many background states. Related discussions can be found in [19].

Applicability. As we saw in the section "Application to a Network with Simultaneous Arrivals," analytic results on the tail asymptotics may not be easy to apply to each specific application because they are not analytically tractable. To fill this gap between theory and application, we have proposed using geometric interpretations instead of analytic formulas. However, this is currently more or less like having numerical tables. We should here make clear what we want to do using tail asymptotics. Once a problem is set up, we might consider solving it using geometric interpretations. There would probably be a systematic way to do this that did not depend on a specific problem. This is also challenging.

Appendix

Proof of Lemma 8.6

Note that $u^2 D_2(u)$ is a polynomial of order 2 at least and order 4 at most. For $k = 1,3$, let c_k be the coefficients of u^k in the polynomial $u^2 D_2(u)$. Then,

$$c_1 = -2(1-p_{00})p_{(-1)0} - 4(p_{(-1)(-1)}p_{01} + p_{(-1)1}p_{0(-1)}) \leq 0,$$
$$c_3 = -2(1-p_{00})p_{10} - 4(p_{1(-1)}p_{01} + p_{11}p_{0(-1)}) \leq 0.$$

Hence, if both $u_1^{(1,\max)}$ and $-u_1^{(1,\max)}$ are the solutions of $u^2 D_2(u) = 0$, then

$$2(c_1 u_1^{(1,\max)} + c_3 (u_1^{(1,\max)})^3) = (u_1^{(1,\max)})^2 (D_2(u_1^{(1,\max)}) - D_2(-u_1^{(1,\max)})) = 0.$$

Since $u_1^{(1,\max)} > 0$, this holds true if and only if $c_1 = c_3 = 0$, which is equivalent to $p_{01} = p_{0(-1)} = p_{(-1)0} = p_{10} = 0$ because $p_{00} = 1$ is impossible. Hence, $u^2 D_2(u) = 0$ has the two solutions $u_1^{(1,\max)}$ and $-u_1^{(1,\max)}$ if and only if (v-a) does not hold. In this case, we have $c_1 = c_3 = 0$, which implies that $u^2 D_2(u)$ is an even function. Since $u^2 D_2(u) = 0$ has only real solutions including $u_1^{(1,\max)}$ by Lemmas 8.4 and 8.5, we complete the proof. □

Proof of Lemma 8.7

By (8.17), we have

$$\sum_{i\in\{-1,0,1\}}\sum_{j\in\{-1,0,1\}} p_{ij}x^iy^j = 1, \qquad \sum_{i\in\{-1,0,1\}}\sum_{j\in\{-1,0,1\}} (-1)^{i+j}p_{ij}x^iy^j = 1.$$

Subtracting both sides of these equations we have

$$p_{10}x + p_{01}y + p_{0(-1)}y^{-1} + p_{(-1)0}x^{-1} = 0.$$

Since x, y are positive, this equation holds true if and only if

$$p_{10} = p_{01} = p_{0(-1)} = p_{(-1)0} = 0.$$

This is the condition that (v-a) does not hold. □

Acknowledgements We are grateful to Mark S. Squillante for encouraging us to complete this work. We are also thankful to the three anonymous referees. This research was supported in part by the Japan Society for the Promotion of Science under Grant No. 21510165.

References

1. Avram, F., Dai, J.G., Hasenbein, J.J.: Explicit solutions for variational problems in the quadrant. Queue. Syst. **37**, 259–289 (2001)
2. Borovkov, A.A., Mogul'skii, A.A.: Large deviations for Markov chains in the positive quadrant. Russ. Math. Surv. **56**, 803–916 (2001)
3. Dai, J.G., Miyazawa, M.: Reflecting Brownian motion in two dimensions: exact asymptotics for the stationary distribution. Stoch. Syst. **1**, 146–208 (2011)
4. Dai, J.G., Miyazawa, M.: Stationary distribution of a two-dimensional SRBM: geometric views and boundary measures. Submitted for publication (arXiv:1110.1791v1) (2011)
5. Fayolle, G., Malyshev, V.A., Menshikov, M.V.: Topics in the Constructive Theory of Countable Markov Chains. Cambridge University Press, Cambridge (1995)
6. Fayolle, G., Iasnogorodski, R., Malyshev, V.: Random Walks in the Quarter-Plane: Algebraic Methods, Boundary Value Problems and Applications. Springer, New York (1999)
7. Flajolet, P., Sedqewick, R.: Analytic Combinatorics. Cambridge University Press, Cambridge (2009)
8. Flatto, L., Hahn, S.: Two parallel queues by arrivals with two demands I. SIAM J. Appl. Math. **44**, 1041–1053 (1984)
9. Flatto, L., McKean, H.P.: Two queues in parallel. Comm. Pure Appl. Math. **30**, 255–263 (1977)
10. Foley, R.D., McDonald, D.R.: Bridges and networks: exact asymptotics. Ann. Appl. Probab. **15**, 542–586 (2005)
11. Foley, R.D., McDonald, D.R.: Large deviations of a modified Jackson network: Stability and rough asymptotics. Ann. Appl. Probab. **15**, 519–541 (2005)
12. Guillemin, F., van Leeuwaarden, J.S.H.: Rare event asymptotics for a random walk in the quarter plane. Queue. Syst. **67**, 1–32 (2011)

13. Kobayashi, M., Miyazawa, M.: Tail asymptotics of the stationary distribution of a two dimensional reflecting random walk with unbounded upward jumps. Submitted for publication (2011)
14. Li, H., Zhao, Y.Q.: Exact tail asymptotics in a priority queue–characterizations of the preemptive model. Queue. Syst. **63**, 355–381 (2009)
15. Li, H., Zhao, Y.Q.: Tail asymptotics for a generalized two-demand queueing models – a kernel method. Queue. Syst. **69**, 77–100 (2011)
16. Li, H., Zhao, Y.Q.: A kernel method for exact tail asymptotics: random walks in the quarter plane. Preprint (2011)
17. Markushevich, A.I.: Theory of Functions, vols. I, II and III, 2nd edn., trans. by R.A. Silverman, reprinted by American Mathematical Society, Providence, Rhode Island (1977)
18. Miyazawa, M.: Tail decay rates in double QBD processes and related reflected random walks. Math. Oper. Res. **34**, 547–575 (2009)
19. Miyazawa, M.: Light tail asymptotics in multidimensional reflecting processes for queueing networks. TOP **19**, 233–299 (2011)
20. Miyazawa, M., Rolski, T.: Exact asymptotics for a Levy-driven tandem queue with an intermediate input. Queue. Syst. **63**, 323–353 (2009)
21. Miyazawa, M., Zhao, Y.Q.: The stationary tail asymptotics in the GI/G/1-type queue with countably many background states. Adv. Appl. Probab. **36**, 1231–1251 (2004)

Chapter 9
Two-Dimensional Fluid Queues with Temporary Assistance

Guy Latouche, Giang T. Nguyen, and Zbigniew Palmowski

Introduction

Stochastic fluid models have a wide range of applications such as water reservoir operational control, industrial and computer engineering, risk analysis, environmental analysis, and telecommunications. In particular, they have been used in telecommunication modeling since the seminal article [3]. With the advent of differentiated services, buffers have, in a very natural way, become multidimensional. To give another example, that of decentralized mobile networks, callers transmit data via each other's equipment, and it is necessary to determine the appropriate fractions of caller capacity, be it buffer space or power, that may be allocated to other users.

In computer processing, a situation where the problem of effective resource sharing can arise is when there are more tasks than schedulers that can process them. Aggarwal et al. [1] consider this problem in the particular setting of two ON–OFF streams of tasks: routine and nonroutine, and one central processing unit (CPU) to serve both streams, one of which is specifically being determined by a workload threshold. The CPU serves routine or nonroutine tasks, depending on whether the amount of workload for the routine tasks is above or below the threshold, respectively. To determine the optimal threshold value that minimizes the weighted sum of the probability of exceeding undesirable workload limits, the authors derive the workload distribution of routine tasks and approximate that of nonroutine tasks. Mahabhashyam et al. [18] extend the resource-sharing model to allow a partial split of the CPU's capacity. More specifically, the CPU serves routine

G. Latouche • G.T. Nguyen (✉)
Université Libre de Bruxelles, Département d'Informatique, CP 212, Boulevard du Triomphe, 1050 Brussels, Belgium
e-mail: latouche@ulb.ac.be; giang.nguyen@adelaide.edu.au

Z. Palmowski
University of Wrocław, Mathematical Institute, 50-384 Wrocław, Poland
e-mail: zbigniew.palmowski@gmail.com

G. Latouche et al. (eds.), *Matrix-Analytic Methods in Stochastic Models*, Springer Proceedings in Mathematics & Statistics 27, DOI 10.1007/978-1-4614-4909-6_9,

tasks when their accumulated workload is above the threshold; according to some predetermined proportion, the CPU serves both routine and nonroutine tasks when the threshold is not exceeded; and the CPU serves nonroutine tasks when there is no routine task left.

We generalize this model further by allowing the input model to better fit an environment where multiple users independently decide when to use the system, thereby allowing for the intensity of the load to vary in time. Specifically, each input stream of fluid is formed by N exponential ON–OFF sources, with $N \geq 1$, and we analyze the model using a two-dimensional stochastic fluid.

A Markov-modulated single-buffer fluid model is a two-dimensional Markov process $\{X(t), \varphi(t) : t \in \mathbb{R}^+\}$, where $X(t)$ is the continuous *level* of the buffer and $\varphi(t)$ is the discrete *phase* of the underlying irreducible Markov chain that governs the rates of change. A practical and well-studied case is *piecewise constant* rates: the fluid is assumed to have a constant rate c_i when $\varphi(t) = i$, for i in a finite state space $\mathcal{S}$. The traditional approach to obtaining performance measures of Markov-modulated single-buffer fluids with piecewise constant rates is to use spectral analysis [3, 13, 17, 19, 23]. Over the last two decades, matrix-analytic methods have attracted a lot of attention as an alternatives and algorithmically effective approach to analyzing these standard fluids [2, 5–8, 10, 11, 21].

The mathematical model we consider is a Markov process $\{X(t), Y(t), \varphi_1(t), \varphi_2(t) : t \in \mathbb{R}^+\}$, where $X(t) \geq 0$ and $Y(t) \geq 0$ represent the levels of buffers 1 and 2, respectively. At a given time $t \geq 0$, the rates of change of buffer 1 depend only on the underlying Markovian phase $\varphi_1(t)$; the rates of change of buffer 2, on the other hand, depend on both $\varphi_2(t)$ and $X(t)$ because, while each buffer receives its own input sources, both share a fixed output capacity c, in a proportion dependent on the level of buffer 1. More specifically, buffer j receives N ON–OFF input sources, with each having exponentially distributed ON and OFF intervals at corresponding rates α_j and β_j, and continuously generates fluid at rate R_j during ON intervals for $j = 1, 2$. When the fluid level $X(t)$ of buffer 1 is above the threshold $x^* > 0$, buffer 1 is allocated the total shared output capacity c, leaving buffer 2 without any; when $0 < X(t) < x^*$, buffer j has output capacity c_j, $c_1 + c_2 = c$; and when $X(t) = 0$, buffer 1 has output capacity $\min\{iR_1, c_1\}$, and buffer 2 $c - \min\{iR_1, c_1\}$, where i is the number of inputs of buffer 1 being on at time t.

The generalization of the number of ON–OFF inputs necessitates modifications in the original rules of output-capacity sharing from Mahabhashyam et al. [18]. When $X(t) = 0$, the policy in the single ON–OFF input model is to allocate the total capacity c to buffer 2. The totality rule is logical when there is only one ON–OFF input for each buffer: buffer 1 is empty only when its input is off, and in that case, buffer 2 can receive the whole output capacity c until the moment the input of buffer 1 is on again. Here, it is possible for buffer 1 to be empty while i inputs are on, for $0 < i \leq \lfloor \frac{c_1}{R_1} \rfloor$. Under these circumstances, assigning the total output capacity c to buffer 2 would immediately cause buffer 1 to try to increase from level 0, consequently grabbing back c_1 amount of output capacity. However, as $i \leq \lfloor \frac{c_1}{R_1} \rfloor$, the output capacity c_1 would be sufficient to empty buffer 1, forcing it to give away the whole output capacity c to buffer 2, etc. Therefore, applying the original totality rule at $X(t) = 0$ for the generalized N ON–OFF input model would potentially lead to inconsistency.

The behavior described above at level 0 for buffer 1 when $0 \leq i \leq \lfloor \frac{c_1}{R_1} \rfloor$ is referred to as being *sticky* [11], a property arising when net rates of the buffer for the same Markovian phase but different levels are different in a particular way that makes it unable to go up or down, thereby remaining stuck at a level until the background Markov chain switches to a nonsticky phase. In our model, by allocating iR_1 output capacity to buffer 1 and $c - iR_1$ to buffer 2 when $X(t) = 0$ and $0 \leq i \leq \lfloor \frac{c_1}{R_1} \rfloor$, we let buffer 1 remain at level zero while eliminating potential uncertainty and utilizing the total output capacity in the most effective way. For the same reason, when $X(t) = x^*$ and $\lceil \frac{c_1}{R_1} \rceil \leq i \leq \lfloor \frac{c}{R_1} \rfloor$, the output capacity is iR_1 for buffer 1 and $c - iR_1$ for buffer 2. While the stickiness, borne in the generalization of the number of ON–OFF inputs, necessitates only slight modifications in the output-capacity allocation policy, it considerably complicates the analysis and numerical computation of performance measures of the model. To deal with this complication, we employ a mixture of tools from both dominant approaches – spectral analysis and matrix-analytic methods.

One may change the system in many ways and still use the same method. For example, in the last part of this chapter, we restrict buffer 1 to a finite size but keep buffer 2 infinite. This affects the analysis of buffer 1, but the analytical expressions for buffer 2 remain unchanged. We take $N = 1$ there for better illustration.

The rest of the paper is organized as follows: in the section "Reference Model," we formulate the model mathematically. Assuming that both buffer sizes are infinite, we derive the marginal probability distribution of buffer 1 in the section "Infinite Buffer 1" and bounds for those of buffer 2 in the section "Analysis for buffer 2." In the section "Finite Buffer 1, with One Input," restricting buffer 1 to a finite size, we determine its marginal probability distribution in the particular case of $N = 1$, thereby providing numerical comparisons to the corresponding results in [18], where buffer 1 is assumed to be infinite.

Reference Model

Consider a four-dimensional Markov process $\{X(t), Y(t), \varphi_1(t), \varphi_2(t) : t \in \mathbb{R}^+\}$, where $X(t) \geq 0$ and $Y(t) \geq 0$ are the levels in buffers 1 and 2, respectively, and for $j = 1, 2$, $\varphi_j(t)$ represents the phase of the background irreducible Markov chain for buffer j with finite state space $\mathcal{S} = \{0, \ldots, N\}$ with $N \geq 1$; state $i \in \mathcal{S}$ indicates that i ON–OFF inputs are on. The generator T_j for $\{\varphi_j(t)\}$ is

$$T_j = \begin{bmatrix} * & N\beta_j & & & \\ \alpha_j & * & (N-1)\beta_j & & \\ & \ddots & \ddots & \ddots & \\ & & (N-1)\alpha_j & * & \beta_j \\ & & & N\alpha_j & * \end{bmatrix},$$

with each diagonal element $*$ defined appropriately such that each row sum of T_j is 0. For $i_1, i_2 \in \mathcal{S}$ we denote by $\dot{x}_{i_1}$ and $\dot{y}_{i_2}$ the respective net rates for buffer 1 in phase i_1 and buffer 2 in phase i_2. For $X(t) > x^*$ and $Y(t) > 0$,

$$\begin{aligned}\dot{x}_{i_1} &= i_1 R_1 - c,\\ \dot{y}_{i_2} &= i_2 R_2;\end{aligned}$$

for $X(t) = x^*$ and $Y(t) > 0$,

$$\begin{aligned}\dot{x}_{i_1} &= 0 \quad \text{for } \left\lceil \frac{c_1}{R_1} \right\rceil \le i_1 \le \left\lfloor \frac{c}{R_1} \right\rfloor,\\ &= i_1 R_1 - c_1 \quad \text{otherwise},\\ \dot{y}_{i_2} &= i_2 R_2 - (c - i_1 R_1) \quad \text{for } \left\lceil \frac{c_1}{R_1} \right\rceil \le i_1 \le \left\lfloor \frac{c}{R_1} \right\rfloor,\\ &= i_2 R_2 - c_2 \quad \text{otherwise};\end{aligned}$$

for $0 < X(t) < x^*$ and $Y(t) > 0$,

$$\begin{aligned}\dot{x}_{i_1} &= i_1 R_1 - c_1,\\ \dot{y}_{i_2} &= i_2 R_2 - c_2;\end{aligned}$$

and for $X(t) = 0$ and $Y(t) > 0$,

$$\begin{aligned}\dot{x}_{i_1} &= 0 \quad \text{for } 0 \le i_1 \le \left\lfloor \frac{c_1}{R_1} \right\rfloor,\\ &= i_1 R_1 - c_1 \quad \text{otherwise},\\ \dot{y}_{i_2} &= i_2 R_2 - (c - i_1 R_1) \quad \text{for } 0 \le i_1 \le \left\lfloor \frac{c_1}{R_1} \right\rfloor,\\ &= i_2 R_2 - c_2 \quad \text{otherwise}.\end{aligned}$$

For $Y(t) = 0$, $\dot{y}_{i_2}$ is the maximum between 0 and the net rate of buffer 2 in $i_2 \in \mathcal{S}$ when $Y(t) > 0$.

We assume that $NR_j > c$, $\frac{c_j}{R_j}, \frac{c}{R_j} \notin \mathbb{N}$, and the system is positive recurrent. The first assumption ensures that for $X(t) > x^*$, the set of states for which the net rates of buffer 1 are positive is nonempty. We impose the second assumption to avoid having states with zero rates for buffer 1 when $X(t) \notin \{0, x^*\}$ and for buffer 2 when $Y(t) \neq 0$. This assumption is purely to simplify some technical details, without any loss of generality, as any single-buffer-fluid model with zero rates can be transformed into a single-buffer-fluid model without zero rates [9]. The third assumption is equivalent to

$$\sum_{i=0}^{N} i R_1 q_i^{(1)} + \sum_{i=0}^{N} i R_2 q_i^{(2)} < c_1 + c_2, \tag{9.1}$$

where $\underline{q}^{(1)}$ and $\underline{q}^{(2)}$ are the stationary probability vectors of T_1 and T_2, respectively. Inequality (9.1) is obvious when considering the stability condition for the equivalent single-buffer-fluid model with a constant output $c_1 + c_2$ and $2N$ exponential ON–OFF inputs, half of which switch on at rate β_1 and switch off at rate α_1, and the other half switch on at rate β_2 and switch off at rate α_2. For $i = 1, 2$ and for $n = 1, \ldots, N$,

$$q_0^{(i)} = \frac{\alpha_i^N}{(\alpha_i + \beta_i)^N},$$

$$q_n^{(i)} = q_0^{(i)} \binom{N}{n} (\beta_i / \alpha_i)^n,$$

which reduces (9.1) to

$$N \frac{R_1 \beta_1}{\alpha_1 + \beta_1} + N \frac{R_2 \beta_2}{\alpha_2 + \beta_2} < c_1 + c_2. \tag{9.2}$$

Infinite Buffer 1

To analyze buffer 1 when $N = 1$, Mahabhashyam et al. [18] consider an equivalent system of two standard single subbuffers, each with a single ON–OFF input, one subbuffer with constant output capacity c_1 and the other with constant output capacity c. Decomposing buffer 1 in this fashion, the authors show that the marginal probability distribution of buffer 1 can be obtained by appropriately combining the average time of going up from x^* and then going down to x^* in Subbuffer 1 and the average time of going down from x^* and then going up to x^* in Subbuffer 2. The authors determine analytic expressions for the former average time by using, from [20], the busy period distribution of a standard single buffer with one exponential ON–OFF input and constant output capacity, and for the latter by establishing a pair of partial differential equations, transformed into ordinary differential equations and then solved by a spectral decomposition technique.

In this chapter, for general $N \geq 1$ we analyze buffer 1 by applying matrix-analytic methods. With this approach, while it is not simple to obtain closed-form expressions for $N \geq 2$, we can obtain various performance measures numerically using fast convergent algorithms (see, most relevantly, [5, 11] and the references therein). The focus of this section is the marginal probability distribution for buffer 1.

We refer to $X(t) = 0$ and $X(t) = x^*$ as *boundaries* $\circ$ and $*$, and $0 < X(t) < x^*$ and $X(t) > x^*$ as *bands* 1 and 2. While T_1 governs the transitions of $\{\varphi_1(t)\}$ for all $X(t) \geq 0$, the rate of buffer 1 in the same phase varies between boundaries and bands. Therefore, we partition $\mathcal{S}$ differently for each boundary and each band. We denote, respectively, by $\mathcal{S}_d^{(\bullet)}$, $\mathcal{S}_s^{(\bullet)}$, and $\mathcal{S}_u^{(\bullet)}$ the sets of states with negative, zero, and positive net rates when buffer 1 is at boundary $\bullet$, for $\bullet \in \{\circ, *\}$, and by $\mathcal{S}_-^{(k)}$ and

$\mathcal{S}_+^{(k)}$ the sets of states with negative and positive net rates when buffer 1 is in band k, for $k = 1, 2$. Then $\mathcal{S} = \mathcal{S}_s^{(\circ)} \cup \mathcal{S}_u^{(\circ)} = \mathcal{S}_-^{(1)} \cup \mathcal{S}_+^{(1)} = \mathcal{S}_d^{(*)} \cup \mathcal{S}_s^{(*)} \cup \mathcal{S}_u^{(*)} = \mathcal{S}_-^{(2)} \cup \mathcal{S}_+^{(2)}$, with

$$\mathcal{S}_s^{(\circ)} = \mathcal{S}_-^{(1)} = \mathcal{S}_d^{(*)} = \mathcal{S}_-^{(2)} = \left\{0, \ldots, \left\lfloor \frac{c_1}{R_1} \right\rfloor \right\},$$

$$\mathcal{S}_u^{(\circ)} = \mathcal{S}_+^{(1)} = \mathcal{S}_+^{(2)} = \left\{ \left\lceil \frac{c_1}{R_1} \right\rceil, \ldots, N \right\},$$

$$\mathcal{S}_s^{(*)} = \left\{ \left\lceil \frac{c_1}{R_1} \right\rceil, \ldots, \left\lfloor \frac{c}{R_1} \right\rfloor \right\}, \mathcal{S}_u^{(*)} = \left\{ \left\lceil \frac{c}{R_1} \right\rceil, \ldots, N \right\}.$$

For each band k, we partition T_1 into submatrices $T_{\ell m}^{(k)}$, of which each element $[T_{\ell m}^{(k)}]_{ij}$ is the transition rate from $i \in \mathcal{S}_\ell^{(k)}$ to $j \in \mathcal{S}_m^{(k)}$, and we denote by $C_\ell^{(k)}$ the diagonal matrix of absolute net rates for $i \in \mathcal{S}_\ell^{(k)}$:

$$C_-^{(1)} = \begin{bmatrix} |-c_1| & & & \\ & |R_1 - c_1| & & \\ & & \ddots & \\ & & & |\lfloor \frac{c_1}{R_1} \rfloor R_1 - c_1| \end{bmatrix},$$

$$C_+^{(1)} = \begin{bmatrix} \lceil \frac{c_1}{R_1} \rceil R_1 - c_1 & & & \\ & (\lceil \frac{c_1}{R_1} \rceil + 1) R_1 - c_1 & & \\ & & \ddots & \\ & & & NR_1 - c_1 \end{bmatrix},$$

$$C_-^{(2)} = \begin{bmatrix} |-c| & & & \\ & |R_1 - c| & & \\ & & \ddots & \\ & & & |\lfloor \frac{c}{R_1} \rfloor R_1 - c| \end{bmatrix},$$

$$C_+^{(2)} = \begin{bmatrix} \lceil \frac{c}{R_1} \rceil R_1 - c & & & \\ & (\lceil \frac{c}{R_1} \rceil + 1) R_1 - c & & \\ & & \ddots & \\ & & & NR_1 - c \end{bmatrix}.$$

We illustrate in Fig. 9.1 the relationships between the large cast of characters. Exploiting Markov-renewal arguments, da Silva Soares and Latouche [11, Theorem 4.2] prove that the stationary density vector of a Markov-modulated, level-dependent, single-buffer-fluid queue can be obtained by properly combining limiting densities from above and below each boundary (when possible) and

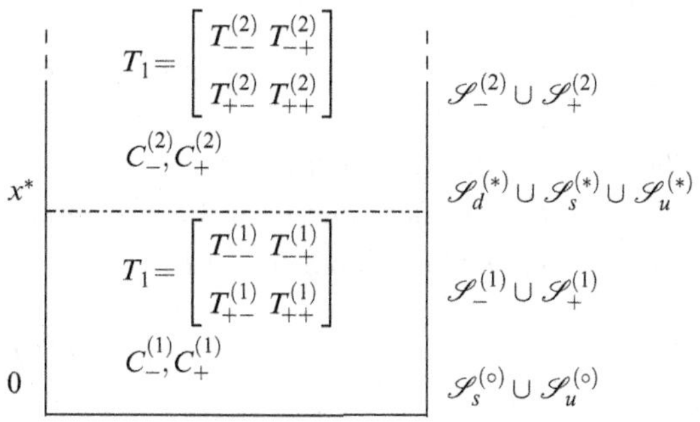

Fig. 9.1 Buffer 1

steady state probability masses at these boundaries. To that effect, we consider the jump chain $\{J_n : n \geq 0\}$ of the process $\{X(t), \varphi_1(t)\}$ restricted to the set of boundary states $\mathcal{B} = \{(\bullet, i) : \bullet \in \{\circ, *\}, i \in \mathcal{S}\}$. We note that this jump chain will also be useful for obtaining bounds on marginal probabilities of buffer 2, as described in the section "Analysis for Buffer 2." By [11, Theorem 4.4], the $(2N+2) \times (2N+2)$ transition matrix Ω of $\{J_n\}$, block-partitioned according to $\mathcal{B} = (\circ, \mathcal{S}^{(\circ)}_u) \cup (\circ, \mathcal{S}^{(\circ)}_s) \cup (*, \mathcal{S}^{(*)}_u) \cup (*, \mathcal{S}^{(*)}_s) \cup (*, \mathcal{S}^{(*)}_d)$, is

$$\Omega = \left[\begin{array}{cc|ccc} \cdot & \Psi^{(\circ)}_{us} & \Lambda^{(\circ,*)}_{uu} & \Lambda^{(\circ,*)}_{us} & \cdot \\ P^{(\circ)}_{su} & P^{(\circ)}_{ss} & \cdot & \cdot & \cdot \\ \hline \cdot & \cdot & \cdot & \Psi^{(*)}_{us} & \Psi^{(*)}_{ud} \\ \cdot & \cdot & P^{(*)}_{su} & P^{(*)}_{ss} & P^{(*)}_{sd} \\ \cdot & \hat{\Lambda}^{(*,\circ)}_{ds} & \hat{\Psi}^{(*)}_{du} & \hat{\Psi}^{(*)}_{ds} & \cdot \end{array}\right], \tag{9.3}$$

where

$\Psi^{(\bullet)}_{um}, \hat{\Psi}^{(*)}_{dm}, \Lambda^{(\circ,*)}_{um}, \hat{\Lambda}^{(*,\circ)}_{dm}$, and $P^{(\bullet)}_{sm}$ denote various *first passage probability* matrices, with

$[\Psi^{(\bullet)}_{um}]_{ij}$ = the probability of returning to $\bullet$ and in $j \in \mathcal{S}^{(\bullet)}_m$, after initially increasing from $\bullet$ and in $i \in \mathcal{S}^{(\bullet)}_u$, while avoiding a higher boundary (if there exists one);

$[\hat{\Psi}^{(*)}_{dm}]_{ij}$ = the probability of returning to x^* and in $j \in \mathcal{S}^{(*)}_m$, after initially decreasing from x^* and in $i \in \mathcal{S}^{(*)}_d$, while avoiding level 0;

$[\Lambda^{(\circ,*)}_{um}]_{ij}$ = the probability of reaching x^* and in $j \in \mathcal{S}^{(*)}_m$, while avoiding level 0 after initially increasing from there in $i \in \mathcal{S}^{\circ}_u$;

$[\hat{\Lambda}^{(*,\circ)}_{ds}]_{ij}$ = probability of reaching level 0 and in $j \in \mathcal{S}^{(\circ)}_s$, while avoiding x^* after initially decreasing from there in $i \in \mathcal{S}^{*}_d$; and

$[P^{(\bullet)}_{sm}]_{ij}$ = the probability of going from $i \in \mathcal{S}^{(\bullet)}_s$ to $j \in \mathcal{S}^{(\bullet)}_m$ in one transition.

The jump chain of the Markov process $\{\varphi_1(\cdot)\}$ has the transition matrix

$$P = I - \Delta^{-1} T_1, \tag{9.4}$$

where Δ is the diagonal matrix with $[\Delta]_i = [T_1]_{ii}$; for the remainder of the paper, we denote by I the identity matrix of the appropriate size. Clearly each of $P_{su}^{(\circ)}$, $P_{ss}^{(\circ)}$, $P_{su}^{(*)}$, $P_{ss}^{(*)}$, and $P_{sd}^{(*)}$ is a submatrix of P:

$$P = \begin{bmatrix} P_{ss}^{(\circ)} & P_{su}^{(\circ)} \\ P_{uu}^{(\circ)} & P_{us}^{(\circ)} \end{bmatrix} = \begin{bmatrix} P_{dd}^{(*)} & P_{ds}^{(*)} & P_{du}^{(*)} \\ P_{sd}^{(*)} & P_{ss}^{(*)} & P_{su}^{(*)} \\ P_{ud}^{(*)} & P_{us}^{(*)} & P_{uu}^{(*)} \end{bmatrix}.$$

The matrices $\Psi_{us}^{(\circ)}$, $[\Lambda_{us}^{(\circ,*)}, \Lambda_{uu}^{(\circ,*)}]$, $\hat{\Lambda}_{ds}^{(*,\circ)}$, and $[\hat{\Psi}_{ds}^{(*)}, \hat{\Psi}_{du}^{(*)}]$ are, respectively, equal to $\Psi_{+-}^{(1)}$, $\Lambda_{++}^{(1)}$, $\hat{\Lambda}_{--}^{(1)}$, and $\hat{\Psi}_{-+}^{(1)}$, the corresponding first passage probability matrices for the level-independent fluid queue $\{M_1(t), \rho_1(t) : t \in \mathbb{R}^+\}$ with finite size x^*, state space $\mathcal{S}_-^{(1)} \cup \mathcal{S}_+^{(1)}$, generator T_1, and rate matrices $C_-^{(1)}$ and $C_+^{(1)}$. By [10, Theorem 5.2],

$$\begin{bmatrix} \Lambda_{++}^{(1)} & \Psi_{+-}^{(1)} \\ \hat{\Psi}_{-+}^{(1)} & \hat{\Lambda}_{--}^{(1)} \end{bmatrix} = \begin{bmatrix} e^{\hat{U}_1 x^*} & \Psi_1 \\ \hat{\Psi}_1 & e^{U_1 x^*} \end{bmatrix} \begin{bmatrix} I & \Psi_1 e^{U_1 x^*} \\ \hat{\Psi}_1 e^{\hat{U}_1 x^*} & I \end{bmatrix}^{-1}, \tag{9.5}$$

where Ψ_1 is the minimum nonnegative solution to the Riccati equation

$$(C_+^{(1)})^{-1} T_{+-}^{(1)} + (C_+^{(1)})^{-1} T_{++}^{(1)} \Psi_1 + \Psi_1 (C_-^{(1)})^{-1} T_{--}^{(1)} + \Psi_1 (C_-^{(1)})^{-1} T_{-+}^{(1)} \Psi_1 = 0, \tag{9.6}$$

$\hat{\Psi}_1$ is the minimum nonnegative solution to the Riccati equation

$$(C_-^{(1)})^{-1} T_{-+}^{(1)} + (C_-^{(1)})^{-1} T_{--}^{(1)} \hat{\Psi}_1 + \hat{\Psi}_1 (C_+^{(1)})^{-1} T_{++}^{(1)} + \hat{\Psi}_1 (C_+^{(1)})^{-1} T_{+-}^{(1)} \hat{\Psi}_1 = 0, \tag{9.7}$$

$$U_1 = (C_-^{(1)})^{-1} T_{--}^{(1)} + (C_-^{(1)})^{-1} T_{-+}^{(1)} \Psi_1, \tag{9.8}$$

and

$$\hat{U}_1 = (C_+^{(1)})^{-1} T_{++}^{(1)} + (C_+^{(1)})^{-1} T_{+-}^{(1)} \hat{\Psi}_1. \tag{9.9}$$

Similarly, $[\Psi_{us}^{(*)}, \Psi_{ud}^{(*)}] = \Psi_2$, which is the first passage probability matrix for the infinite level-independent fluid queue $\{M_2(t), \rho_2(t) : t \in \mathbb{R}^+\}$ with state space $\mathcal{S}_-^{(2)} \cup \mathcal{S}_+^{(2)}$, generator T_1, and the rate matrices $C_-^{(2)}$ and $C_+^{(2)}$. By [22], the matrix Ψ_2 is the minimum nonnegative solution to the Riccati equation

$$(C_+^{(2)})^{-1}T_{+-}^{(2)} + (C_+^{(2)})^{-1}T_{++}^{(2)}\Psi_2 + \Psi_2(C_-^{(2)})^{-1}T_{--}^{(2)} + \Psi_2(C_-^{(2)})^{-1}T_{-+}^{(2)}\Psi_2 = 0. \quad (9.10)$$

Applying fast convergent algorithms described in [4, 9], we can solve Riccati Eqs. (9.6), (9.7), and (9.10) to obtain Ψ_1, $\hat{\Psi}_1$, and Ψ_2 and, consequently, Ω.

We denote by $\underline{m} = [\underline{p}_s^{(\circ)}, \underline{p}_s^{(*)}]$ the probability mass vector of buffer 1 at the set of boundary sticky states $\mathcal{K} = \{(\bullet, \zeta) : \bullet \in \{\circ, *\}, \zeta \in \mathcal{S}_s^{(\bullet)}\}$, and we define $\mathcal{E}^{(*)} = \{(*, \zeta) : \zeta \in \mathcal{S}_u^{(*)} \cup \mathcal{S}_d^{(*)}\}$ and $\mathcal{E}^{(\circ)} = \{(\circ, \zeta) : \zeta \in \mathcal{S}_u^{(\circ)}\}$. Note that $\mathcal{K} = \mathcal{B} - \{\mathcal{E}^{(*)} \cup \mathcal{E}^{(\circ)}\}$. Proceeding in two steps, we write the transition matrix $\Omega^{(*)}$ of the censored fluid queue on $\{\mathcal{B} - \mathcal{E}^{(*)}\}$ as

$$\Omega^{(*)} = \begin{bmatrix} \cdot & \Psi_{us}^{(\circ)} & \Lambda_{us}^{(\circ,*)} \\ P_{su}^{(\circ)} & P_{ss}^{(\circ)} & \cdot \\ \cdot & \cdot & P_{ss}^{(*)} \end{bmatrix} + \begin{bmatrix} \Lambda_{uu}^{(\circ,*)} & \cdot \\ \cdot & \cdot \\ P_{su}^{(*)} & P_{sd}^{(*)} \end{bmatrix} \begin{bmatrix} I & -\Psi_{ud}^{(*)} \\ -\hat{\Psi}_{ud}^{(*)} & I \end{bmatrix}^{-1} \begin{bmatrix} \cdot & \cdot & \Psi_{us}^{(*)} \\ \cdot & \hat{\Lambda}_{ds}^{(*,\circ)} & \hat{\Psi}_{ds}^{(*)} \end{bmatrix}$$

and find that the transition matrix of the censored fluid queue on $\mathcal{K}$ is

$$\Omega^{(\circ)} = \Omega_{\mathcal{K}\mathcal{K}}^{(*)} + \Omega_{\mathcal{K}\mathcal{E}^{(\circ)}}^{(*)} \left\{ I - \Omega_{\mathcal{E}^{(\circ)}\mathcal{E}^{(\circ)}}^{(*)} \right\}^{-1} \Omega_{\mathcal{E}^{(\circ)}\mathcal{K}}^{(*)}, \quad (9.11)$$

and its generator matrix is

$$\Theta = \Delta^{(s)}(I - \Omega^{(\circ)}),$$

where $\Delta^{(s)}$ is the diagonal matrix with $[\Delta^{(s)}]_{ii} = [T_1]_{ii}$ for $0 \le i \le \lfloor \frac{c}{R_1} \rfloor$. By [11, Theorems 4.5 and 4.2],

$$\underline{m} = \kappa[\underline{x}_s^{(\circ)}, \underline{x}_s^{(*)}], \quad (9.12)$$

and the density vector $\underline{\pi}(x)$ of buffer 1 is

$$\begin{aligned} \underline{\pi}(x) &= \kappa \underline{y}_1(x) \quad \text{for } 0 < x < x^*, & (9.13) \\ &= \kappa \underline{y}_2(x) \quad \text{for } x > x^*, & (9.14) \end{aligned}$$

where

$$\kappa = \left\{ [\underline{x}_s^{(\circ)}, \underline{x}_s^{(*)}]\underline{1} + \int_0^{x^*} \underline{y}_1(x)\underline{1}\mathrm{d}x + \int_{x^*}^{\infty} \underline{y}_2(x)\underline{1}\mathrm{d}x \right\}^{-1},$$

the vector $[\underline{x}_s^{(\circ)}, \underline{x}_s^{(*)}]$ is a solution of $[\underline{x}_s^{(\circ)}, \underline{x}_s^{(*)}]\Theta = \underline{0}$,

$$\underline{y}_2(x) = \{\underline{u}C_+^{(2)}N_+^{(2)}(x - x^*)\}(C^{(2)})^{-1}, \quad (9.15)$$

$$\underline{y}_1(x) = \{\underline{x}_s^{(\circ)}T_{-+}^{(1)}N_+^{(1)}(0, x) + \underline{d}C_-^{(1)}N_-^{(1)}(x^*, x)\}(C^{(1)})^{-1}, \quad (9.16)$$

the vectors $\underline{u}$ and $\underline{d}$ are the solution of

$$\underline{d} = \{\underline{x}_s^{(*)}T_{s-}^{(2)} + \underline{u}C_+^{(2)}\Psi_2\}(C_-^{(1)})^{-1}, \quad (9.17)$$

$$\underline{u} = \{\underline{x}_s^{(*)}T_{s+}^{(2)} + \underline{x}_s^{(\circ)}T_{-+}^{(1)}\Lambda_{uu}^{(\circ,*)} + \underline{d}C_-^{(1)}\hat{\Psi}_{-+}^{(1)}\}(C_+^{(2)})^{-1}, \quad (9.18)$$

and

$N_+^{(2)}(w)$ is the matrix of the expected number of visits to $w > 0$ in a phase of $\mathcal{S}_+^{(2)}$, while avoiding 0 after initially increasing from there, for the infinite fluid queue $\{M_2(t), \rho_2(t)\}$;

$N_-^{(1)}(x^*, w)$ is the matrix of the expected number of visits to $w < x^*$ in a phase of $\mathcal{S}_-^{(1)}$, after initially decreasing from x^* and while avoiding both x^* and 0, for the finite fluid queue $\{M_1(t), \rho_1(t)\}$; and

$N_+^{(1)}(0, w)$ is the matrix of the expected number of visits to $w < x^*$ in a phase of $\mathcal{S}_+^{(1)}$, after initially increasing from 0 and while avoiding both 0 and x^*, for $\{M_1(t), \rho_1(t)\}$.

By [21, Theorems 2.1 and 2.2],

$$N_+^{(2)}(w) = \mathrm{e}^{K_2 w}, \tag{9.19}$$

with

$$K_2 = (C_+^{(2)})^{-1} T_{++}^{(2)} + \Psi_2 (C_-^{(2)})^{-1} T_{-+}^{(2)},$$

and by [10, Lemma 4.1],

$$\begin{bmatrix} N_+^{(1)}(0,w) \\ N_-^{(1)}(x^*,w) \end{bmatrix} = \begin{bmatrix} I & \mathrm{e}^{K_1 x^*}\Psi_1 \\ \mathrm{e}^{\hat{K}_1 x^*}\hat{\Psi}_1 & I \end{bmatrix}^{-1} \begin{bmatrix} \mathrm{e}^{K_1 w} & \mathrm{e}^{K_1 w}\Psi_1 \\ \mathrm{e}^{\hat{K}_1 (x^*-w)}\hat{\Psi}_1 & \mathrm{e}^{\hat{K}_1 (x^*-w)} \end{bmatrix},$$

with

$$\begin{aligned} K_1 &= (C_+^{(1)})^{-1} T_{++}^{(1)} + \Psi_1 (C_-^{(1)})^{-1} T_{-+}^{(1)}, \\ \hat{K}_1 &= (C_-^{(1)})^{-1} T_{--}^{(1)} + \hat{\Psi}_1 (C_+^{(1)})^{-1} T_{+-}^{(1)}. \end{aligned}$$

Therefore, the marginal distribution function of buffer 1 is

$$\begin{aligned} \lim_{t\to\infty} P(X(t) \le x) &= \underline{p}_s^{(\circ)} \underline{1} + \int_0^x \underline{\pi}(x)\mathrm{d}x \quad \text{for } 0 < x < x^*, \\ &= [\underline{p}_s^{(\circ)}, \underline{p}_s^{(*)}]\underline{1} + \int_0^x \underline{\pi}(x)\mathrm{d}x \quad \text{for } x \ge x^*. \end{aligned}$$

Analysis for Buffer 2

Deriving the marginal probability distribution for buffer 2 is not easy. Since its output capacity is dependent on $X(t)$, when analyzed as a standalone process, $\{Y(t), \varphi_2(t) : t \in \mathbb{R}^+\}$, buffer 2 does not enjoy the Markovian property of

$\{X(t), \varphi_1(t) : t \in \mathbb{R}^+\}$. Gautam et al. [15] give bounds for the stationary distribution of fluid models with semi-Markov inputs and constant outputs. To apply these results, we first need to transform buffer 2 into an equivalent fluid queue with semi-Markov inputs and a constant output. We achieve the transformation by employing a *compensating source*, a concept developed by Elwalid and Mitra [12] and extended in Mahabhashyam et al. [18]. The role of a compensating source is to add the exact amount of input for maintaining a constant output, c in our case, while keeping all the while the fluid level the same as that of the original, output-varying, buffer.

Consider a virtual fluid queue $\{Z(t), A(t), \varphi_2(t) : t \in \mathbb{R}^+\}$ that has N exponential ON–OFF inputs and one independent compensating source. Here, $Z(t) \geq 0$ is the level, $A(t)$ is the semi-Markov process that drives the compensating source, and $\varphi_2(t)$ is the irreducible Markov chain controlling ON–OFF inputs, with state space $\mathcal{S}$ and generator T_2. The semi-Markov process $A(t)$ has state space $\mathcal{B}$, and the set of boundary states for the jump chain $\{J_n\}$ defined in the section "Infinite Buffer 1" for the analysis of buffer 1, as the output capacity of buffer 2, and consequently the compensating source, changes each time $X(t)$ is in a boundary state. Specifically, the input rates $\dot{a}_{\bullet,i}$ of the compensating source are

$$\begin{aligned} \dot{a}_{\bullet,i} &= iR_1 \;\text{ for } (\bullet,i) \in (\circ, \mathcal{S}_s^{(\circ)}) \cup (*, \mathcal{S}_s^{(*)}), \\ &= c_1 \;\text{ for } (\bullet,i) \in (\circ, \mathcal{S}_u^{(\circ)}) \cup (*, \mathcal{S}_d^{(*)}), \\ &= c \;\;\text{ for } (\bullet,i) \in (*, \mathcal{S}_u^{(*)}). \end{aligned}$$

Let S_n be the time of the nth jump epoch in $A(t)$, B_n the state of $A(t)$ immediately after the nth jump, and $\Omega(t)$ the kernel of $A(t)$, where

$$[\Omega(t)]_{ij} = P(S_1 \leq t, B_1 = j | B_0 = i).$$

It is clear that $\Omega(\infty) = \Omega$, the transition matrix of the jump chain $\{J_n\}$, given by (9.3). We denote by $\widetilde{\Omega}(s)$ the matrix of Laplace–Stieltjes transforms of S_1, and, in general, by $\widetilde{D}(s)$ and $\bar{D}(s)$ the respective LST counterparts of the submatrices D and $\hat{D}$ of Ω. The matrices $\widetilde{P}_{su}^{(\circ)}(s)$, $\widetilde{P}_{ss}^{(\circ)}(s)$, $\widetilde{P}_{su}^{(*)}(s)$, $\widetilde{P}_{ss}^{(*)}(s)$, and $\widetilde{P}_{sd}^{(*)}(s)$ are submatrices of $\widetilde{P}(s)$, where

$$\widetilde{P}(s) = (sI - \Delta)^{-1}(T_1 - \Delta). \tag{9.20}$$

To obtain the remaining submatrices of $\widetilde{\Omega}(s)$, we follow an analysis analogous to that described in the section "Infinite Buffer 1." The matrices $\widetilde{\Psi}_{us}^{(\circ)}(s)$, $[\widetilde{\Lambda}_{us}^{(\circ,*)}(s), \widetilde{\Lambda}_{uu}^{(\circ,*)}(s)]$, $\bar{\Lambda}_{ds}^{(*,\circ)}(s)$, and $[\bar{\Psi}_{ds}^{(*)}(s), \bar{\Psi}_{du}^{(*)}(s)]$ are equal to $\widetilde{\Psi}_{+-}^{(1)}(s)$, $\widetilde{\Lambda}_{++}^{(1)}(s)$, $\bar{\Lambda}_{--}^{(1)}(s)$, and $\bar{\Psi}_{-+}^{(1)}(s)$, the corresponding matrices of the LSTs of *first passage times* for $\{M_1(t), \rho_1(t)\}$. By [8, Theorem 3], for s such that $\mathrm{Re}(s) > 0$,

$$\begin{bmatrix} \widetilde{\Lambda}_{++}^{(1)}(s) & \widetilde{\Psi}_{+-}^{(1)}(s) \\ \bar{\Psi}_{-+}^{(1)}(s) & \bar{\Lambda}_{--}^{(1)}(s) \end{bmatrix} = \begin{bmatrix} \mathrm{e}^{\bar{U}_1(s)x^*} & \widetilde{\Psi}_1(s) \\ \bar{\Psi}_1(s) & \mathrm{e}^{\widetilde{U}_1(s)x^*} \end{bmatrix} \begin{bmatrix} I & \widetilde{\Psi}_1(s)\mathrm{e}^{\widetilde{U}_1(s)x^*} \\ \bar{\Psi}_1(s)\mathrm{e}^{\bar{U}_1(s)x^*} & I \end{bmatrix}^{-1},$$

where $\widetilde{\Psi}_1(s)$ is the minimum nonnegative solution to the Riccati equation

$$(C_+^{(1)})^{-1}(T_{+-}^{(1)} - sI) + (C_+^{(1)})^{-1}(T_{++}^{(1)} - sI)\widetilde{\Psi}_1(s)$$
$$+\widetilde{\Psi}_1(s)(C_-^{(1)})^{-1}(T_{--}^{(1)} - sI) + \widetilde{\Psi}_1(s)(C_-^{(1)})^{-1}(T_{-+}^{(1)} - sI)\widetilde{\Psi}_1(s) = 0, \quad (9.21)$$

$\bar{\Psi}_1(s)$ is the minimum nonnegative solution to the Riccati equation

$$(C_-^{(1)})^{-1}(T_{-+}^{(1)} - sI) + (C_-^{(1)})^{-1}(T_{--}^{(1)} - sI)\bar{\Psi}_1(s)$$
$$+\bar{\Psi}_1(s)(C_+^{(1)})^{-1}(T_{++}^{(1)} - sI) + \bar{\Psi}_1(s)(C_+^{(1)})^{-1}(T_{+-}^{(1)} - sI)\bar{\Psi}_1(s) = 0, \quad (9.22)$$
$$\widetilde{U}_1(s) = (C_-^{(1)})^{-1}(T_{--}^{(1)} - sI) + (C_-^{(1)})^{-1}(T_{-+}^{(1)} - sI)\widetilde{\Psi}_1(s),$$

and

$$\bar{U}_1(s) = (C_+^{(1)})^{-1}(T_{++}^{(1)} - sI) + (C_+^{(1)})^{-1}(T_{+-}^{(1)} - sI)\bar{\Psi}_1(s).$$

Similarly, $[\widetilde{\Psi}_{us}^{(*)}(s), \widetilde{\Psi}_{ud}^{(*)}(s)] = \widetilde{\Psi}_2(s)$, which is the matrix of the LST of first passage times for $\{M_2(t), \rho_2(t)\}$. By [6, Theorem 1], $\widetilde{\Psi}_2(s)$ is the minimum nonnegative solution to the Riccati equation

$$(C_+^{(2)})^{-1}(T_{+-}^{(2)} - sI) + (C_+^{(2)})^{-1}(T_{++}^{(2)} - sI)\widetilde{\Psi}_2(s)$$
$$+\widetilde{\Psi}_2(s)(C_-^{(2)})^{-1}(T_{--}^{(2)} - sI) + \widetilde{\Psi}_2(s)(C_-^{(2)})^{-1}(T_{-+}^{(2)} - sI)\widetilde{\Psi}_2(s) = 0. \quad (9.23)$$

Bean et al. [7] give efficient algorithms for solving (9.21)–(9.23) to obtain $\widetilde{\Psi}_1(s)$, $\bar{\Psi}_1(s)$ and $\widetilde{\Psi}_2(s)$, and consequently $\widetilde{\Omega}(s)$.

Before we state the bounds for buffer 2, we need to define *effective bandwidths* and *failure rate functions*. For $v > 0$, the *effective bandwidth* $\mathsf{eb}(v)$ of an input that generates $F(t)$ amount of fluid at time t is defined to be

$$\mathsf{eb}(v) = \lim_{t\to\infty} \frac{1}{vt} \log E[\mathrm{e}^{vF(t)}]$$

(see, for example, [13, 16]). By [3, 13], the effective bandwidth $\mathsf{eb}_\mathrm{e}(v)$ of a single exponential ON–OFF source for fixed v is

$$\mathsf{eb}_\mathrm{e}(v) = \frac{R_2 v - \alpha_2 - \beta_2 + \sqrt{(R_2 v - \alpha_2 - \beta_2)^2 + 4\beta_2 R_2 v}}{2v}. \quad (9.24)$$

To obtain the effective bandwidth $\mathsf{eb}_\mathrm{c}(v)$ for the compensating source, we begin by defining $\Phi(v,u)$ to be the matrix with submatrices

$$[\Phi(v,u)]_{(\bullet,i),(\bar{\bullet},i')} = [\widetilde{\Omega}(v(u - \dot{a}_{\bullet,i}))]_{(\bullet,i),(\bar{\bullet},i')}. \quad (9.25)$$

Denote by $\chi(D)$ the maximal real eigenvalue of a matrix D; then, by [15, Sects. 4 and 5], the effective bandwidth $\mathsf{eb}_c(v)$ for fixed v is the unique positive solution to the equation

$$\chi(\Phi(v, \mathsf{eb}_c(v))) = 1. \tag{9.26}$$

With these, we define η to be the minimum positive solution to

$$\mathsf{eb}_c(\eta) + N\mathsf{eb}_e(\eta) = c. \tag{9.27}$$

The existence of such η is guaranteed by the facts [14, Sect. 2.2.2] that $\mathsf{eb}_c(v)$ and $\mathsf{eb}_e(v)$ are both increasing functions with respect to v and that for any given $v > 0$,

$$0 \leq \mathsf{eb}_c(v) \leq c,$$

and

$$\lim_{v \to 0} \mathsf{eb}_c(v) = 0 \quad \text{and} \quad \lim_{v \to \infty} \mathsf{eb}_c(v) = c.$$

For fixed v, we can solve (9.26) using fixed point iteration, as $\chi(\Phi(v,u))$ is a decreasing function with respect to u [14, Sect. 2.2.3], and solve (9.27) using bisection.

For $i \in \mathcal{B}$, we denote by τ_i the expected sojourn time of $A(t)$ in i

$$\tau_i = -\sum_{j \in \mathcal{B}} [\widetilde{\Omega}'(0)]_{ij}, \tag{9.28}$$

by $\underline{p}$ the vector with elements

$$p_i = \frac{\omega_i \tau_i}{\sum\limits_{j \in \mathcal{B}} \omega_i \tau_j}, \tag{9.29}$$

where $\underline{\omega}$ is the stationary vector associated with Ω ($\underline{\omega}\Omega = \underline{1}$, $\underline{\omega}\underline{1} = 1$), and by $\underline{h}$ the left eigenvector of $\Phi(\eta, \mathsf{eb}_c(\eta))$ corresponding to the eigenvalue one. Now, we are ready to define $\Xi_{\max}(i,j)$ and $\Xi_{\min}(i,j)$ as follows:

$$\Xi_{\min}(i,j) = \tfrac{h_i \tau_i}{p_i} \inf_x f_{ij}(x),$$
$$\Xi_{\max}(i,j) = \tfrac{h_i \tau_i}{p_i} \sup_x f_{ij}(x),$$

where

$$f_{ij}(x) = \frac{\displaystyle\int_x^{\infty} e^{\eta(\dot{a}_i - \mathsf{eb}_c(\eta))y} d[\Omega(y)]_{ij}}{e^{\eta(\dot{a}_i - \mathsf{eb}_c(\eta))x} \{[\Omega]_{ij} - [\Omega(x)]_{ij}\}}. \tag{9.30}$$

Applying [15, Theorems 6 and 7] and then simplifying using [14, Sect. 4.2.4], we obtain the following result.

Theorem 9.1. *For $x > 0$,*

$$K_* \mathrm{e}^{-\eta x} \leq \lim_{t \to \infty} P(Y(t) > x) \leq K^* \mathrm{e}^{-\eta x}, \tag{9.31}$$

where

$$K_* = \frac{\left[\frac{R_2}{\mathsf{eb}_\mathrm{e}(\eta)\alpha_2}\right]^N H_c}{\max\limits_{s,(i,j)} D(s)\Xi_{\max}(i,j)}$$

and

$$K^* = \frac{\left[\frac{R_2}{\mathsf{eb}_\mathrm{e}(\eta)\alpha_2}\right]^N H_c}{\min\limits_{s,(i,j)} D(s)\Xi_{\min}(i,j)},$$

with $i, j \in \mathcal{B}$ and $1 \leq s \leq N$ such that $\dot{a}_i + sR_2 > c$ and $[\Omega]_{i,j} > 0$, where

$$H_c = \sum_{i \in \mathcal{B}} \left[\frac{h_i}{\eta(\dot{a}_i - \mathsf{eb}_\mathrm{c}(\eta))}\right] \left[\sum_{j \in \mathcal{B}} [\Phi(\eta, \mathsf{eb}_c(\eta))]_{ij} - 1\right],$$

$$D(s) = \left[\frac{\alpha_2 + \beta_2}{\alpha_2 \beta_2}\right]^s \left[\frac{(\alpha_2 + \beta_2)(R_2 - \mathsf{eb}_\mathrm{e}(\eta))}{\mathsf{eb}_e(\eta)\alpha_2^2}\right]^{(N-s)}.$$

For $i, j \in \mathcal{B}$ the *failure rate function* $\lambda_{ij}(x)$ of the compensating source is

$$\lambda_{ij}(x) = \frac{[\Omega'(x)]_{ij}}{[\Omega]_{ij} - [\Omega(x)]_{ij}}. \tag{9.32}$$

The function $[\Omega(x)]_{ij}$ is said to be increasing (IFR) if $\lambda_{ij}(x)$ is an increasing function of x and decreasing (DFR) if $\lambda_{ij}(x)$ is a decreasing function of x. In cases where $[\Omega(x)]_{ij}$ is either IFR or DFR, $\Xi_{\max}(i,j)$ and $\Xi_{\min}(i,j)$ are given in Table 9.1. For a sticky state $i \in (\circ, \mathcal{S}_s^{(\circ)}) \cup (*, \mathcal{S}_s^{(*)})$, $[\Omega(x)]_{ij}$ has a constant failure rate λ_{ij}, and $\Xi_{\min}(i,j) = \Xi_{\max}(i,j)$.

When $[\Omega(x)]_{ij}$ is neither IFR nor DFR, $\Xi_{\max}(ij)$ and $\Xi_{\min}(ij)$ may be estimated by numerical computation. The LST of the numerator of $f_{ij}(\cdot)$ is $-\widetilde{\Omega}(s - \eta(\dot{a}_i - \mathsf{eb}_\mathrm{c}(\eta)))$; hence, both the numerator and the denominator of $f_{ij}(x)$ are obtainable by numerical inversion of $\widetilde{\Omega}(\cdot)$.

Finite Buffer 1, with One Input

In this section, we determine the marginal probability distribution of buffer 1 in the particular case $N = 1$, with an added assumption that it has finite size $V > x^*$. Our aim is to illustrate the difference in distributions of the finite buffer 1, as derived

Table 9.1 $\Xi_{\max}(i,j)$ and $\Xi_{\min}(i,j)$ in simple cases

	$\Xi_{\max}(i,j)$	$\Xi_{\min}(i,j)$
IFR, $\dot{a}_i > \mathsf{eb}_c(\eta)$, or DFR, $\dot{a}_i \leq \mathsf{eb}_c(\eta)$	$\frac{[\Phi(\eta,\mathsf{eb}_c(\eta))]_{ij}\tau_i h_i}{[\Omega]_{ij}p_i}$	$\frac{\tau_i h_i \lambda_{ij}(\infty)}{p_i(\lambda_{ij}(\infty)-\eta(\dot{a}_i-\mathsf{eb}_c(\eta)))}$
IFR, $\dot{a}_i \leq \mathsf{eb}_c(\eta)$ or DFR, $\dot{a}_i > \mathsf{eb}_c(\eta)$	$\frac{\tau_i h_i \lambda_{ij}(\infty)}{p_i(\lambda_{ij}(\infty)-\eta(\dot{a}_i-\mathsf{eb}_c(\eta)))}$	$\frac{[\Phi(\eta,\mathsf{eb}_c(\eta))]_{ij}\tau_i h_i}{[\Omega]_{ij}p_i}$

here, and of the infinite buffer 1, as in [18]. As mentioned in the introduction, while the analysis in this section can be extended in a straightforward manner to the general case $N \geq 1$, we specifically consider the case $N = 1$ to better illustrate the analytic approach. We only carry out the analysis for buffer 1, as the expressions for buffer 2 remain the same.

The assumptions of the reference model, stated in the section "Reference Model," become $R_j > c$, for $j = 1,2$, and $\beta_1(R_1 - c) < \alpha_1 - c$. The imposed finiteness leads to a third boundary $X(t) = V$, in addition to the two boundaries $X(t) = 0$ and $X(t) = x^*$, and the second band becomes $x^* < X(t) < V$. All state spaces are simplified significantly:

$$\mathcal{S}_s^{(\circ)} = \mathcal{S}_-^{(1)} = \mathcal{S}_d^{(*)} = \mathcal{S}_-^{(2)} = \{0\},$$

$$\mathcal{S}_u^{(\circ)} = \mathcal{S}_+^{(1)} = \mathcal{S}_u^{(*)} = \mathcal{S}_+^{(2)} = \{1\}.$$

While the set $\mathcal{S}_s^{(*)}$ of sticky states at x^* is empty, there is a new sticky state at V, that is, $\mathcal{S}_s^{(V)} = \{1\}$ and $\mathcal{S}_d^{(V)} = \{0\}$. The generator matrix T_1 is

$$T_1 = \begin{bmatrix} T_{--}^{(1)} & T_{-+}^{(1)} \\ T_{+-}^{(1)} & T_{++}^{(1)} \end{bmatrix} = \begin{bmatrix} T_{--}^{(2)} & T_{-+}^{(2)} \\ T_{+-}^{(2)} & T_{++}^{(2)} \end{bmatrix} = \begin{bmatrix} -\beta_1 & \beta_1 \\ \alpha_1 & -\alpha_1 \end{bmatrix},$$

and the rate matrices are now scalars:

$$C_+^{(1)} = R_1 - c_1, \qquad C_-^{(1)} = c_1,$$

$$C_+^{(2)} = R_1 - c, \qquad C_-^{(2)} = c.$$

By [11, Theorem 4.4], the jump chain $\{J_n : n \geq 0\}$ of the process $\{X(t), \varphi_1(t)\}$ restricted to the set of boundary states $\mathcal{B} = \{(\bullet, i) : \bullet \in \{\circ, *, V\}, i = \{1,2\}\}$ has a transition matrix

$$\Omega = \left[\begin{array}{cc|cc|cc} \cdot & \Psi_{us}^{(\circ)} & \Lambda_{uu}^{(\circ,*)} & \cdot & \cdot & \cdot \\ 1 & \cdot & \cdot & \cdot & \cdot & \cdot \\ \hline \cdot & \cdot & \cdot & \Psi_{ud}^{(*)} & \Lambda_{us}^{(*,V)} & \cdot \\ \cdot & \hat{\Lambda}_{ds}^{(*,\circ)} & \hat{\Psi}_{du}^{(*)} & \cdot & \cdot & \cdot \\ \hline \cdot & \cdot & \cdot & \cdot & \cdot & 1 \\ \cdot & \cdot & \cdot & \hat{\Lambda}_{dd}^{(V,*)} & \hat{\Psi}_{ds}^{(V)} & \cdot \end{array}\right], \tag{9.33}$$

where $\Psi_{us}^{(\circ)}$, $\Lambda_{uu}^{(\circ,*)}$, $\hat{\Lambda}_{ds}^{(*,\circ)}$, and $\hat{\Psi}_{du}^{(*)}$ are the solutions of (9.5). Here, (9.6) and (9.7) reduce to scalar quadratic equations, from which one easily obtains the minimal solutions

$$\Psi_1 = 1, \qquad \hat{\Psi}_1 = \frac{\beta_1(R_1 - c_1)}{\alpha_1 c_1}. \tag{9.34}$$

Substituting (9.34) into (9.8) and (9.9) leads to

$$U_1 = 0, \qquad \hat{U}_1 = \frac{-\alpha_1 c_1 + \beta_1(R_1 - c_1)}{c_1(R_1 - c_1)}. \tag{9.35}$$

Then, substituting (9.35) into (9.5) gives us

$$\Lambda_{uu}^{(\circ,*)} = \Lambda_{++}^{(1)} = \frac{1 - \hat{\Psi}_1}{\mathrm{e}^{-\hat{U}_1 x^*} - \hat{\Psi}_1}, \tag{9.36}$$

$$\Psi_{us}^{(\circ)} = \Psi_{+-}^{(1)} = \frac{1 - \mathrm{e}^{\hat{U}_1 x^*}}{1 - \hat{\Psi}_1 \mathrm{e}^{\hat{U}_1 x^*}}, \tag{9.37}$$

$$\hat{\Psi}_{du}^{(*)} = \hat{\Psi}_{-+}^{(1)} = \frac{\hat{\Psi}_1 - \hat{\Psi}_1 \mathrm{e}^{\hat{U}_1 x^*}}{1 - \hat{\Psi}_1 \mathrm{e}^{\hat{U}_1 x^*}}, \tag{9.38}$$

$$\hat{\Lambda}_{ds}^{(*,\circ)} = \hat{\Lambda}_{--}^{(1)} = \frac{1 - \hat{\Psi}_1}{1 - \hat{\Psi}_1 \mathrm{e}^{\hat{U}_1 x^*}}. \tag{9.39}$$

Since the second band is now finite, we follow the same steps as for the first band and find that the matrices $\Psi_{ud}^{(*)}$, $\Lambda_{us}^{(*,V)}$, $\hat{\Lambda}_{dd}^{(V,*)}$, and $\hat{\Psi}_{ds}^{(V)}$ are equal to $\Psi_{+-}^{(2)}$, $\Lambda_{++}^{(2)}$, $\hat{\Lambda}_{--}^{(2)}$, and $\hat{\Psi}_{-+}^{(2)}$, the corresponding first passage probability matrices for the level-independent fluid queue $\{M_2(t), \rho_2(t) : t \in \mathbb{R}^+\}$ with finite size $V - x^*$, state space $\mathcal{S}_-^{(2)} \cup \mathcal{S}_+^{(2)}$, generator T_1, and rates $C_-^{(2)}$ and $C_+^{(2)}$. By [10, Theorem 5.2],

$$\begin{bmatrix} \Lambda_{++}^{(2)} & \Psi_{+-}^{(2)} \\ \hat{\Psi}_{-+}^{(2)} & \hat{\Lambda}_{--}^{(2)} \end{bmatrix} = \begin{bmatrix} \mathrm{e}^{\hat{U}_2(V-x^*)} & \Psi_2 \\ \hat{\Psi}_2 & \mathrm{e}^{U_2(V-x^*)} \end{bmatrix} \begin{bmatrix} 1 & \Psi_2 \mathrm{e}^{U_2(V-x^*)} \\ \hat{\Psi}_2 \mathrm{e}^{\hat{U}_2(V-x^*)} & 1 \end{bmatrix}^{-1},$$

where

$$\Psi_2 = 1, \qquad \hat{\Psi}_2 = \frac{\beta_1(R_1 - c)}{\alpha_1 c}, \tag{9.40}$$

$$U_2 = 0, \qquad \hat{U}_2 = \frac{-\alpha_1 c + \beta_1(R_1 - c)}{c(R_1 - c)}. \tag{9.41}$$

Substituting (9.40) and (9.41) into (9.40) gives us

$$\Lambda_{us}^{(*,V)} = \Lambda_{++}^{(2)} = \frac{1 - \hat{\Psi}_2}{\mathrm{e}^{-\hat{U}_2(V-x^*)} - \hat{\Psi}_2}, \tag{9.42}$$

$$\Psi_{ud}^{(*)} = \Psi_{+-}^{(2)} = \frac{1 - \mathrm{e}^{\hat{U}_2(V-x^*)}}{1 - \hat{\Psi}_2 \mathrm{e}^{\hat{U}_2(V-x^*)}}, \tag{9.43}$$

$$\hat{\Psi}_{ds}^{(V)} = \hat{\Psi}_{-+}^{(2)} = \frac{\hat{\Psi}_2 - \hat{\Psi}_2 \mathrm{e}^{\hat{U}_2(V-x^*)}}{1 - \hat{\Psi}_2 \mathrm{e}^{\hat{U}_2(V-x^*)}}, \tag{9.44}$$

$$\hat{\Lambda}_{dd}^{(V,*)} = \hat{\Lambda}_{--}^{(2)} = \frac{1 - \hat{\Psi}_2}{1 - \hat{\Psi}_2 \mathrm{e}^{\hat{U}_2(V-x^*)}}. \tag{9.45}$$

Together, Eqs. (9.36)–(9.39) and (9.42)–(9.45) complete the transition matrix Ω, specified in (9.33), of the jump chain $\{J_n\}$ on the set $\mathcal{B}$ of boundary states. The set $\mathcal{K}$ of sticky states is $\{(\circ,0),(V,1)\}$. Straightforward but tedious calculations show that the jump chain on $\mathcal{K}$ has the transition matrix

$$\Omega^{(\circ)} = \begin{bmatrix} \Psi_{us}^{(\circ)} + \dfrac{\Lambda_{uu}^{(\circ,*)}\Psi_{ud}^{(*)}\hat{\Lambda}_{ds}^{(*,\circ)}}{1 - \Psi_{ud}^{(*)}\hat{\Psi}_{du}^{(*)}} & \dfrac{\Lambda_{uu}^{(\circ,*)}\Lambda_{us}^{(*,\circ)}}{1 - \Psi_{ud}^{(*)}\hat{\Psi}_{du}^{(*)}} \\[2ex] \dfrac{\hat{\Lambda}_{dd}^{(V,*)}\hat{\Lambda}_{ds}^{(*,\circ)}}{1 - \Psi_{ud}^{(*)}\hat{\Psi}_{du}^{(*)}} & \hat{\Psi}_{ds}^{(V)} + \dfrac{\hat{\Lambda}_{dd}^{(V,*)}\hat{\Psi}_{du}^{(*)}\Lambda_{us}^{(*,V)}}{1 - \Psi_{ud}^{(*)}\hat{\Psi}_{du}^{(*)}} \end{bmatrix}$$

and, consequently, the generator matrix

$$\Theta = \begin{bmatrix} -\beta_1 & \\ & -\alpha_1 \end{bmatrix} (I - \Omega^{(\circ)}).$$

A solution of $[x_s^{(\circ)}, x_s^{(V)}]\Theta = \underline{0}$ is

$$x_s^{(\circ)} = 1,$$

$$x_s^{(V)} = -\frac{\beta_1(1 - [\Omega^{(\circ)}]_{11})}{\alpha_1[\Omega^{(\circ)}]_{21}}.$$

By [11, Theorem 4.5], the probability mass vector $\underline{m} = [p_s^{(\circ)}, p_s^{(V)}]$ of buffer 1 at $\mathcal{K}$ is given by $\underline{m} = \kappa[x_s^{(\circ)}, x_s^{(*)}]$, with

$$\kappa = \left\{1 + \frac{\alpha_1(1-[\Omega^{(\circ)}]_{11})}{\beta_1[\Omega^{(\circ)}]_{22}} + \int_0^{x^*} \underline{y}_1(x)\underline{1}\mathrm{d}x + \int_{x^*}^{V} \underline{y}_2(x)\underline{1}\mathrm{d}x\right\}^{-1},$$

and

$$\underline{y}_1(x) = \{\beta_1 N_+^{(1)}(0,x) + c_1\gamma_1 N_-^{(1)}(x^*,x)\}(C^{(1)})^{-1},$$
$$\underline{y}_2(x) = \{(R_1-c)\gamma_2 N_+^{(2)}(0,x-x^*) + c\gamma_3 N_-^{(2)}(V-x^*,x-x^*)\}(C^{(2)})^{-1};$$

the vectors γ_1, γ_2, and γ_3 are the solution of the system

$$\begin{aligned}
\gamma_3 &= x_s^{(V)} T_{+-}^{(2)} (C_-^{(2)})^{-1} \\
&= \frac{\alpha_1}{c} x_s^{(V)}, \\
\gamma_2 &= \left\{x_s^{(\circ)} T_{-+}^{(1)} \Lambda_{++}^{(\circ,*)} + \gamma_1 C_-^{(1)} \hat{\Psi}_{-+}^{(1)}\right\} (C_+^{(2)})^{-1} \\
&= \frac{1}{R_1-c}\left\{\beta_1 \Lambda_{++}^{(\circ,*)} + c_1\gamma_1 \hat{\Psi}_{-+}^{(1)}\right\}, \\
\gamma_1 &= \left\{\gamma_2 C_+^{(2)} \Psi_{+-}^{(2)} + \gamma_3 C_-^{(2)} \hat{\Lambda}_{--}^{(V,*)}\right\} (C_-^{(1)})^{-1} \\
&= \frac{1}{c_1}\left\{(R_1-c)\gamma_2 \Psi_{+-}^{(2)} + \alpha_1 x_s^{(V)} \hat{\Lambda}_{--}^{(V,*)}\right\}.
\end{aligned}$$

Solving for γ_1 and γ_2 leads to

$$\begin{aligned}
\begin{bmatrix}\gamma_1 \\ \gamma_2\end{bmatrix} &= \begin{bmatrix} \dfrac{c_1\hat{\Psi}_{-+}^{(1)}}{R_1-c} & -1 \\ 1 & -\dfrac{(R_1-c)\Psi_{+-}^{(2)}}{c_1}\end{bmatrix}^{-1} \begin{bmatrix} -\dfrac{\beta_1\Lambda_{++}^{(\circ,*)}}{R_1-c} \\ \dfrac{\alpha_1 x_s^{(V)}\hat{\Lambda}_{--}^{(V,*)}}{c_1}\end{bmatrix} \\
&= \frac{1}{1-\hat{\Psi}_{-+}^{(1)}\Psi_{+-}^{(2)}} \begin{bmatrix} \dfrac{\beta_1}{c_1}\Psi_{+-}^{(2)}\Lambda_{++}^{(\circ,*)} + \dfrac{\alpha_1 x_s^{(V)}\hat{\Lambda}_{--}^{(V,*)}}{c_1} \\ \dfrac{\beta_1}{R_1-c}\Lambda_{++}^{(\circ,*)} + \dfrac{\alpha_1 x_s^{(V)}}{R_1-c}\hat{\Psi}_{-+}^{(1)}\hat{\Lambda}_{--}^{(V,*)}\end{bmatrix}.
\end{aligned}$$

By [10, Lemma 4.1],

$$\begin{bmatrix} N_+^{(1)}(0,x) \\ N_-^{(1)}(x^*,x) \end{bmatrix} = \begin{bmatrix} 1 & e^{K_1 x^*} \\ e^{\hat{K}_1 x^*}\hat{\Psi}_1 & 1 \end{bmatrix}^{-1} \begin{bmatrix} e^{K_1 x} & e^{K_1 x} \\ e^{\hat{K}_1 (x^*-x)}\hat{\Psi}_1 & e^{\hat{K}_1 (x^*-x)} \end{bmatrix}, \tag{9.46}$$

with

$$K_1 = -\frac{\alpha_1}{R_1 - c_1} + \frac{\beta_1}{c_1}, \qquad \hat{K}_1 = 0.$$

Consequently,

$$N_+^{(1)}(0,x) = \frac{1}{1-\hat{\Psi}_1 e^{K_1 x^*}} \left[e^{K_1 x} - \hat{\Psi}_1 e^{K_1 x^*}, e^{K_1 x} - e^{K_1 x^*} \right]$$

and

$$N_-^{(1)}(x^*,x) = \frac{1}{1-\hat{\Psi}_1 e^{K_1 x^*}} \left[-\hat{\Psi}_1 e^{K_1 x} + \hat{\Psi}_1, -\hat{\Psi}_1 e^{K_1 x} + 1 \right].$$

Similarly, by [10, Lemma 4.1] again,

$$\begin{bmatrix} N_+^{(2)}(0,x-x^*) \\ N_-^{(2)}(V-x^*,x-x^*) \end{bmatrix} = \begin{bmatrix} 1 & e^{K_2 (V-x^*)} \\ e^{\hat{K}_2 (V-x^*)}\hat{\Psi}_2 & 1 \end{bmatrix}^{-1} \begin{bmatrix} e^{K_2 (x-x^*)} & e^{K_2 (x-x^*)} \\ e^{\hat{K}_2 (V-x)}\hat{\Psi}_2 & e^{\hat{K}_2 (V-x)} \end{bmatrix},$$

with

$$K_2 = -\frac{\alpha_1}{R_1 - c} + \frac{\beta_1}{c}, \qquad \hat{K}_2 = 0.$$

Consequently,

$$N_+^{(2)}(0,x-x^*) = \frac{e^{-K_2 x^*}}{1-\hat{\Psi}_2 e^{K_2 (V-x^*))}} \left[e^{K_2 x} - \hat{\Psi}_2 e^{K_2 V}, e^{K_2 x} - e^{K_2 V} \right]$$

and

$$N_-^{(2)}(V-x^*,x-x^*) = \frac{1}{1-\hat{\Psi}_2 e^{K_2 (V-x^*)}} \left[-\hat{\Psi}_2 e^{K_2 (x-x^*)} + \hat{\Psi}_2, -\hat{\Psi}_2 e^{K_2 (x-x^*)} + 1 \right].$$

The density vector $\underline{\pi}(x)$ of buffer 1 is

$$\begin{aligned} \underline{\pi}(x) &= \kappa \underline{y}_1(x) \quad \text{for } 0 < x < x^*, \\ &= \kappa \underline{y}_2(x) \quad \text{for } x^* < x < V. \end{aligned}$$

As an illustration, we consider Scenarios A, E, and F from [18, Table 1] to compare marginal probabilities for buffer 1 in the finite and infinite cases. In all three scenarios, $R_1 = 12.48$, $\alpha_1 = 11$, $\beta_1 = 1, x^* = 1.5$, and $c = 2.6$. For Scenario A, $c_1 = 1.6$ and $c_2 = 1$; for Scenario B, $c_1 = 1.19$ and $c_2 = 1.41$; and for Scenario C,

Table 9.2 $\lim_{t\to\infty} P(X(t) > x^*)$

V	∞	3.5	6	20
A	0.1706	0.1411	0.1660	0.1706
E	0.1942	0.1615	0.1891	0.1942
F	0.3501	0.3009	0.3426	0.3501

Table 9.3 $\lim_{t\to\infty} P(X(t) > 3)$

V	∞	3.5	6	20
A	0.0572	0.0237	0.0519	0.0572
E	0.0651	0.0271	0.0592	0.0651
F	0.1173	0.0505	0.1072	0.1173

$c_1 = 0.2$ and $c_2 = 2.4$. In Tables 9.2 and 9.3, the values of $\lim_{t\to\infty} P(X(t) > x^*)$ and of $\lim_{t\to\infty} P(X(t) > 3)$ for $V = \infty$ are taken from the last and second columns of [18, Table 2], respectively. It is clear that the marginal probabilities for buffer 1 differ between the infinite and finite cases and that these differences quickly tend to zero as V tends to infinity.

Acknowledgements This work was subsidized by the ARC Grant AUWB-08/13-ULB 5 financed by the Ministère de la Communauté Française de Belgique. The second author also gratefully acknowledges the hospitality of the Mathematical Institute at the University of Wrocław, where part of this work was done.

References

1. Aggarwal, V., Gautam, N., Kumara, S.R.T., Greaves, M.: Stochastic fluid flow models for determining optimal switching thresholds with an applicaiton to agent task scheduling. Perfom. Eval. **59**(1), 19–46 (2004)
2. Akar, N., Sohraby, K.: Inifinite- and finite-buffer Markov fluid queues: a unified analysis. J. Appl. Probab. **41**, 557–569 (2004)
3. Anick, D., Mitra, D., Sondhi, M.M.: Stochastic theory of a data handling system with multiple sources. Bell Syst. Tech. J. **61**(8), 1871–1894 (1982)
4. Bean, N., O'Reilly, M.M., Taylor, P.G.: Algorithms for return probabilities for stochastic fluid flows. Stoch. Models **21**(1), 149–184 (2005)
5. Bean, N.G., O'Reilly, M.M.: Performance measures of a multi-layer Markovian fluid model. Ann. Oper. Res. **160**(1), 99–120 (2008)
6. Bean, N.G., O'Reilly, M.M., Taylor, P.G.: Hitting probabilities and hitting times for stochastic fluid flows. Stoch. Process. Appl. **115**, 1530–1556 (2005)
7. Bean, N.G., O'Reilly, M.M., Taylor, P.G.: Algorithms for the Laplace–Stieltjes transforms of first return times for stochastic fluid flows. Methodol. Comput. Appl. Probab. **10**, 381–408 (2008)
8. Bean, N.G., O'Reilly, M.M., Taylor, P.G.: Hitting probabilities and hitting times for stochastic fluid flows: the bounded model. Probab. Eng. Inf. Sci. **32**(1), 121–147 (2009)
9. da Silva Soares, A., Latouche, G.: Further results on the similarity between fluid queues and QBDs. In: Latouche, G., Taylor, P.G., (eds) Proceedings of the 4th International Conference on Matrix-Analytic Methods. Matrix-Analytic Methods: Theory and Applications, pp. 89–106, World Scientific, Singapore (2002)

10. da Silva Soares, A., Latouche, G.: Matrix-analytic methods for fluid queues with finite buffers. Perform. Eval. **63**, 295–314 (2006)
11. da Silva Soares, A., Latouche, G.: Fluid queues with level dependent evolution. Eur. J. Oper. Res. **196**, 1041–1048 (2009)
12. Elwalid, A.I., Mitra, D.: Analysis, approximations and admission control of a multi-service multiplexing system with priorities. In: Proceedings of INFOCOM'95, vol. 2, pp. 463–472. IEEE Computer Society, New York (1995)
13. Elwalid, A.I., Mitra, D.: Effective bandwidth of general Markovian traffic sources and admission control of high speed networks. IEEE Trans. Netw. **1**(3), 329–343 (1993)
14. Gautam, N., Quality of Service for Multi-class Traffic in High-speed Networks, Ph.D. Thesis, Dept. of OR, University of North Carolina, Chapel Hill, (1997)
15. Gautam, N., Kulkani, V.G., Palmowski, Z., Rolski, T.: Bounds for fluid models driven by semi-Markov inputs. Probab. Eng. Inf. Sci. **13**, 429–475 (1999)
16. Kelly, F.P.: Notes on Effective Bandwidths, vol. 4, pp. 141–168. Oxford University Press, Oxford (1996)
17. Kosten, L.: Stochastic theory of a multi-entry buffer, part 1. Delft Prog. Rep. F **1**(1), 10–18 (1974)
18. Mahabhashyam, S.R., Gautam, N., Kumara, S.R.T.: Resource-sharing queueing systems with fluid-flow traffic. Oper. Res. **56**(3), 728–744 (2008)
19. Mitra, D.: Stochastic theory of a fluid model of producers and consumers coupled by a buffer. Adv. Appl. Probab. **20**, 646–676 (1988)
20. Narayanan, A., Kulkani, V.G.: First passage times in fluid models with an application to two-priority fluid systems. In: IEEE International Computer Performance and Dependability Symposium, pp. 166–175 (1996)
21. Ramaswami, V.: Matrix analytic methods for stochastic fluid flows. In: Smith, D., Hey, P. (eds.) Proceedings of the 16th International Teletraffic Congress, Teletraffic Engineering in a Competitive World, pp. 1019–1030. Elsevier Science B. V. NY
22. Roger, L.C.G.: Fluid models in queueing theory and Wiener-Hopf factorization of Markov chains. Adv. Appl. Probab. **4**, 390–413 (1994)
23. Stern, T.E., Elwalid, A.I.: Analysis of separable Markov-modulated rate models for information-handling systems. Adv. Appl. Probab. **23**, 105–139 (1991)

Chapter 10
A Fluid Introduction to Brownian Motion and Stochastic Integration

Vaidyanathan Ramaswami

This chapter develops an approach to numerical simulation and integration of stochastic differential equations (SDEs) using elementary linear fluid flows defined on finite state Markov chains. It serves as an introductory tutorial on Brownian motion and stochastic integration as well as a vehicle for introducing for the first time my ideas based on fluid flow models. I believe that recent advances in algorithmic methods for stochastic fluid flow models offer the opportunity to develop a class of new algorithms of value to various application areas. To a student my approach may somewhat ease the transition from simple discrete state space processes to diffusions on continuous state spaces. Being somewhat pedagogical in nature, I shall include a brief introduction to various intermediate concepts, eschew attempts to obtain the highest level of generality that can be achieved for integrands and integrators, and keep measure-theoretic formalism to a minimum. The advanced reader may skip the first few sections and go directly to the section "Fluid Approximation to Brownian Motion."

Preliminaries

I begin with some preliminary facts needed to approximate stochastic processes. First is the notion of stochastic process convergence, also known as weak convergence of stochastic processes, and certain tools related to it. For a detailed discussion of stochastic process convergence, see Billingsley [6] and Whitt [31].

Throughout the discussion I assume that (S,m) denotes a complete, separable metric space (Polish space) and consider random elements taking values in S.

V. Ramaswami (✉)
AT&T Labs, Research,
180 Park Avenue, Florham Park, NJ 07932, USA
e-mail: vram@research.att.com

G. Latouche et al. (eds.), *Matrix-Analytic Methods in Stochastic Models*, Springer Proceedings in Mathematics & Statistics 27, DOI 10.1007/978-1-4614-4909-6_10,

A random element may be a random variable, random vector, or an indexed set of random variables (a stochastic process). A highly relevant example of a random element for us is a real-valued stochastic process $\{X(t) : t \in I\}$ indexed by a compact interval $I \subset [0,\infty)$ and having continuous sample paths. Here I may take S to be the set $C(I)$ of all continuous functions on I along with the uniform metric $m(f,g) = \sup_{t\in I} |f(t) - g(t)|$.

Given a random element X on a probability space $(\Omega,\mathcal{F},P)$ taking values in S, recall that it induces a probability measure P_X on the Borel sets of (S,m) through the rule $P_X(A) = P[X^{-1}(A)]$. Using a customary [31] abuse of terminology, I shall say "the random element X on (S,m)" to really denote a random element on a probability space $(\Omega,\mathcal{F},P)$ taking values in S. Also, when I consider a measure μ on $(S,\mathcal{B}_S)$, where $\mathcal{B}_S$ is the sigma field of Borel sets of S, I shall simply say that μ is a measure on (S,m), once again an abuse of terminology consistent with that used by Whitt [31].

Definition 10.1. A sequence of probability measures $\{P_n : n \geq 1\}$ on (S,m) converges weakly to a probability measure P (I write $P_n \Rightarrow P$) if $\int_S f dP_n \to \int_S f dP$ for all real-valued functions f on (S,m) that are continuous and bounded. A sequence of random elements $\{X_n : n \geq 1\}$ on (S,m) converges weakly to a random element X on (S,m), and I write $X_n \Rightarrow X$ if the corresponding probability measures $P_n X_n^{-1} \Rightarrow PX^{-1}$ on (S,m).

Remark 10.1. As defined previously, weak convergence of random elements is simply a statement about certain probability measures on the space (S,m) and says nothing about the random elements themselves. Indeed, the underlying random elements may in fact be defined even on different probability spaces.!

As is widely known [8,9], for real-valued random variables, weak convergence is equivalent to the weak convergence of the variables' distribution functions; that is, in the case of random variables, $X_n \Rightarrow X$ iff $PX_n^{-1}(-\infty,x] \to PX^{-1}(-\infty,x]$ at all continuity points x of the distribution function $F(x) = PX^{-1}(-\infty,x]$. A similar result holds for finite-dimensional random vectors in terms of joint distribution functions. However, when one is dealing with more general random elements X_n, as would be the case when each X_n is a stochastic process on (S,m), some additional conditions are needed besides weak convergence of finite-dimensional distribution functions; see Whitt [31], Example 11.6.1. Therein lies one essential complexity in dealing with stochastic process convergence, which is needed to assert the weak convergence not only of the coordinate random variables or a finite number of them jointly, but of all bounded, continuous functionals of the processes. (Later I will see that a stochastic integral is one such functional.) A key notion needed in that context is that of tightness. For details and proofs of the results below, see Billingsley [6].

Definition 10.2. A set $\mathcal{P}$ of probability measures on (S,m) is said to be tight if for every $\varepsilon > 0$ there exists a compact subset $K \subset S$ such that $P(K) > 1 - \varepsilon$ for all $P \in \mathcal{P}$. A set of random elements on (S,m) is tight if the corresponding probability measures on (S,m) induced by them is tight.

Stochastic processes indexed by a compact interval and with continuous sample paths are much easier to deal with than more general ones. For them, an important tool in proving convergence is provided by the next result.

Theorem 10.1. *Let I be a closed, bounded interval of R^1. A sequence of stochastic processes $\{X_n(t) : t \in I\}$ with continuous sample paths converges to a stochastic process $\{X(t) : t \in I\}$ with continuous sample paths iff the sequence $\{X_n\}$ is tight and all finite-dimensional distributions of $\{X_n\}$ converge to the corresponding ones of X weakly, i.e., $(X_n(t_1),\ldots,X_n(t_k)) \Rightarrow (X(t_1),\ldots,X(t_k))$ for all $(t_1,\ldots,t_k) \in I^k$, $k \geq 1$.*

A moment criterion for tightness is provided by the next theorem and is often easy to check.

Theorem 10.2. *A sequence of stochastic processes $\{X_n\}$ indexed by a compact interval $I = [0,T] \subset R^1$ and with continuous sample paths is tight if the sequence $\{X_n(0)\}$ is tight and there exist constants $\gamma > 0$ and $\alpha > 1$ and a nondecreasing continuous function g on I such that*

$$E[|X_n(t) - X_n(s)|^\gamma \leq |g(t) - g(s)|^\alpha, \text{ for all } s,t \in I \text{ with } s \leq t.$$

As noted previously, weak convergence of stochastic processes is a property of the probability laws governing them and sheds no light on their path behavior. This has some major implications for the stochastic integrals I wish to define later. If one depended only on weak convergence as a basis for defining an integral, one would run into difficulty in using a simulated integral computed from a path of the approximating process as an estimate of the path of the integral from the limit process. Coming to one's aid in such situations is an important result called the Skorohod representation theorem [26].

Theorem 10.3. *If the random elements $\{X_n : n \geq 1\}$ and X_0 on a separable metric space (S,m) are such that $X_n \Rightarrow X_0$, then there exist random elements $\{\tilde{X}_n : n \geq 0\}$, all defined on a common probability space $(\Omega,\mathcal{F},P)$, such that the finite-dimensional distributions of $\{X_n\}_0^\infty$ coincide with the corresponding ones of $\{\tilde{X}_n\}_0^\infty$, and furthermore $\tilde{X}_n \to \tilde{X}_0$ almost surely (a.s.).*

Once again, note that the Skorohod tepresentation theorem only asserts the existence of a probability space and versions of the processes on that space such that convergence is almost sure. In specific instances, one may actually want to demonstrate such a probability space and use that construction for simulation purposes when one needs to understand the path behavior of functionals of the limiting process based on simulations of the approximants. For a majority of application purposes, however, one is often interested only in measures that depend on the probability law of the functional, and then it suffices to know that such a construction does indeed exist. Thus, for many discussions I may assume that I am working with versions defined on a common probability space.

Brownian Motion

I begin with a definition.

Definition 10.3. A stochastic process $\{W(t) : t \geq 0\}$ defined on a probability space $(\Omega, \mathcal{F}, P)$ is a (standard) Brownian motion if (a) for each $\omega \in \Omega$ its sample path $W(t,\omega)$, $t \geq 0$ is a continuous function of t and (b) for each pair $t \geq 0$, $s > 0$, the increment $W(t+s) - W(t)$ is independent of the history $\{W(u) : 0 \leq u \leq t\}$ and is normally distributed with mean zero and variance s.

Assuming the existence of Brownian motion, its many elementary properties such as the following ones are fairly easy to establish.

P1. For each $t > 0$, $W(t)$ is a normal random variable with mean 0 and variance t.
P2. For $0 < t_1 < \cdots < t_n$ the joint distribution of $(W(t_1), \ldots, W(t_n))$ is multivariate normal with mean 0 and variance covariance matrix Σ whose (i,j)th element is given by $\sigma(t_i,t_j) = t_i$ for $i \leq j$.
P3. For all $t,s \geq 0$, $E[W(t+s) \mid W(u),\ 0 \leq u \leq t] = W(t)$. That is, $W(\cdot)$ is a martingale.
P4. $W(\cdot)$ is a Markov process. That is, for all $0 < t_1 < \cdots < t_n$ and $t > 0$ I have for all Borel sets B_k

$$
\begin{aligned}
&P[W(t+t_1) \in B_1, \ldots, W(t+t_n) \in B_n) \mid W(u), 0 \leq u \leq t] \\
&= P[W(t+t_1) \in B_1, \ldots, W(t+t_n) \in B_n \mid W(t)].
\end{aligned}
$$

With a little additional effort one can actually prove the strong Markov property, namely, that in **P4** I may replace t by a stopping time τ, where the random time τ is a stopping time for $W(\cdot)$ if for each $t \geq 0$ the event $[\tau \leq t]$ is in the history sigma field induced by $\{W(u) : 0 \leq u \leq t\}$; from a practical perspective, τ is a stopping time if for all t the occurrence or nonoccurrence of the event $[\tau \leq t]$ can be determined from the values of $W(\cdot)$ in the interval $[0,t]$.

By simple computations, one can also determine various first passage time distributions and exit time distributions of Brownian motion; see Breiman [8], Chung [9] or Borodin & Salminen [7].

The demonstration of the existence, however, is not straightforward, although many diverse approaches do exist in the literature of which the one by Wiener [32] appears to be the earliest. The one by Ciesielski [10], cited in [24], also obtains Brownian motion as a random Fourier series. A classical approach due to Donsker [11] (see Whitt [31] for illustrations through simulation) proceeds by considering a scaled random walk on the integers by demonstrating that the scaled random walk converges in the stochastic process sense to a process with finite-dimensional distributions as in **P2** and then by appealing to the Skorohod representation theorem. In this approach, one specifically considers the random walk defined by an independent and identically distributed sequence of random variables $\{X_n\}$ each assuming values ± 1 with equal probability 1/2, i.e., the partial sums

$S_n = \sum_{k=1}^n X_k$, and one constructs a scaled process in the interval $[0,1]$ by a linear interpolation using the formula

$$W_n(t) = \frac{1}{\sqrt{n}}[S_{u(t)} + (nt - u(t))\{S_{1+u(t)} - S_{u(t)}\}], \text{ where } u(t) = \lfloor nt \rfloor, \ t > 0,$$

with $W_n(0) = 0$. Equivalently, I could consider a random walk that at time intervals of length $1/n$ takes equiprobable steps of size $\pm 1/\sqrt{n}$, generate a continuous path by linear interpolation, and take that as the path of $W_n(\cdot)$. Either way, one demonstrates that $W_n(\cdot) \Rightarrow W(\cdot)$ on $[0,t]$; see Breiman [8]. An appeal to Skorohod completes the existence proof. See Szabados [29] for a construction of this approximation scheme on a common probability space that allows one to bypass an appeal to Skorohod to get a.s. (locally uniform) convergence. Szabados and Shékely [30] indeed use this construction to provide an elementary approach to stochastic integrals with respect to Brownian motion. The random walk approximation appears to be a popular method used in computational finance [19, 25].

Fluid Approximation to Brownian Motion

I begin with the consideration of a linear stochastic fluid flow process as in [1,2]. The fluid level starts at 0 at time 0 and is modulated by a continuous-time Markov chain on the state space $\{1,2,3,4\}$ with initial probability vector $\alpha = (1/4, 1/4, 1/4, 1/4)$ and infinitesimal generator

$$Q_\lambda = \begin{bmatrix} -\lambda & \frac{\lambda}{2} & \frac{\lambda}{2} & 0 \\ \frac{\lambda}{2} & -\lambda & 0 & \frac{\lambda}{2} \\ \frac{\lambda}{2} & 0 & -\lambda & \frac{\lambda}{2} \\ 0 & \frac{\lambda}{2} & \frac{\lambda}{2} & -\lambda \end{bmatrix}. \tag{10.1}$$

The rates of change of the fluid in the environmental states (phases) $\{1, \cdots, 4\}$ are given by the vector $\sqrt{\frac{\lambda}{2}}(1, 1, -1, -1)$.

The process $\mathcal{F}_\lambda = \{(F_\lambda(t), J_\lambda(t)), \ t \geq 0\}$, composed of the fluid level $F_\lambda(t)$ and the phase $J_\lambda(t)$ at time $t+$, constitutes a continuous-time Markov process and can be analyzed by a variety of methods. This is the model for a fluid buffer that on successive intervals whose lengths are distributed independently as exponential random variables with mean $1/\lambda$ increases or decreases at a rate $\sqrt{\lambda/2}$ with equal chance independently of its behavior in past intervals. I will prove soon that as $\lambda \to \infty$, the fluid level process $F_\lambda(\cdot) \Rightarrow W(\cdot)$, which is standard Brownian motion, although the phase process itself has no limit.

Remark 10.2. Let τ be exponentially distributed with mean $1/\lambda$. Observe that $F_\lambda(\tau)$ has mean 0 and variance $1/\lambda$, whence the scaling I have used is indeed such that $\mathrm{Var}(F_\lambda(\tau))/E[\tau] = 1$, consistent with the rate of growth of the variance I wish to achieve for Brownian motion. The scaling of time and space used here is quite different in spirit from the standard setup using Donsker's theorem [11]. Unlike the latter, which works on a deterministic partitioning of time, the preceding scheme builds on a stochastic discretization of time with Poisson processes. For more on stochastic discretization in the context of fluid flows, see [1].

Theorem 10.4. *The (marginal) probability law of $F_\lambda(\cdot)$ is the same as that of the fluid in the fluid process starting at 0 and modulated by the continuous-time Markov chain with initial probability vector $\tilde{\alpha} = (1/2, 1/2)$ and infinitesimal generator*

$$\tilde{Q}_\lambda = \begin{bmatrix} -\lambda/2 & \lambda/2 \\ \lambda/2 & -\lambda/2 \end{bmatrix}, \tag{10.2}$$

with fluid rates $\pm\sqrt{\lambda/2}$.

Proof. The result follows, and I note that any interval of continuous ascent or continuous descent of $F_\lambda(\cdot)$ is exponentially distributed with mean $2/\lambda$ by virtue of the fact that it is a random sum of successive convolutions of the exponential distribution with mean $1/\lambda$, wherein the number of terms in the sum is geometrically distributed with parameter 1/2. □

Remark 10.3. In the simple binomial random walk setting, I cannot use an alternating process like the one obtained in the foregoing theorem since the piecewise linear process there would return to zero at every even step and the process would fail to develop. The reduction in dimensionality in the fluid case is useful for computations since all quantities related to the fluid model of Theorem 10.4 can be obtained from a pair of scalar kernels [1]. From a computational point of view, I do not have to deal with a "binomial tree" but a simple two-state process at each step for the phases.

Lemma 10.1. *As $\lambda \to \infty$, the distribution of $F_\lambda(t)$ converges to that of $N(0,t)$, the normal distribution with mean zero and variance t.*

Proof. For $-\infty < s < \infty$ and $w > 0$ define

$$\phi_\lambda(s,t) = E[\mathrm{e}^{sF_\lambda(t)}] \text{ and}$$

$$\tilde{\phi}_\lambda(s,w) = \int_0^\infty \mathrm{e}^{-wt}\phi(s,t)\,\mathrm{d}t.$$

A simple probability argument, conditioning on the first transition epoch of the phase process in $\mathcal{F}_\lambda$, gives

$$\begin{aligned}\phi_\lambda(s,t) &= \mathrm{e}^{-\lambda t}\frac{1}{2}\left(\mathrm{e}^{s\sqrt{\frac{\lambda}{2}}t} + \mathrm{e}^{-s\sqrt{\frac{\lambda}{2}}t}\right) \\ &\quad + \frac{1}{2}\lambda \int_0^t \mathrm{e}^{-\lambda u}\left(\mathrm{e}^{s\sqrt{\frac{\lambda}{2}}u} + \mathrm{e}^{-s\sqrt{\frac{\lambda}{2}}u}\right)\phi(s,t-u)\,\mathrm{d}u.\end{aligned} \tag{10.3}$$

From this one easily obtains after some algebra

$$\tilde{\phi}_\lambda(s,w) = \frac{1+w/\lambda}{w - s^2/2 + w^2/\lambda}. \tag{10.4}$$

Clearly, as $\lambda \to \infty$, $\tilde{\phi}_\lambda(s,w) \to 1/[w - s^2/2]$. I recognize the limit to be the Laplace transform of the function $\mathrm{e}^{s^2 t/2}$, which is indeed the moment-generating function of $N(0,t)$, and the proof is complete. □

Remark 10.4. As one would anticipate from familiar theory [12] based on scaled random walk approximations, I observe that the rate of convergence in the foregoing analysis is $O(1/\lambda)$.

Figure 10.1 shows the results of 10,000 simulations of the fluid model with different values of λ. For a fluid level at $t = 1$, I have shown histograms and QQnorm plots computed using the statistical package R on a Dell laptop.

Corollary 10.1. *For any n-tuple $0 < t_1 < t_2 < \cdots < t_n$, the joint distribution of the increments $F_\lambda(t_1), F_\lambda(t_2) - F_\lambda(t_1), \ldots, F_\lambda(t_n) - F_\lambda(t_{n-1})$ weakly converges as $\lambda \to \infty$ to that of a set of n independent normal random variables with mean 0 and variances $t_1, t_2 - t_1, \ldots, t_n - t_{n-1}$. In other words, as $\lambda \to \infty$, the finite-dimensional distributions of the process $\mathcal{F}_\lambda$ converge to those of the standard Brownian motion.*

Theorem 10.5. *Let I be a compact interval of $[0,\infty)$. As $\lambda \to \infty$, the marginal process $\{F_\lambda(t) : t \in I\} \Rightarrow \{W(t) : t \in I\}$, where $W(\cdot)$ is standard Brownian motion.*

Proof. For $t, w > 0$ let

$$m_{\lambda,2k}(t) = E[(F_\lambda(t))^{2k}], \text{ and } \tilde{m}_{\lambda,2k}(w) = \int_0^\infty \mathrm{e}^{-wt} m_{\lambda,2k}(t)\,\mathrm{d}t.$$

Rewriting Eq. (10.4) as

$$(w^2 - \lambda s^2/2 + \lambda w)\tilde{\phi}_\lambda(s,w) = \lambda + w,$$

differentiating both sides of that equation twice and four times respectively, and setting $s = 0$, one easily obtains that

$$\tilde{m}_{\lambda,2}(w) = \frac{1}{w^2}\frac{\lambda}{\lambda + w}$$

and

$$\tilde{m}_{\lambda,4}(w) = \frac{6}{w^3}\left(\frac{\lambda}{\lambda + w}\right)^2.$$

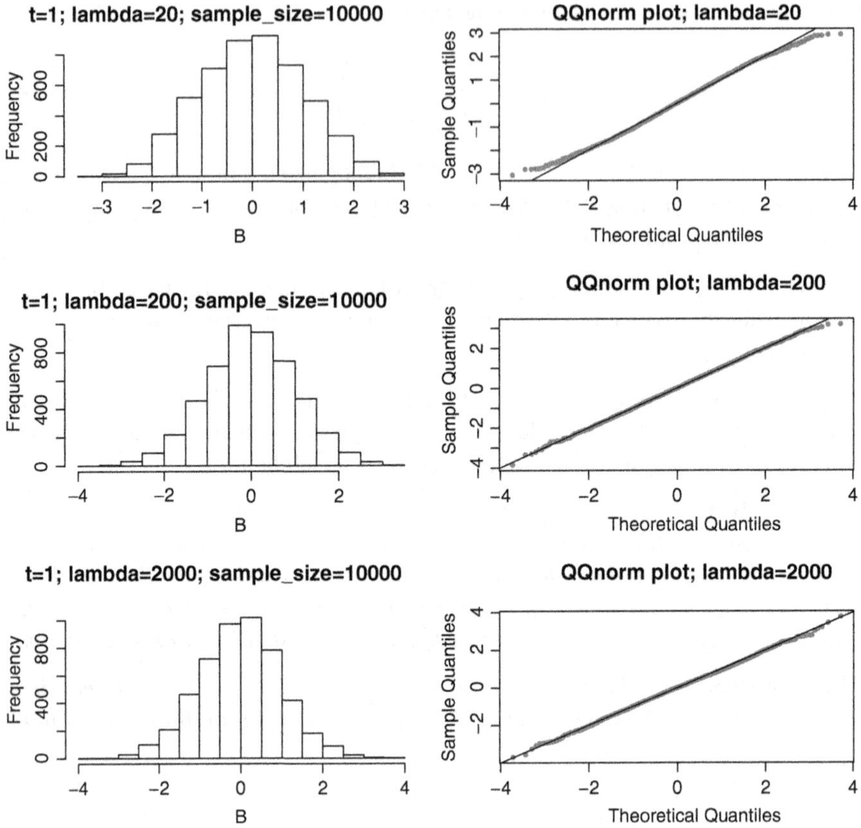

Fig. 10.1 Fluid level at $t=1$

Inverting the preceding Laplace transform I get

$$m_{\lambda,4}(t) = 6\int_0^t \lambda^2 e^{-\lambda u} u(t-u)^2 \, du \le 6t^2 \int_0^t \lambda^2 e^{-\lambda u} u \, du \le 6t^2.$$

From this it follows easily that

$$E[|F_\lambda(t) - F_\lambda(s)|^4] \le 6|t-s|^2 \le 6(t^2+s^2). \tag{10.5}$$

Since $F_\lambda(0) \equiv 0$ a.s., the set of random variables $\{F_\lambda(0)\}$ is trivially tight in R, and now, along with inequality (10.5), I have by Theorem 10.2 that the family of processes $\{F_\lambda(\cdot)\}$ is tight on any compact interval; here, the conditions of Theorem 10.2 are satisfied with $\gamma = 4$, $g(t) = \sqrt{6}t$, and $\alpha = 2$. Now, in light of Corollary 10.1, an appeal to Theorem 10.1 completes the proof. □

The Stochastic Integral

Throughout this discussion, I assume that the fluid processes and the Brownian motion have all been defined on a common probability space $(\Omega, \mathcal{A}, P)$. Having a modest goal of not aiming at the greatest generality, I consider a real-valued random function $f(t, W(t))$ on $[0, \infty) \times \Omega$ that (a) is jointly measurable with respect to both coordinates (t, ω), (b) has continuous second partial derivatives, and, furthermore, (c) is adapted to the family of sigma fields $\{\mathcal{W}(t) : t \geq 0\}$, where $\mathcal{W}(t)$ is the history sigma field generated by $\{W(u) : 0 \leq u \leq t\}$. I now define the two types of integrals, respectively due to Ito [13] and Stratonovich [28]. For a nice summary of Ito integration, refer to Steele [27], and for a discussion of both Ito and Stratonovich integrals, refer to Oksendal [15, 16, 20].

Definition 10.4. The Ito integral $\int_0^t f(u, W(u))\,\mathrm{d}W(u)$ is defined as the random variable obtained as the limit

$$
\begin{aligned}
I(t) &= \int_0^t f(u, W(u))\,\mathrm{d}W(u) \\
&= \lim_{\lambda \to \infty} \sum_{k=1}^{N_\lambda(t)} \left[f(t_{k-1}, F_\lambda(t_{k-1}))\right]\left[F_\lambda(t_k) - F_\lambda(t_{k-1})\right] \\
&\quad + f(t_{N_\lambda(t)}, F_\lambda(t_{N_\lambda(t)}))[F_\lambda(t) - F_\lambda(t_{N_\lambda(t)})],
\end{aligned} \tag{10.6}
$$

where $N_\lambda(t)$ is the number of points in $[0, t]$ of the Poisson process underlying the fluid flow model, $t_0 = 0$, and t_i, $i \geq 1$, are the epochs of events in that Poisson process.

Definition 10.5. The Stratonovich integral $\int_0^t f(u, W(u)) \circ \mathrm{d}W(u)$ is defined as the random variable obtained as the limit

$$
\begin{aligned}
S(t) &= \int_0^t f(u, W(u)) \circ \mathrm{d}W(u) \\
&= \lim_{\lambda \to \infty} \sum_{k=1}^{N_\lambda(t)} [f(t_{k-1}^*, F_\lambda(t_{k-1}^*))][F_\lambda(t_k) - F_\lambda(t_{k-1})] \\
&\quad + f(t_{N_\lambda(t)}^*, F_\lambda(t_{N_\lambda(t)}^*))[F_\lambda(t) - F_\lambda(t_{N_\lambda(t)})],
\end{aligned} \tag{10.7}
$$

where

$$t_k^* = 0.5[t_{k-1} + t_k] \text{ for } k < N_\lambda(t) \text{ and } t_{N_\lambda(t)}^* = 0.5(t_{N_\lambda(t)} + t).$$

Remark 10.5. (a) The existence of the limits in (10.6) and (10.7) is trivial to establish for step functions; for $f \geq 0$ I proceed by approximating f using a sequence of step functions, and for a general f by considering f^+ and f^-, the positive and negative parts of f. (b) The Ito integral uses the value of the function f at the lower end point, whereas the Stratonovich integral uses the value at the midpoint of each partitioning interval. The two integrals are not equal, and that exemplifies a major difference between stochastic integrals and ordinary Riemann–Stieltjes integrals.

Remark 10.6. The classical approach to stochastic integrals is based on approximating the integrand by step functions and to take the integral as the limit of those for the approximating step functions with respect to Brownian motion. As opposed to this, my approach is based on approximating Brownian motion itself by a sequence of Markovian fluid flows and considering the limit of the integrals with respect to the fluid flows.

Remark 10.7. A useful theoretical construct yielding a sequence of fluid approximations on a common probability space converging to the standard Brownian motion is as follows. Here, having defined an approximation to the Brownian motion in terms of a Poisson process with a parameter λ, a new Poisson process with the same parameter λ is superimposed with it, and up and down marks for the fluid are defined at these new points through a fair coin toss while retaining the up and down marks for the original points; the rates of the fluid should of course be adjusted for the fluid flow based on the superposition. This ensures that the sequence of Poisson processes that I use in building the fluid flow approximations to the standard Brownian motion are all defined on a common probability space and are nested. With this construction, the convergence on the right-hand side of Eqs. (10.6) and (10.7) can be strengthened to convergence in probability; I omit the details.

Remark 10.8. Based on the foregoing discussion, in practice one chooses a large value of λ and approximates the value of the integral through the sums appearing in (10.6) and (10.7).

The next result asserts that through the preceding definitions, I do obtain two stochastic processes on $(\Omega, \mathcal{A}, P)$ with continuous sample paths.

Theorem 10.6. *For each $T < \infty$, $\{I(u) : 0 \leq u \leq T\}$ and $\{S(u) : 0 \leq u \leq T\}$ are stochastic processes on $(\Omega, \mathcal{A}, P)$ with continuous sample paths. In addition, the Ito integral process $I(\cdot)$ is a martingale adapted to the filtration $\mathcal{W}(\cdot)$; that is, for each Borel set B, the event $[I(t) \in B]$ is a member of $\mathcal{W}(t)$, and*

$$E[I(t+s)|I(u),\ 0 \leq u \leq t] = I(t),\ \ \textit{for all}\ 0 \leq t \leq t+s \leq t\ \ a.s.$$

Proof. My assumptions on f entail that it is a.s. bounded over any compact interval. All the asserted results are now easy to establish using standard arguments based on the dominated convergence theorem. I omit the details.

Remark 10.9. The special properties of the Ito integral noted previously make it particularly relevant for applications in mathematical finance; see Shreve [25]. There, if one models the value of an asset as a Brownian motion and f as the position of the asset held, it is natural to assume that f is adapted to the price process (buy/sell decisions are based only on knowledge of the history), and the value process $I(\cdot)$ is then also adapted to the price process.

Example 10.1. It is widely known [and fairly straightforward to prove from the definitions in (10.6) and (10.7)] that

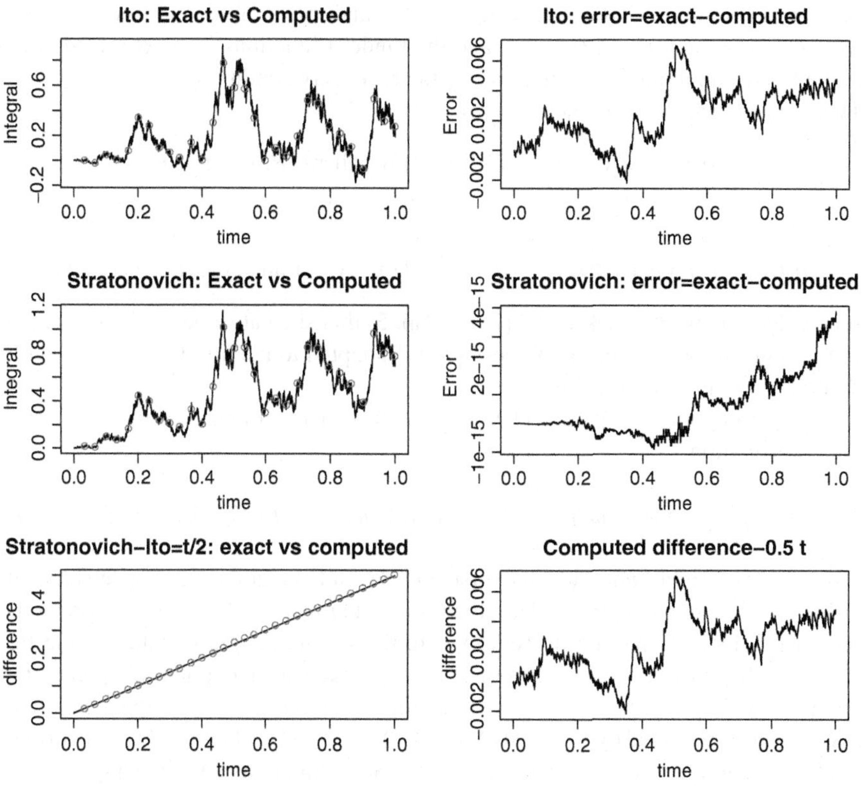

Fig. 10.2 Numerical computations $\int W\,\mathrm{d}W$

$$\int_0^t W(u)\,\mathrm{d}W(u) = \frac{1}{2}(W(t))^2 - t/2, \text{ and } \int_0^t W(u) \circ \mathrm{d}W(u) = (W(t))^2/2.$$

In Fig. 10.2 below, I show a simulation result for the foregoing explicit results for the integrals and their approximations computed as the sums in (10.6) and (10.7) with $\lambda = 10{,}000$ at a set of points in $[0,1]$; note that the differences are indeed negligible.

Stochastic Differential Equations

Consider the SDE

$$\mathrm{d}X(t) = b(t,X(t))\,\mathrm{d}t + \sigma(t,X(t))\,\mathrm{d}W(t),\ 0 \le t \le T < \infty, \tag{10.8}$$

with initial condition $X(0) = Z$, where Z is a random variable independent of the Brownian motion with $E(Z^2) < \infty$, and the random functions b and σ are adapted to the Brownian motion $W(\cdot)$ and satisfy the regularity conditions:
There exist constants C, D such that (a)

$$|b(t,x)| + |\sigma(t,x)| \leq C(1+|x|) \text{ for all } x \in R,\ 0 \leq t \leq T;$$

and (b)

$$|b(t,x) - b(t,y)| + |\sigma(t,x) - \sigma(t,y)| \leq D|x-y| \text{ for all } x,y \in R,\ 0 \leq t \leq T.$$

It is widely known (see Oksendal [20], Chap. 5) that such an equation has a unique solution given by a stochastic process $\{X(t)\}$ adapted to the filtration generated by Z and $W(\cdot)$ with $E[\int_0^T X(t)^2 \mathrm{d}t] < \infty$.

In the foregoing setup, Eq. (10.8) is taken as a mnemonic for the stochastic integral equation

$$X(t) = \int_0^t b(u, X(u))\,\mathrm{d}u + \int_0^t \sigma(u, X(u))\mathrm{d}W(u), \quad 0 \leq t \leq T, \quad X(0) = Z, \tag{10.9}$$

where on the right-hand side the first and second integrals are respectively an ordinary Riemann integral and a stochastic integral as defined in the previous section. (I shall assume the Ito setup where the second integral is taken in the Ito sense; the Stratonovich case is similar.) From a practical standpoint, the preceding integral equation asserts that in a small time interval $(t, t+\Delta t]$, the increment $\Delta X(t)$ is approximated by $b(t, X(t))\Delta t + \sigma(t, X(t))N(\Delta t)$, where $N(\Delta t)$ is a normal random variable with mean 0 and variance Δt. Such equations arise in many areas of applied probability such as population dynamics, time series, mathematical finance, hydrology, signal processing, biomedical engineering, etc. For a friendly discussion of SDEs with examples from a diverse set of application areas, I refer the reader to Higham [12] and the book by Kloeden and Platten [17], who also provide an extensive set of numerical recipes for solving SDEs. For a treatment of stochastic integrals at a high level of generality, refer to Protter [21].

The approximation of a stochastic integral through a scheme based on linear fluid flows immediately provides a technique for solving SDEs numerically. Without belaboring the reader with a lot of equations that essentially repeat my construction of the stochastic integral, I wish to just note that with my scheme of evaluating the integral, what I get is roughly equivalent to the Euler–Maruyama scheme [17]. Higher-order schemes providing higher accuracy can of course be developed as in [14, 17].

Figure 10.3 presents a set of simulations of the exact and computed solutions of a set of SDEs taken from Kloeden and Platten [17] for which explicit results are available. Note that the two terms compared in each example are computed in an entirely different manner; while the "exact" result is computed using the explicit formula, which involves only the time point t, the "approximation" is computed using the sums in Eqs. (10.6) and (10.7), which are functionals of the entire path up to time t.

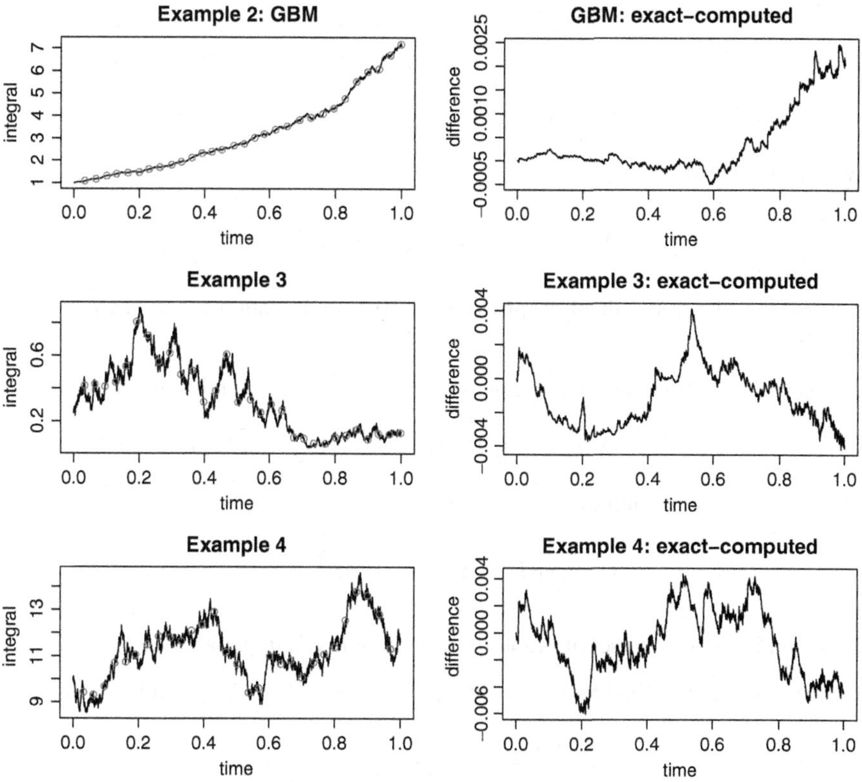

Fig. 10.3 Example SDEs solved using fluid flows

I am not proposing that one should simulate the stochastic integral via the fluid approximation; in the one-dimensional case, there is little to be gained compared to the familiar binomial scheme in terms of complexity. These examples are presented only to confirm my assertion that a fluid-based approximation can indeed be developed and works well. Given that much progress has been made in characterizing time-dependent distributions of fluid flow models, this opens up an interesting possibility of using those results for computing various characteristics of stochastic integrals using them. From a theoretical standpoint, it is also worth examining whether the approximation involved here of Brownian motion based on fluids that are continuous-time martingales offers any advantages. I emphasize that for stochastic integrals based on a multidimensional Brownian motion, my approach has definite advantages relative to the binomial schemes in that, unlike the latter in which one quickly encounters the curse of dimensionality, the dimensionality of the phase space remains the same at each stochastic discretization epoch.

The following examples are presented.

Example 10.2. This example is of the geometric Brownian motion defined by

$$\mathrm{d}X(t) = X(t)[\mu\,\mathrm{d}t + \sigma\,\mathrm{d}W(t)],\ \ X(0) = 1,$$

which has the solution

$$X(t) = \exp\left[(\mu - \frac{\sigma^2}{2})t + \sigma W(t)\right].$$

I showed in the first pair of graphs of Fig. 10.3 a simulation for the case $\mu = 2$, $\sigma = 0.2$.

Example 10.3. This SDE is Example 4.38 on p. 123 of [17] and is given by

$$\mathrm{d}X(t) = -X(t)[2\ln X(t) + 1]\,\mathrm{d}t - 2X(t)\sqrt{-\ln X_t}\,\mathrm{d}W(t)$$

and has the explicit solution

$$X(t) = \exp\left(-(\ W(t) + \sqrt{-\ln X(0)}\ )^2\right).$$

I showed in the second set of graphs of Fig. 10.3 this example with $X(0) = 0.25$.

Example 10.4. The final SDE example is Example 4.41 from [17]. It is given by

$$\mathrm{d}X(t) = \frac{1}{3}[X(t)]^{1/3}\,\mathrm{d}t + [X(t)]^{2/3}\,\mathrm{d}W(t)$$

and has the solution

$$X(t) = \left([X(0)]^{1/3} + \frac{1}{3}W(t)\right)^3.$$

I chose the value $X(0) = 10$ and showed this example in the last pair of graphs of Fig. 10.3.

Markov-Modulated Fluid Models

I note that the process governed by the stochastic differential equation

$$\mathrm{d}X(t) = \mu_{J(t)}\,\mathrm{d}t + \sigma_{J(t)}\,\mathrm{d}W(t),$$

where $\{J(t) : t \geq 0\}$ is an m-state continuous-time Markov chain, is indeed the Markov-modulated Brownian motion (MMBM) considered by Asmussen [5]. Thus, I may treat it as a stochastic integral and apply techniques for SDEs.

But it is also obvious that my approximation to the Brownian motion with a linear fluid flow will allow us to approximate MMBM itself by a Markovian linear fluid flow modulated by a $2m$-state continuous-time Markov chain. I can combine the fast algorithms now available for the transient analysis of linear fluid flow models to obtain approximate transient results for the MMBM numerically; I shall explore these in a forthcoming work.

For the linear Markovian fluid flow, Ramaswami [22] established some strong connections with quasi-birth-and-death processes (QBDs) to obtain [1, 3, 4, 23] powerful algorithms based on matrix-geometric methods. Readers familiar with that development, and particularly [1], will recognize immediately that a similar connection exists for MMBM with QBDs. Reasoning in way that is almost identical to that in Sect. 2 of [1], one obtains MMBM as the limit of reward processes defined on a sequence of QBDs. The detailed construction is almost identical to that in Sect. 2 of [1], with the only change being that the spatial scaling needs to be effected through Poisson processes at a rate of $\sqrt{2n\lambda}$ while retaining the semi-Markov structure with $\exp(n\lambda)$ for interevent times along the time axis. For the reflected MMBM case, this analogy is mapped immediately to familiar functional central limit theorems for Markov-modulated queues [31].

Concluding Remarks

I have identified an approach to approximating Brownian motion with a Markov-modulated linear fluid flow model and shown how it may be used to compute stochastic integrals and to solve stochastic differential equations. The approach appears to have many interesting features:

- The approach is elementary, and many computations and proofs follow easily from results on stochastic fluid flows.
- As opposed to a discretization using binomials, there is no tree to deal with. At each point in time I only have to track the state of a two-state phase process. That is a significant gain relative to binomial-tree-based methods.
- It offers the potential to exploit recent advances in the area of fluid flow models such as the quadratically convergent algorithm of Ahn and Ramaswami [3] to obtain powerful algorithms to compute the distribution and various expectations of stochastic integrals with respect to Brownian motion.
- If one computes the integral with respect to the fluid model, then the SDE really reduces to a random ordinary differential equation (RODE) which is simpler; for a detailed discussion of RODEs, refer to [14].

A particularly interesting aspect of the fluid approach as compared with binomial trees based on discretization is its ability to handle dimensionality. For an N-dimensional Brownian motion, the binomial tree approach with K discretization points over the time interval would require a very large number of nodes that grows exponentially with N. However, the two-state fluid approximation of each Brownian component entails a dimensionality of only $N+1$ by keeping track of the number of processes in phase 1 at any time. This aspect could be particularly useful in many financial models like multifactor interest-rate models.

I have initiated two extensions of the present work: (a) developing algorithms for the transient analysis of second-order fluid flows as in Asmussen [5]; this model is really an MMBM where the drift and diffusion coefficients depend on the state of an environmental continuous-time Markov chain; and (b) approximation of fractional Brownian motion through the introduction of correlations among successive exponential intervals.

I consider my work as opening the door to many interesting research problems. Besides those concerning a careful analysis of discretization errors and stability issues, there are interesting questions related to the development of higher-order schemes.

It is worth noting that instead of an exponential distribution to generate the random partitions of time, I could have used other distributions. As long as these are in the domain of attraction [8] of the normal law, I get approximations to Brownian motion. Now, an interesting question is what might be gained by considering, say, a phase-type distribution. I conjecture that the order of convergence would remain the same, but the closer the distribution is to a normal, the more the constant appearing in the convergence order will be improved.

I wish to point out that my work on fluid flows can be extended to Markov renewal process modulated fluids; see Latouche and Takine [18] for a steady-state analysis of such a model. Such fluid models would offer an approach to stochastic integrals with respect to stable stochastic processes through appropriate scaling. This aspect is particularly interesting as a potential candidate for the numerical analysis of heavy-tailed and self-similar systems modeled as SDEs.

Finally, there is also the challenging multidimensional case.

References

1. Ahn, S., Ramaswami, V.: Transient analysis of fluid flow models via stochastic coupling to a queue. Stoch. Models **20**(1), 71–101 (2004)
2. Ahn, S., Ramaswami, V.: Transient analysis of fluid flow models via elementary level crossing argument. Stoch. Models **22**(1), 129–148 (2006)
3. Ahn, S., Ramaswami, V.: Efficient algorithms for transient analysis of stochastic fluid flow models. J. Appl. Probab. **42**(2), 531–549 (2005)
4. Ahn, S., Badescu, A.L, Ramaswami, V.: Transient analysis of finite buffer fluid flows and risk models with a dividend barrier. QUESTA **55**, 207–222 (2007)
5. Asmussen, S: Stationary distributions of fluid flow models with or without Brownian noise. Stoch. Models **11**, 1–20 (1995)
6. Billingsley, P.: Convergence of Probability Measures, 2nd edn. Wiley, New York (1999)
7. Borodin, A.N., Salminen, P.: Handbook of Brownian Motion - Facts and Formulae. Birkhäuser, Basel, Boston and Berlin (1996)
8. Breiman, L.: Probability. Addison-Wesley, Reading, MA (1968)
9. Chung, K.L.: A Course in Probability Theory. Harcourt, Brace & World (1968)

10. Ciesielski, Z.: Hölder conditions for realisations of Gaussian processes. Trans. Am. Math. Soc. **99**, 403–413 (1961)
11. Donsker, M.: An invariance principle for certain probability limit theorems. Mem. Am. Math. Soc. **6**, 1–10 (1951)
12. Higham, D.: An algorithmic introduction to numerical simulation of stochastic differential equations. SIAM Rev. **43**(3), 525–546 (2001)
13. Itô, K.: Stochastic Integral. Proc. Imper. Acad. Tokyo **20**, 519–524 (1944)
14. Jentzen, A., Kloeden, P.E.: Pathwise Taylor schemes for random ordinary differential equations. BIT Numer. Math. **49**, 113–140 (2009)
15. Karandikar, R.: Pathwise solutions of stochastic differential equations. Sankhya A **43**(2), 121–132 (1981)
16. Karatzas, I., Shreve, S.E.: Brownian Motion and Stochastic Calculus. Springer, New York (1991)
17. Kloeden, P.E., Platten, E.: Numerical Solution of Stochastic Differential Equations. Springer, Berlin (1995)
18. Latouche, G., Takine, T.: Markov renewal fluid queues. J. Appl. Probab. **41**, 746–757 (2004)
19. Nelson, D., Ramaswamy, K.: Simple binomial processes as diffusion approximations in financial models. Rev. Financ. Stud. **3**, (3), 393–430 (1990)
20. Øksendal, B.: Stochastic Differential Equations: An Introduction with Applications. Springer, Berlin (2003)
21. Protter, P.: Stochastic Integration and Differential Equations. Springer, New York (1990)
22. Ramaswami, V.: Matrix analytic methods for stochastic fluid flows. In: Smith, D., Key, P. (eds.) Teletraffic Engineering in a Competitive World. Proceedings of the 16th International Teletraffic Congress, pp. 1019–1030. Elsevier, New York (1999)
23. Ramaswami, V.: Passage times in fluid models with application to risk processes. Meth. Comput. Appl. Probab. **8**, 497–515 (2006)
24. Rogers, L.C.G., Williams, D.: Diffusions, Markov Processes and Martingales, vol. I. Cambridge University Press, Cambridge (2000)
25. Shreve, S.E.: Stochastic Calculus for Finance, vol. 2. Continuous Time Models. Springer, Berlin (2004)
26. Skorohod, A.V.: Limit theorems for stochastic processes. Theor. Probab. Appl. **1**, 261–290 (1956)
27. Steele, M.J.: Ito calculus. In: Teugels, J., Sundt, B. (eds.) The Encyclopedia of Actuarial Science. Wiley, New York (2004)
28. Stratonovich, R.L.: A new representation for stochastic integrals and equations. J. SIAM Contr. **4**(2), 362–371 (1966)
29. Szabados, T.: An elementary approach to the Wiener process and stochastic integrals. Stud. Sci. Math. Hung. **41**, 101–126 (2004)
30. Szabados, T., Székely, B.: Stochastic integration based on simple, symmetric random walks. J. Theor. Probab. **22**, 203–219 (2009)
31. Whitt, W.: Stochastic Process Limits. Springer, Berlin (2002)
32. Wiener, N.: Un problème de probabilités énombrables. Bull. Soc. Math. Fr. **52**, 569–568 (1924)

Chapter 11
Impact of Dampening Demand Variability in a Production/Inventory System with Multiple Retailers

B. Van Houdt and J.F. Pérez

Introduction

Consider a two-echelon supply chain consisting of a single retailer and a single manufacturer, where the retailer places an order for a batch of items with the manufacturer at regular time instants, i.e., the time between two orders is fixed and denoted by r. The manufacturer may be regarded as a single-server queue that produces these items and delivers them to the retailer as soon as a full order is finished. The retailer sells the items and maintains an inventory on hand to meet customer demand. When the customer demand exceeds the current inventory on hand, only part of the demand is immediately fulfilled and the remaining items are delivered as soon as new items become available at the retailer. Hence, items are backlogged instead of being lost (i.e., there are no *lost sales*). We assume that the manufacturer does not maintain an inventory but simply produces items whenever an order arrives, i.e., it operates on a *make-to-order* basis.

A key performance measure in such a system is the *fill rate*, which is a measure of the proportion of customer demand that can be met without any delay. To guarantee a certain fill rate, it is important to determine the size of the orders placed at regular time instants. This size will depend on the current *inventory position*, defined as the inventory on hand plus the number of items on order minus the number of backlogged items. The rule that determines the order size is termed the *replenishment* rule. A well-studied replenishment rule exists in ordering an amount

B. Van Houdt (✉)
Department of Mathematics and Computer Science, University of Antwerp – IBBT, Middelheimlaan 1, 2020 Antwerp, Belgium
e-mail: benny.vanhoudt@ua.ac.be

J.F. Pérez
Department of Electrical and Electronics Engineering, Universidad de los Andes, Cra 1 No. 18A-12, Bogotá, Colombia
e-mail: jf.perez33@uniandes.edu.co

G. Latouche et al. (eds.), *Matrix-Analytic Methods in Stochastic Models*, Springer Proceedings in Mathematics & Statistics 27, DOI 10.1007/978-1-4614-4909-6_11,

such that the inventory position is raised after each order to some fixed position S, called the *base-stock level*. This basically means that at regular time instants, you simply order the amount of items sold since the last order instant. As a result, the order policy of the retailer is called an (R,S) policy.

A common approach in the analysis of such a policy is to assume an *exogenous* lead time, which means that the time required to *deliver* an order is independent of the size of the current order and independent of the lead time of previous orders. In Boute et al. [3] studied the (R,S) policy with *endogenous* lead times, meaning the lead times depend on the order size and consecutive lead times are correlated. Their results indicate that exogenous lead times result in a severe underestimation of the required inventory on hand, as expected.

When the lead times are endogenous, it is clear that a high variability in the order sizes comes at a cost because this increases the variability of the arrival process at the manufacturer and therefore increases the lead times. As a result, replenishment rules that smooth the order pattern at the retailer were studied by Boute et al. [4], and it was shown that the retailer could reduce the upstream demand variability without having to increase the safety stock (much) to maintain customer service at the same target level. Moreover, on many occasions the retailer could even decrease the safety stock somewhat when the orders were smoothed. This is clearly advantageous for both the retailer and the manufacturer. The manufacturer receives a less variable order pattern and the retailer can decrease the safety stock while maintaining the same fill rate, so that a cooperative surplus is realized.

In this chapter we analyze the same set of replenishment rules as in Boute et al. [4], but now we look at a two-echelon supply chain consisting of one manufacturer and two retailers, where either both, one, or neither of the retailers uses a smoothing rule. The main issue that we wish to address therefore consists of studying whether all parties can still benefit when the orders are smoothed and, moreover, who benefits most.

As in Boute et al. [4], one of the key steps in the analysis of this supply chain system will exist in setting up a GI/M/1-type Markov chain [8] that has only two nonzero blocks, denoted by A_0 and A_d. However, as opposed to Boute et al. [4], the size of these blocks often prohibits us from storing them in main (or secondary) memory. This implies that iteratively computing the dense R matrix, used to express the matrix geometric steady state vector of the GI/M/1-type Markov chain, using one of the existing methods such as functional iterations or cyclic reduction [1] is no longer possible/efficient. Instead, we will rely on the specific structure of the matrices A_0 and A_d and make use of numerical methods typically used to solve large finite Markov chains, such as the shuffling algorithm [5], Kronecker products, the power method, the Gauss–Seidel iteration, and GMRES [10].

Model Description

We consider a two-echelon supply chain with two retailers and a single manufacturer, where both retailers maintain their own inventory. Every period, both retailers review their customer demand. If there is enough on-hand inventory available at a retailer, the demand is immediately satisfied. If not, the shortage is backlogged. To maintain an appropriate amount of inventory on hand, both retailers place a replenishment order with the manufacturer at the end of every period. The manufacturer does not hold a finished-goods inventory but produces the orders on a make-to-order basis. The manufacturer's production system is characterized by a single-server queueing model that sequentially processes the orders, which require stochastic processing times. Once the complete replenishment order of both retailers is produced, the manufacturer replenishes both inventories. Hence, the order in which the two orders are produced is irrelevant because shipping only occurs when both orders are ready.

The time from the moment an order is placed to the moment that it replenishes a retailer's inventory is the *replenishment lead time* T_{r}. The queueing process at the manufacturer clearly implies that the retailer's replenishment lead times are stochastic and correlated with the order quantity. The sequence of events in a period is as follows. The retailer first receives goods from the manufacturer, then he observes and satisfies customer demand, and, finally, he places a replenishment order with the manufacturer. The following additional assumptions are made.

1. Customer demand during a period for retailer i is independently and identically distributed (i.i.d.) over time according to an arbitrary, finite, discrete distribution $D^{(i)}$ with a maximum of $m_D^{(i)}$, for $i = 1$ and 2. The demand at one retailer is also assumed to be independent of the demand at the other retailer. For further use, denote $m_D = m_D^{(1)} + m_D^{(2)}$.
2. The order quantity $O_t^{(i)}$ of retailer i during period t is determined by the retailer's replenishment rule and influences the variability in the orders placed with the manufacturer. Possible replenishment rules are discussed in the next section.
3. The replenishment orders are processed by a single FIFO server. This excludes the possibility of order crossovers. When the server is busy, new orders join a queue of unprocessed orders.
4. The orders placed during period t are delivered when both orders have been produced.
5. Orders consist of multiple items, and the production time of a single item is i.i.d. according to a discrete-time phase type (PH) distribution with representation (α, U). For further use, we define $u^* = e - Ue$, with e a column vector of ones.

The PH distribution is determined using the matching procedure presented in [4], which matches the first two moments of the production time using an order 2 representation, meaning the matrix U is a 2×2 matrix and α a size 2 row vector, even if the squared coefficient of variation is small by exploiting the scaling factor as in [2]. This implies that the length of a time slot is chosen as half of the mean production time of an item. In other words, the mean production time of an item is two time slots, while the length of a period is denoted as d time slots, where d is assumed to be an integer.

The time from the moment the order arrives at the production queue to the point that the production of the entire batch is finished is the *production* lead time or response time, denoted by T_p. Note that the production lead time is not necessarily an integer number of periods. Since in our inventory model events occur on a discrete-time basis with a time unit equal to one period, the replenishment lead time T_r is expressed in terms of an integer number of periods. For instance, suppose that the retailer places an order at the *end* of period t, and it turns out that the production lead time is 1.4 periods. This order quantity will be added to the inventory in period $t+2$ and, due to our sequence of events, can be used to satisfy demand in period $t+2$. As such, we state that the replenishment lead time T_r is $\lfloor T_\mathrm{p} \rfloor$ periods, i.e., one period in our example.

Replenishment Rules

The retailers considered in this chapter apply an (r,S) policy with or without smoothing, meaning, among other things, that they place an order at the end of each period. Without smoothing, the order size is such that the inventory position IP, defined as the on-hand inventory plus the number of items on order minus the backlogged items, equals some fixed S after the order is placed. In other words, the size of the order O_t at the end of period t simply equals the demand D_t observed during period t.

If smoothing is applied with parameter $0 < \beta < 1$, then we do not order the difference between S and IP but instead only order β times $S - IP$. As will become clear subsequently, this does not imply that fewer items are ordered in the long run; it simply means that some items will be ordered at a later time. As shown in [4], this rule is equivalent to stating that the size of the order at the end of period t, denoted O_t, is given by

$$O_t = (1-\beta)O_{t-1} + \beta D_t,$$

where D_t is the demand observed by a retailer in period t. Hence, setting $\beta = 1$ implies that we do not smooth. This equation also shows that the mean order size is still equal to the mean demand size $E[D]$. It is also easy to show [4] that the variance of the order size $\mathrm{Var}[O]$ equals

$$\frac{\beta}{(2-\beta)}\mathrm{Var}[D],$$

meaning the variance decreases to zero as β approaches zero, where $\mathrm{Var}[D]$ is the variance in the demand. It is also possible to consider β values between 1 and 2, but this would amplify the variability instead of dampening it.

The key question that our analytical model will answer is how to select the base-stock level S such that the fill rate, a measure of the proportion of demand that can be immediately delivered from the inventory on hand, defined as

$$1 - \frac{\text{expected number of backlogged items}}{\text{expected demand}},$$

is sufficiently high. The level S is typically expressed using the *safety stock* SS, defined as the average net stock just before a replenishment arrives (where the net stock equals the inventory on hand minus the number of backlogged items). For a retailer that smooths with parameter β, S, and SS are related as follows [4]:

$$S = \mathrm{SS} + (E[T_{\mathrm{r}}]+1)E[D] + \frac{1-\beta}{\beta}E[D],$$

where $E[T_{\mathrm{r}}]$ is the mean replenishment lead time. Thus, a good policy will result in a smaller safety stock SS, which implies a lower average storage cost for the retailer.

Markov Chain

Both Markov chains developed in this section are a generalization of the Markov chain introduced in [4] for a system with a single retailer. The numerical method to attain a stationary probability vector, discussed in the section "Numerical Solution," is, however, very different.

Henceforth we will express all our variables in time slots, where the length of a single slot equals half of the mean production time, i.e., $\alpha(I-U)^{-1}e/2$, and orders are placed by both retailers every d time slots. Hence, the order size of retailer i at the end of period t is now written as $O_{td}^{(i)}$ and

$$O_{td}^{(i)} = (1-\beta_i)O_{(t-1)d}^{(i)} + \beta_i D^{(i)},$$

where β_i is the smoothing parameter of retailer i for $i = 1,2$. As the order size must be an integer, the integer amount ordered, $O_{td}^{(i*)}$, will equal $\lceil O_{td}^{(i)} \rceil$ with probability $O_{td}^{(i)} - \lfloor O_{td}^{(i)} \rfloor$ and $\lfloor O_{td}^{(i)} \rfloor$ with probability $\lceil O_{td}^{(i)} \rceil - O_{td}^{(i)}$ in case $O_{td}^{(i)}$ is not an integer. This guarantees that $E[O_{td}^{(i*)}] = E[O_{td}^{(i)}] = E[D^{(i)}]$.

A joint order O_{td}^{*} of both retailers placed at time td equals $O_{td}^{(1*)} + O_{td}^{(2*)}$. Recall that both these orders are only delivered by the manufacturer when a joint order has been produced. Next, define the following random variables:

- t_n: the time of the nth observation point, defined as the nth time slot during which the server is busy;
- $a(n)$: the arrival time of the joint order in service at time t_n;
- B_n: the age of the joint order in service at time t_n, expressed in time slots, i.e., $B_n = t_n - a(n)$;
- C_n: the number of items of the joint order in service that still need to start or complete service at time t_n;
- S_n: the service phase at time t_n.

All events such as arrivals, transfers from the waiting line to the server, and service completions are assumed to occur at instants immediately after the discrete time epochs. This implies that the age of an order in service at some time epoch t_n is at least 1. We start by introducing the Markov chain for the case where both retailers smooth.

Both Retailers Smooth

It is clear that the stochastic process $(B_n, C_n, O^{(1)}_{a(n)}, O^{(2)}_{a(n)}, S_n)_{n\geq 0}$ forms a discrete-time Markov process on the state space $\mathbb{N}_0 \times \{(c,x_1,x_2)|c \in \{1,\ldots,m_D\}, 1 \leq x_i \leq m_D^{(i)}, i \in \{1,2\}\} \times \{1,2\}$, as the PH service requires only two phases. Note that the process makes use of the order quantities $O^{(i)}_{a(n)}$ instead of the integer values $O^{(i*)}_{a(n)}$. Given that these order quantities are real numbers, the Markov process $(B_n, C_n, O^{(1)}_{a(n)}, O^{(2)}_{a(n)}, S_n)_{n\geq 0}$ has a continuous state space, which makes it very hard to find its steady state vector.

Therefore, instead of keeping track of $O^{(i)}_{a(n)}$ in an exact manner, we will round it in a probabilistic way to the nearest multiple of $1/g$, where $g \geq 1$ is an integer termed the *granularity* of the system. Clearly, the larger g is, the better the approximation will be. Hence, we approximate the foregoing Markov process by the Markov chain $(B_n, C_n, O^{g,(1)}_{a(n)}, O^{g,(2)}_{a(n)}, S_n)_{n\geq 0}$ on the discrete state space $\mathbb{N}_0 \times \{(c,x_1,x_2)|c \in \{1,\ldots,m_D\}, x_i \in \mathbb{S}_g^{(i)}, i \in \{1,2\}\} \times \{1,2\}$, where $\mathbb{S}_g^{(i)} = \{1, 1+1/g, 1+2/g, \ldots, m_D^{(i)}\}$ and the quantity $O^{g,(i)}_{td}$ evolves as follows. Let

$$x = (1-\beta_i)O^{g,(i)}_{(t-1)d} + \beta_i D^{(i)};$$

then $O^{g,(i)}_{td} = x$ if $x \in \mathbb{S}^{(i)}$; otherwise it equals $\lceil x \rceil_g$ with probability $g(x - \lfloor x \rfloor_g)$, or $\lfloor x \rfloor_g$ with probability $g(\lceil x \rceil_g - x)$, where $\lceil x \rceil_g$ ($\lfloor x \rfloor_g$) rounds up (down) to the nearest element in $\mathbb{S}_g^{(i)}$. Notice that, by induction, we have $E[O^{g,(i)}_{td}] = E[D^{(i)}]$. Using this probabilistic rounding, we can easily compute the conditional probabilities $P[O^{g,(i)}_{td} = q'|O^{g,(i)}_{(t-1)d} = q]$, which we denote by $p_g^{(i)}(q,q')$, from $D^{(i)}$ (see [4, Eq. (12)] for details).

The transition matrix P_g of the Markov chain $(B_n, C_n, O^{(1)}_{a(n)}, O^{(2)}_{a(n)}, S_n)_{n\geq 0}$ is a GI/M/1-type Markov chain [8] with the following structure:

$$P_g = \begin{bmatrix} A_d & A_0 & & & \\ \vdots & & \ddots & & \\ A_d & & & A_0 & \\ & A_d & & & A_0 \\ & & \ddots & & & \ddots \end{bmatrix},$$

as B_n either increases by one if the same joint order remains in service or decreases by $d-1$ if a joint order is completed. Hence, there are d occurrences of A_d on the first block column. The size m of the square matrices A_0 and A_d is $2m_D m_g$, with $m_g = \prod_{i=1}^2 (m_D^{(i)} g - g + 1)$, which is typically such that we cannot store the matrices A_0 and A_d in memory. Although we can eliminate close to 50% of the states by removing the transient states with $C_n > \lceil O^{(1)}_{a(n)} \rceil + \lceil O^{(2)}_{a(n)} \rceil$, the size m remains problematic, and this would slow down the numerical solution method presented in the section "Numerical Solution." A more detailed discussion of the structure of A_0 and A_d is given in the section "Fast Multiplication."

One Retailer Smooths

Assume without loss of generality that retailer 1 smooths, while retailer 2 does not, i.e., $\beta_1 < 1$ and $\beta_2 = 1$. In this case we can also rely on the Markov chain defined previously, but now there is no longer a need to keep track of $O^{g,(2)}_{a(n)}$, as the orders of retailer 2 are distributed according to $D^{(2)}$. This not only simplifies the transition probabilities but also considerably reduces the time and memory requirements of the numerical solution method introduced in the section "Numerical Solution." Although storing the matrices A_0 and A_d in memory may no longer be problematic, a numerical approach as presented in the next section outperforms the more traditional approach, which relies on computing the rate matrix R [8] by a considerable margin.

Numerical Solution

The objective of this section is to introduce a numerical method to compute the steady state distribution of the Markov chain introduced in the section "Both Retailers Smooth" by avoiding the need to store matrices A_0 and A_d.

Fast Multiplication

To multiply the vector $x = (x_0, x_1, \ldots)$ by P_g, where x_i is a length $m = 2m_D m_g$ vector, without storing matrix A_0 or A_d, we will write P_g as the sum of $P^{(0)}_g + P^{(d)}_g =$

$$\begin{bmatrix} A_0 & & & \\ & \ddots & & \\ & & A_0 & \\ & & & \ddots \end{bmatrix} + \begin{bmatrix} A_d & & \\ \vdots & & \\ A_d & & \\ & \ddots & \end{bmatrix}$$

and compute xP_g as $xP^{(0)}_g + xP^{(d)}_g$. To express the time complexity of these multiplications, assume $x_i = 0$ for $i \geq n$ for some n (as will be the case in the next subsection).

The matrix A_0 corresponds to the case where the same joint order remains in service, meaning C_n either remains the same or decreases by one. Due to the order of the random variables, matrix A_0 is a bidiagonal block Toeplitz matrix, with blocks of size $2m_g$. The block appearing on the main diagonal equals $I \otimes U$, as the production of the same item continues in this case. The block below the main diagonal is $I \otimes u^* \alpha$, as the item is finished, but at least one item of the joint order still needs to be produced. Hence,

$$A_0 = \begin{bmatrix} I \otimes U & & & \\ I \otimes u^* \alpha & I \otimes U & & \\ & \ddots & \ddots & \\ & & I \otimes u^* \alpha & I \otimes U \end{bmatrix},$$

where I is the size m_g unity matrix and we have m_D blocks $I \otimes U$ on the main diagonal. As the PH representation is of order 2 (even in the case of low variability), we can multiply x by $P_g^{(0)}$ in $O(mn)$ time.

When multiplying by A_d, we first argue that A_d can be written as

$$A_d = (e_1 \otimes (I \otimes u^*))(W_1 \otimes W_2)(Y \otimes \alpha),$$

where e_1 is a size m_D column vector that equals one in its first entry and zero elsewhere, W_i is a square matrix of size $m_D^{(i)} g - g + 1$, and Y is an $m_g \times m_g m_D$ matrix. To understand this decomposition, we split the transition into four steps. First, a service completion of an order must occur, meaning C_n must equal one and the item in service must be completed. Thus, the matrix $(e_1 \otimes (I \otimes u^*)$ describes this step. Next, in step 2, we determine the new order size for each retailer based on the previous order size (using the granularity g). Let the (q,q')th entry of W_i equal $p_g^{(i)}(q,q')$ (as defined in the section "Both Retailers Smooth") for $i = 1,2$. As each retailer determines its next order size independently, $W_1 \otimes W_2$ captures step 2. To complete the transition, we need to determine the joint *integer* order size given the individual *granularity* g order sizes of both retailers (in step 3) and the initial service phase of the first item of the joint order (in step 4). Step 4 is clearly determined by α, while step 3 corresponds to the matrix Y. A row of the matrix Y contains either 1, 2, or 4 nonzero entries (depending on whether the row corresponds to a case where both, one, or none of the granularity g orders are integers).

Thus, when $x = (x_0, x_1, \ldots)$ is multiplied by $P_g^{(d)}$, each of the vectors x_i is first reduced to a length m_g vector in $O(nm_g)$ time because of $(e_1 \otimes (I \otimes u^*))$. A multiplication by $W_1 \otimes W_2$ is done in two steps. First we multiply by $(I \otimes W_2)$, which can be trivially done in $O((m_D^{(2)} g)^2 m_D^{(1)} g) = O(m_g m_D^{(2)} g)$ for each vector, followed by multiplication by $(W_1 \otimes I)$. This latter multiplication can be rewritten as a multiplication by $(I \otimes W_1)$ using the shuffle algorithm[5]. Hence, it can also be done in $O(m_g m_D^{(1)} g)$. Due to its sparse structure, a multiplication by Y can be implemented in $O(m_g)$ time. In conclusion, the overall time required to multiply x by $P_g^{(d)}$ can be written as $O(nm_g(m_D^{(1)} + m_D^{(2)})g) = O(nmg)$ and the time needed to multiply x by P_g is therefore also $O(nmg)$. In practice, for g small, the multiplication by $P_g^{(0)}$ is more time demanding than the multiplication by $P_g^{(d)}$, and a considerable percentage of the time is also spent on allocating memory.

Power Method, Gauss–Seidel Iteration, and GMRES

To determine the steady state probability vector of the transition matrix P_g we rely on the fast matrix multiplication between a vector x and P_g introduced previously.

When this fast multiplication technique is combined with the power method, we basically start with some initial vector $x(0)$ and define $x(k+1) = x(k)P_g$ until the infinity norm of $x(k+1) - x(k)$ is smaller than some predefined ε_1 (e.g., $\varepsilon_1 = 10^{-8}$). If we start from an empty system, then $x(0)$ has only one nonzero component $x_0(0)$ of length m and $x(k)$ has $k+1$ nonzero components $x_0(k)$ to $x_k(k)$. Whenever some of the last components are smaller than some predefined ε_2, we reduce the length of $x(k)$ (by adding these components to the last component larger than ε_2). Notice that introducing ε_2 is not exactly equivalent to a truncation of the Markov chain at some predefined level N. Instead we dynamically truncate the vector x during the computation, and its length may still vary over time. The impact of both ε_1 used by the stopping criteria and ε_2 used by the dynamic truncation will be examined in the section "Computation Times and Accuracy." Both these parameters will be used in a similar manner for the other iterative schemes as well.

When applying the *forward* Gauss–Seidel iteration [9], we compute $x(k+1)$ from $x(k)$ by solving the linear system

$$x(k+1)(I - P_g^{(0)}) = x(k)P_g^{(d)},$$

which can be done efficiently using forward substitution as $(I - P_g^{(0)})$ is upper triangular. If x is an arbitrary stochastic vector, we initialize $x(0)$ such that it solves $x(0)(I - P_g^{(0)}) = x$. As indicated in [9], this Gauss–Seidel iteration is equivalent to a preconditioned power method if we use $(I - P_g^{(0)})$ as the preconditioning matrix M. Notice that we can benefit from the fast multiplications discussed in the previous section when computing $x(k)P_g^{(d)}$ as well as during the forward substitution phase.

The GMRES method [10] computes an approximate solution of the linear system $(I - P'_g)x = 0$ by finding a vector $x(1)$ that minimizes $\|(I - P'_g)x\|_2$ over the set $x(0) + \mathcal{K}(I - P'_g, r_0, n)$. Here r_0 is the residual of an initial solution $x(0)$: $r_0 = -(I - P'_g)x(0)$; $\mathcal{K}(I - P'_g, r_0, n)$ is the Krylov subspace, i.e., the subspace spanned by the vectors $\{r_0, (I - P'_g)r_0, \ldots, (I - P'_g)^{n-1}r_0\}$; and n is the dimension of the Krylov subspace [6]. To do this, GMRES relies on the Arnoldi iteration to find an orthonormal basis V_n for the Krylov subspace such that $V'_n(I - P'_g)V_n = H_n$, where H_n is an upper Hessenberg matrix of size n. Once V_n and H_n are obtained, a vector y_n is found such that $J(y) = \|\beta e_1 - \tilde{H}_n y\|_2$ is minimized. Here β is the two-norm of r_0, e_1 is the first column of the identity matrix, and $\tilde{H}_n$ is an $(n+1) \times n$ matrix whose first n rows are identical to H_n and whose last row has one nonzero element that also results from the Arnoldi iteration. A new approximate solution $x(1)$ is computed as $x(1) = x(0) + V_n y_n$. The process is then repeated with $x(1)$ as $x(0)$ until the difference between two consecutive solutions is less than some predefined ε. Although this algorithm is defined to solve linear systems of the type $Ax = b$, with A nonsingular, it can also be used to solve homogeneous systems with A singular, as is the case with Markov chains [11].

The GMRES algorithm also benefits from the fast multiplication discussed in the previous section. To find the residual r_0 at each iteration, we need to compute the product $(I - P'_g)x(0) = x(0) - P'_g x(0)$. Also, for the Arnoldi process we need to determine the vectors $v_j = (I - P'_g)^{j-1} r_0$, which are computed iteratively, and require $n-1$ products of the type $(I - P'_g)v_{j-1} = v_{j-1} - P'_g v_{j-1}$. As with the power method, when analyzing several scenarios we can use the final approximate solution of one scenario as the starting solution for the next one to speed up convergence.

Safety Stock

The required safety stock SS_i for each retailer to guarantee a certain fill rate is one of the main performance measures of this supply chain problem. The derivation for the case where both retailers smooth is nearly identical to the one presented in [4] and is mainly included for reasons of completeness. As indicated in the section "Replenishment Rules," computing SS_i is equivalent to determining the base stock S_i provided that we know the mean replenishment lead time $E[T_r]$ (which equals the floor of the production lead time T_p). The production lead time distribution T_p is easy to obtain from the steady state probability vector π of P_g as follows. First define the length-$2m_g$ vectors $\pi_b(c)$ as the steady state probabilities of being in a state with $B_n = b$ and $C_n = c$. Then the probability of having a production lead time of b slots equals

$$P[T_p = b] = \rho \pi_b(1)(e \otimes u^*)/(1/d)$$

for $b > 0$, where $\rho = 2(E[D^{(1)}] + E[D^{(2)}])/d$ is the load at the manufacturer and $1/d$ the arrival rate of the joint orders.

The fill rate is defined as $1 - E[(-NS)^+]/E[D]$, where NS is the net stock (i.e., inventory on hand minus backlog) and $x^+ = \max\{0, x\}$. Hence, $E[(-NS)^+]$ is the expected number of backlogged items. As in [4, Sect. 5.1], we can show that

$$NS_i = S_i + \sum_{j=1}^{k} D^{(i)} + O_k^{(i)}/\beta, \tag{11.1}$$

where k is the age, expressed in periods, of the joint order in production at the manufacturer at the end of a period, and this joint order contains $O_k^{(i)}$ items for retailer i for $i = 1,2$. If $k = 0$, meaning the last order left the queue before the end of the period, then $O_k^{(i)}$ is the number of items ordered by retailer i in the next joint order. Thus, the key step in determining the required base-stock value S_i consists in computing the joint probabilities $p_{k,q}^{(i)}$ of having an order of age kd in service when a period ends and the order in service contains q items for retailer i for $i = 1,2$, $k \geq 0$, and $q \in \{1, \ldots, m_D^{(i)}\}$.

These joint probabilities can be readily obtained from the steady state of the Markov chain introduced in the section "Both Retailers Smooth" as

$$p_{k,q}^{(i)} = \rho d \pi_{kd}^{(i)}(q) e$$

for $k > 0$, where $\pi_b^{(i)}(q)$ is the steady state vector for the states with $B_n = b$ and $O_{a(n)}^{g,(i)} = q$. For $k = 0$, we note that an order finds the queue empty upon arrival if the previous order had a lead time of at most $d-1$, yielding

$$p_{0,q}^{(i)} = \rho d \sum_{b=1}^{d-1} \sum_{q_1,q_2,s} \pi_b(1,q_1,q_2,s) u_s^* p_g(q_i,q),$$

where $\pi_b(c,q_1,q_2,s)$ is the steady state probability of state (b,c,q_1,q_2,s).

If we wish to compute the joint probabilities $p_{k,q}^{(2)}$ from the Markov chain $(B_n, C_n, O_{a(n)}^{g,(1)}, S_n)_{n\geq 0}$ in case only the first retailer smooths, then things are somewhat more involved when $k > 0$. For $k = 0$ we clearly have

$$p_{0,q}^{(2)} = P[T_{\mathrm{p}} < d] P[D^{(2)} = q].$$

For $k > 0$ we start by computing $p_w(q_1,x)$, the probability that an order consisting of q_1 items for retailer 1 has a waiting time of $x > 0$ slots. As the waiting time x of an order with $x > 0$ equals the lead time of the previous order minus the interarrival time d, we find

$$p_w(q_1,x) = \frac{\rho d}{\pi(q)} \sum_{q,s} \pi_{x+d}(1,q,s) u_s^* p_g(q,q_1),$$

where $\pi_b(c,q,s)$ is the steady state probability of state (b,c,q,s) and $\pi(q)$ is the probability that an arbitrary order contains q items for retailer 1.

Next, we determine the probabilities $p_o(q_1,q_2,y)$ that an arbitrary joint order consists of q_i items for retailer i and its production time equals y time slots. These probabilities are readily obtained from $p_g(q,q')$ and (α, U). Then

$$p_a(q_1,q_2,x) = \sum_{y\geq x} \frac{p_o(q_1,q_2,y)}{2(E[D^{(1)}] + E[D^{(2)}])}$$

is the probability that we find a joint order consisting of q_i items for retailer i in service at an arbitrary moment when the server is busy, while the service of this joint order started x time slots ago. Taking the convolution over x between $p_w(q_1,x)$ and $p_a(q_1,q_2,x)$ and summing over q_1 gives us the probability that the order in service has an age of x time slots and consists of q_2 items for retailer 2, given that we observe the system when the server is busy. From these probabilities the joint probabilities $p_{k,q}^{(2)}$ are readily found.

We can also compute the probabilities $p_{k,q}^{(2)}$ from the Markov chain in the section "Both Retailers Smooth" by setting $\beta_2 = 1$, but this approach requires more time and considerably more memory. As required, the numerical experiments indicated a perfect agreement between both approaches.

Numerical Examples

In this section we illustrate the effect of smoothing on the performance of the production/inventory system. We focus on the safety stock as the main measure of performance and consider various scenarios for the demand distribution, the load, and the smoothing parameters β_1 and β_2. The required safety stock in all the numerical examples guarantees a fill rate of 0.98.

For the demand we consider three different distributions; let us call the three associated random variables X, Y, and Z, respectively. X is defined as $X = 1 + \hat{X}$, where $\hat{X}$ is a binomial distribution with parameters $N - 1$ and $p = 1/2$. Thus, X takes vales on the set $\{1,\ldots,N\}$. The expected value and variance of X are $E[X] = (N+1)/2$ and $\text{Var}(X) = (N-1)/4$. The second random variable Y is uniformly distributed between 1 and N, and its expected value and variance are $E[Y] = (N+1)/2$ and $\text{Var}(Y) = (N^2-1)/12$. The last random variable is defined as $P(Z = k) = (1+\alpha)P(Y = k) - \alpha P(X = k)$ for $k = 1,\ldots,N$. As a result, Z has a U-shaped probability mass function, with $E[Z] = (N+1)/2$ and $\text{Var}(Z) = (N^2 - 1 + \alpha(N^2 - 3N + 2))/12$. Clearly, for Z to be a proper random variable, the value of α must be such that $P(Z = k) \geq 0$ for all k. In our experiments we set $N = 10$, for which α can take values up to roughly 0.68. We choose $\alpha = 0.6$ to make Z highly variable. With this setup, $\text{Var}(X) = 2.25$, $\text{Var}(Y) = 8.25$, and $\text{Var}(Z) = 8.25 + 6\alpha = 11.85$. Also, if we set the maximum demand size to $N = 10$, then the size of the square matrices A_0 and A_D ranges from 4,000 (for $g = 1$) to 84,640 (for $g = 5$).

As mentioned previously, the mean production time is set equal to 2, and for the experiments in this section the standard deviation is also set to 2. The load is set by adjusting d, the number of slots between two orders placed by the retailers. In our setup we choose d from the set $\{40, 34, 29, 26\}$, which generates loads of roughly $\{0.55, 0.65, 0.76, 0.85\}$, respectively. We will start by looking at the case where both retailers use the same value of the smoothing parameters β_1 and β_2. Afterward we consider the case where these parameters may differ. However, before we generate any numerical results let us first evaluate the impact of discretizing the state space (that is, the impact of the granularity g) as well as the parameters ε_1 and ε_2 used in the stopping criteria and dynamic truncation of the state space, respectively.

Computation Times and Accuracy

We start by looking at the accuracy and computation times required to obtain the results in the paper with the power, Gauss–Seidel, and GMRES methods when $g = 1$ (even though larger g values are needed for small β values as indicated below). Table 11.1 shows the residual error of the steady state vector, that is, the norm of $xP_g - x$, as well as the accuracy of SS and $E[T_r]$ for both the power and Gauss–Seidel methods when compared against a solution obtained with $\varepsilon_1 = \varepsilon_2 = 10^{-14}$

Table 11.1 Accuracy and computation times of the power and Gauss–Seidel methods for $\varepsilon_2 = 10^{-9}$

	Power			Gauss–Seidel		
ε_1	10^{-6}	10^{-7}	10^{-8}	10^{-6}	10^{-7}	10^{-8}
Residual error	1.76E−5	1.68E−6	1.85E−7	2.14E−6	2.38E−7	2.56E−8
SS	0.64%	0.03%	0.01%	1.10%	0.17%	0.02%
$E[T_r]$	0.11%	0.02%	0.00%	0.63%	0.09%	0.01%
Time (s)	31	54	79	1.7	3.0	4.4
Iteration	804	1,207	1,636	21	34	49

(by the power method). The table also lists the computation times and the required number of iterations. Table 11.2 provides the same data for the GMRES method, where the size of the Krylov subspace was set equal to 1, 3, and 5. These results correspond to the example where the demand follows a binomial distribution, the load $\rho = 0.85$ (which is the most demanding among the four loads considered), and both retailers smooth with $\beta_1 = \beta_2 = 0.8$. All the experiments were run on a PC with four cores at 2.93 GHz and 4 GB of RAM. We observe that, for the same ε_1, the Gauss–Seidel method is far superior to both the power method and GMRES, as it requires substantially less time than the power method and has an accuracy similar to that of the power method. This can be explained by the fact that the Markov chain characterized by P_g typically makes many consecutive upward transitions according to A_0 followed by an occasional downward jump using A_d.

The accuracy of GMRES is quite poor for larger ε_1 values and is far worse than the power or Gauss–Seidel method. As ε_1 decreases, the difference in accuracy between GMRES and the other methods becomes smaller (and eventually negligible). GMRES is faster than the power method for $\varepsilon_1 = 10^{-6}$ and when n is 1 or 3, but, as indicated above, the accuracy of GMRES is poor in these cases. As stated in the section "Power Method, Gauss–Seidel Iteration, and GMRES," the Gauss–Seidel method may be regarded as a preconditioned power method where the preconditioning matrix M is equal to $(I - P_g^{(0)})$. In principle we can use the same preconditioning for GMRES, which should improve the performance of GMRES significantly. However, as GMRES is typically inferior to the power method, it seems unlikely that we can do better than the Gauss–Seidel method using $(I - P_g^{(0)})$ as a preconditioning matrix.

Next, let us have a look at the impact of the granularity of g on the results for the Gauss–Seidel method only, as the other methods are too time consuming for larger g values. We let g vary from 1 to 5 for a load $\rho = 0.85$, while $\varepsilon_1 = 10^{-8}$ and $\varepsilon_2 = 10^{-9}$. Figure 11.1 depicts the required safety stock SS as a function of β for the three demand distributions discussed previously. These results indicate that for β close to one, letting $g = 1$ suffices; however, for smaller β values setting $g = 1$ may lead to a serious overestimation of the required safety stock. Thus, to guarantee an acceptable accuracy for smaller β values, we generated all the subsequent results with $g = 5$ (and $\varepsilon_1 = 10^{-8}$ and $\varepsilon_2 = 10^{-9}$).

Table 11.2 Accuracy and computation times of GMRES for $\varepsilon_2 = 10^{-9}$

	GMRES ($n = 1$)			GMRES ($n = 3$)			GMRES ($n = 5$)		
ε_1	10^{-6}	10^{-7}	10^{-8}	10^{-6}	10^{-7}	10^{-8}	10^{-6}	10^{-7}	10^{-8}
Residual error	4.18E−5	5.31E−6	1.16E−7	2.79E−6	2.47E−7	2.48E−8	1.42E−6	1.33E−7	1.46E−8
SS	15.40%	8.01%	0.92%	9.70%	1.95%	0.29%	6.80%	1.23%	0.19%
$E[T_r]$	10.63%	4.70%	0.48%	6.86%	1.07%	0.14%	4.44%	0.64%	0.09%
Time (s)	15	34	105	21	64	290	38	170	446
Iteration	186	261	797	89	341	301	61	120	190

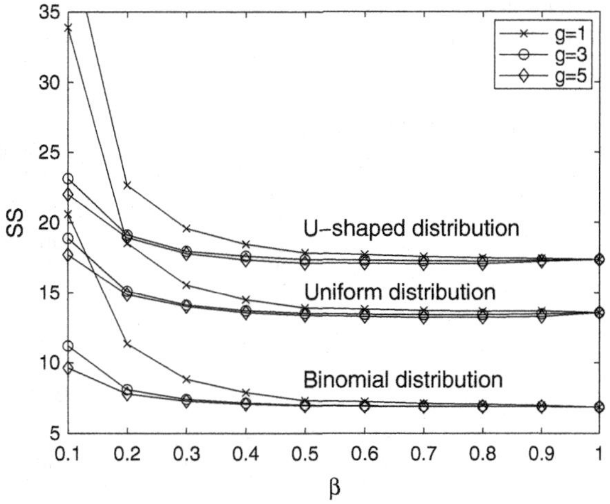

Fig. 11.1 Safety stock as a function of β

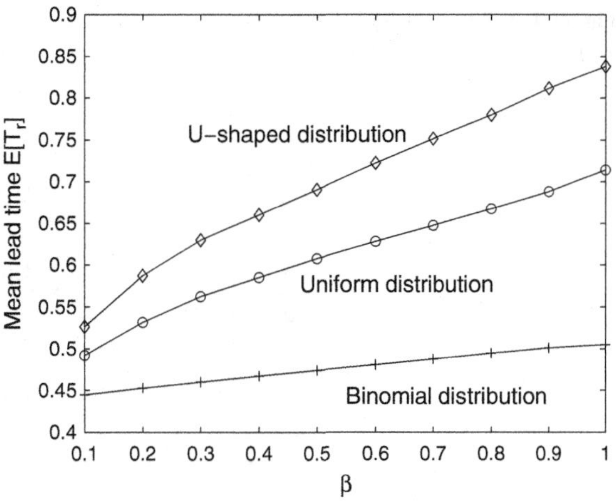

Fig. 11.2 Mean lead time $E[T_r]$ vs. $\beta = \beta_1 = \beta_2$ for $\rho = 0.85$

Finally, we would like to mention that a significant amount of the computation time is devoted to allocating memory, due to the large sizes of the vectors, e.g., the size of the final vector x in Fig. 11.1, for $\beta = 0.8$ and the binomial distribution, is 732,000 (for $g = 1$) and 15,065,920 (for $g = 5$). Since GMRES computes n large vectors, it is more significantly affected by the memory allocation delay. Also, the computation times of all the methods are highly influenced by the system parameters, especially by the load ρ and the variance of the demand and processing times. Larger values for these parameters imply longer computation times and larger memory requirements.

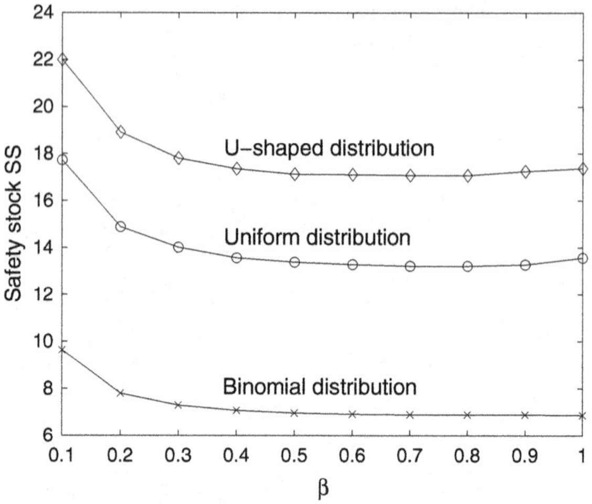

Fig. 11.3 Safety stock SS vs. $\beta = \beta_1 = \beta_2$ for $\rho = 0.85$

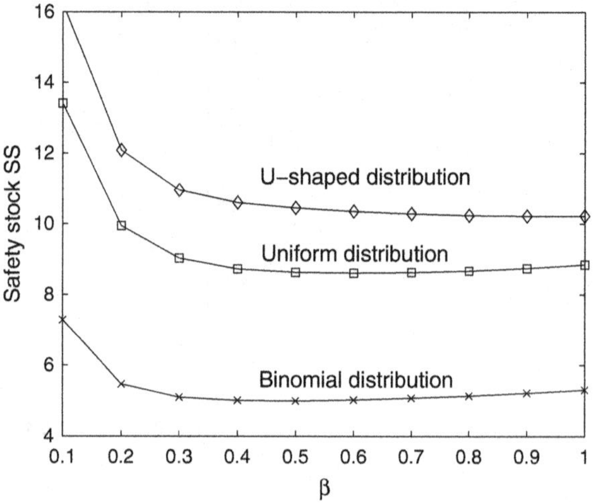

Fig. 11.4 Safety stock SS vs. $\beta = \beta_1 = \beta_2$ for $\rho = 0.65$

Homogeneous Smoothing

We start by looking at a system facing a load of $\rho = 0.85$, and we consider values of $\beta = \beta_1 = \beta_2$ in the set $\{0.1, 0.2, \ldots, 1\}$ and the three different demands described previously. The results are included in Fig. 11.2, where we observe that the mean replenishment lead time increases as a function of β, meaning both retailers benefit from smoothing with respect to the replenishment time. As expected, the lead time reduction increases with the variability of the demand distribution. This reduction in the lead time is key in understanding the effect of β on the safety stock.

Figure 11.3 depicts the corresponding safety stock to guarantee a fill rate of 0.98. The results indicate that unless β is small, the required safety stock does not increase

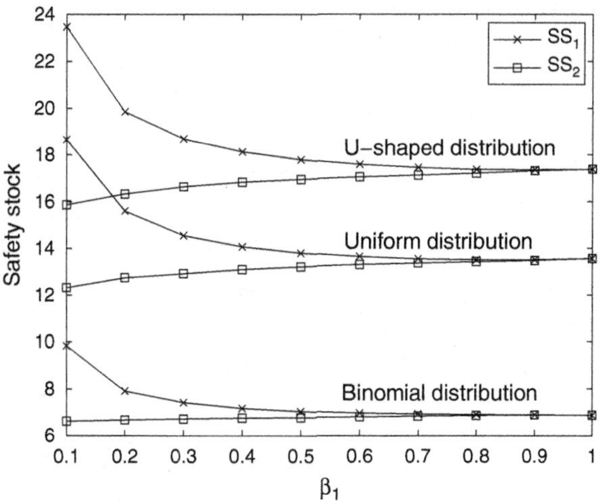

Fig. 11.5 Safety stock SS vs. $\beta_1 - \beta_2 = 1$ and $\rho = 0.85$

much as β decreases, meaning both retailers can perform a considerable amount of smoothing without the need to increase their SS much. Note that, as β decreases, the response of the retailer to a sudden increase in the demand tends to become slower, which intuitively should result in an increased SS. However, the decrease in the lead time (partially) compensates the slower response. When β becomes too small, the reduction in the lead time is insufficient to avoid a significant increase in the SS. Actually, when β decreases, starting in $\beta = 1$, the SS initially even decreases slightly in the case of more variable demand.

Similar results were obtained for lower load scenarios as well, and the corresponding plot for an approximate load of $\rho = 0.65$ (i.e., for $d = 29$) is given in Fig. 11.4. These results and insights are similar in nature to the single-retailer case [4].

Heterogeneous Smoothing

We start by considering the scenario where only one retailer smooths, say retailer 1. Thus, we assume that β_2 is fixed and equal to one, while β_1 changes. As expected, the mean lead time can be shown to decrease as β_1 decreases. Figure 11.5 depicts the safety stock of both retailers as a function of β_1 for $\rho = 0.85$. The results indicate that the safety stock SS_1 of retailer 1 behaves very similar to the homogeneous case (it is a fraction larger to be precise). Thus, the retailer can still smooth his demand considerably without affecting his safety stock too much. The safety stock of the second retailer SS_2, on the other hand, decreases slightly as β_1 decreases. This can be understood by noting that the second retailer also benefits from the reduced lead time while being more reactive to a sudden increase in the demand than retailer 1 (as $\beta_2 = 1$).

Fig. 11.6 Safety stock SS vs. $\beta_1 - \beta_2 = 1$ and $\rho = 0.65$

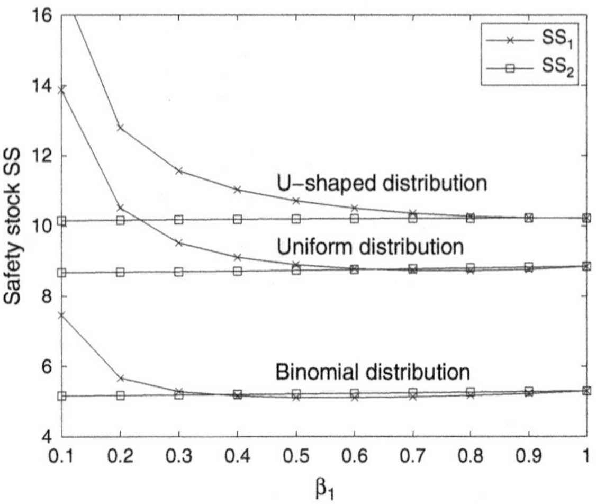

Fig. 11.7 Base stock S vs. $\beta_1 - \beta_2 = 1$ and $\rho = 0.65$

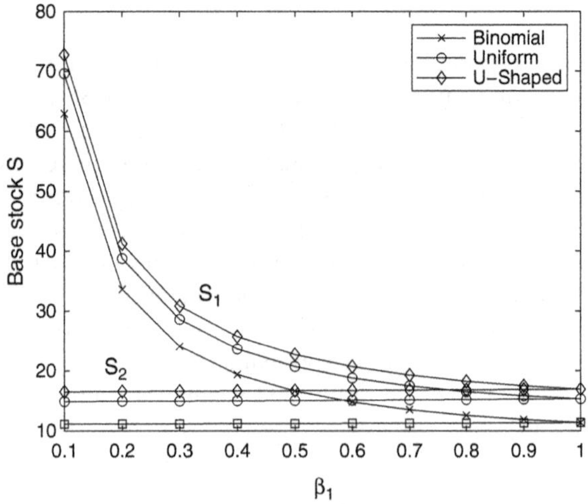

In Fig. 11.6 we consider the same example, but with a reduced load $\rho = 0.65$. In this case we observe a more remarkable result: the safety stock SS_1 of retailer 1 first decreases and is even below the safety stock SS_2 of retailer 2 for some β_1 values. This may seem counterintuitive at first as both retailers benefit from the reduction in lead time, while the second is still more reactive. To understand this, consider Eq. (11.1) for the net stock distribution NS of retailer i. The last term $O_k^{(i)}/\beta$ is clearly larger on average for retailer 1, but $O_k^{(1)}$ is less variable than $O_k^{(2)}$ as the orders of retailer one are smoothed. Thus, if S_1 is chosen larger than S_2 to compensate for the larger average of $O_k^{(i)}/\beta$, the lower variability of $O_k^{(i)}$ might indeed result in a less variable net stock (for β sufficiently close to one) and therefore in a smaller safety stock as well. Figure 11.7 shows that this is exactly what happens: S_1 decreases, while S_2 increases as a function of β_1.

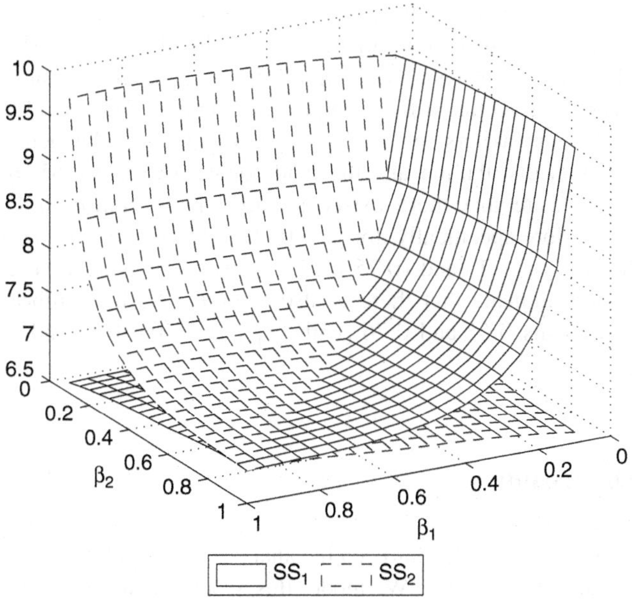

Fig. 11.8 Safety stock SS_1 and SS_2 vs. β_1 and $\beta_2 = 1 - \rho = 0.85$, binomial demand distribution

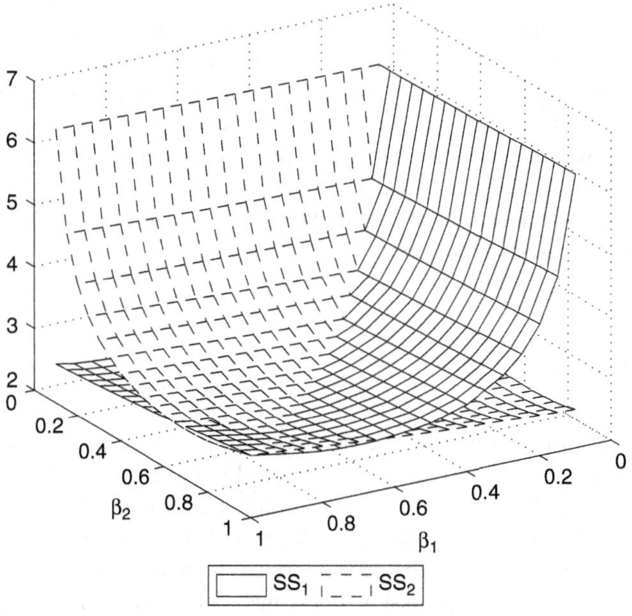

Fig. 11.9 Safety stock SS_1 and SS_2 vs. β_1 and $\beta_2 = 1 - \rho = 0.55$, binomial demand distribution

If we consider the selection of β_1 and β_2 in a game theoretic setting, where the objective of retailer i exists in minimizing SS_i, it is already clear from Fig. 11.6 that $(\beta_1,\beta_2) = (1,1)$ is not always a Nash equilibrium,[1] as retailer 1 can decrease his safety stock SS_1 by selecting a β_1 less than one. Figures 11.8 and 11.9 depict the safety stock of both retailers for $\beta_1,\beta_2 \in \{0.1,0.15,0.2\ldots,1\}$ when the demand follows a binomial distribution and the load equals 0.85 and 0.55, respectively. These results indicate that there exists a unique Nash equilibrium (β_1,β_2) in these scenarios. More specifically, for $\rho = 0.85$ and 0.55 the Nash equilibrium is located in $(\beta_1,\beta_2) = (1,1)$ and $(0.75,0.75)$, respectively. Numerical experiments not depicted here indicate that there is also a unique Nash equilibrium (β_1,β_2) when the load equals 0.65 and 0.76 [with $(\beta_1,\beta_2) = (0.5,0.5)$ and $(0.85,0.85)$, respectively].

Further Discussion

The main focus of this chapter has been the analysis of a supply chain with a single manufacturer and two retailers. We model this system as a GI/M/1-type Markov chain with blocks whose size is large enough to make the computation of the (dense) matrix R, with traditional algorithms, infeasible. To overcome this issue, we propose the use of numerical methods, such as the power method, Gauss–Seidel, and GMRES, to compute the stationary probability vector of the chain. As these methods rely heavily on vector-matrix multiplications, we exploited the structure of the transition-matrix blocks to perform these multiplications efficiently. Clearly, the same approach can be used to analyze other systems modeled as a structured Markov chain whose blocks are large and possess an inner structure that can be exploited to perform the vector-matrix multiplications. In this section, we conclude the paper with two fairly arbitrary examples of other systems that can be analyzed with the approach used in the paper.

An Edge Router

Edge routers provide access to core networks from service providers as well as carrier networks. For instance, edge routers are located at the edge of an Internet service provider (ISP) network, connecting multiple users to the ISP's core network. Therefore, the edge router typically has multiple low-speed interfaces (connected to the users) and one (or a few) high-speed interfaces (connected to the core network). Given the difference in transmission rates, the router may collect multiple, say

[1] A common strategy is called a Nash equilibrium if neither player can improve his objective by deviating from the common strategy.

b, packets arriving from the low-speed interfaces into a single packet to forward through the high-speed channel. Assuming the router always collects b user-generated packets into one packet for high-speed transmission, we can model the number of user-generated packets in the system (buffered and in transmission) as a (continuous-time) GI/M/1-type Markov chain with only three nonzero blocks.

Let $N(t)$, $S(t)$, and $J(t)$ be, respectively, the number of packets, the service phase of the packet in transmission, and the phase of the arrival process at time t. The service time distribution is a continuous PH distribution with parameters (α, T), and the packet arrival process is a Markovian arrival process (MAP) with parameters (D_0, D_1) [7]. The process $X(t) = \{(N(t), S(t), J(t)), t \geq 0\}$ is thus a continuous-time Markov chain with generator matrix

$$Q = \begin{bmatrix} B_1 & B_0 & & & & & & \\ & A_1 & A_0 & & & & & \\ & & A_1 & A_0 & & & & \\ & & & \ddots & \ddots & & & \\ B_{b+1} & & & & A_1 & A_0 & & \\ & A_{b+1} & & & & A_1 & A_0 & \\ & & A_{b+1} & & & & A_1 & A_0 \\ & & & \ddots & & & \ddots & \ddots \end{bmatrix},$$

where $A_0 = D_1 \otimes I$, $A_1 = D_0 \oplus T$, $A_{b+1} = I \otimes t\alpha$, $t = -Te$, and $\oplus$ stands for Kronecker sum [7]. Notice that a packet in service is actually a bundle of b user-generated packets. Assuming the transmission time of the latter packets follows a PH distribution with parameters (β, S), the parameters α and T are given by

$$\alpha = \begin{bmatrix} \beta \; 0 \dots 0 \end{bmatrix}, \text{ and } T = \begin{bmatrix} S & s\beta & & & \\ & S & s\beta & & \\ & & \ddots & \ddots & \\ & & & S & s\beta \\ & & & & S \end{bmatrix},$$

where $s = -Se$. Letting m_s and m_a be the size of the matrices S and D_0, respectively, the block size is $m = bm_s m_a$.

As mentioned previously, the edge router receives packets from many, say n, low-speed interfaces. If the traffic incoming through interface j is modeled as a MAP with parameters (C_0^j, C_1^j), then the total incoming traffic is the superposition of these n MAPs. Thus, D_0 and D_1 are given by

$$D_0 = \oplus_{j=1}^n C_0^j \text{ and } D_1 = \oplus_{j=1}^n C_1^j.$$

If the size of each of the C_0^j matrices is m_u, then the block size is $m = bm_s m_u^n$. As a result, the block size grows linearly with b, the number of user-generated packets per forwarded packet, and exponentially with n, the number of sources. It is clear, then, that the block size can be very large for rather limited values of b and, particularly, n. For instance, with $m_s = m_u = 2$, $b = 10$, and $n = 16$, the block size is over a million. This model is therefore well-suited to be analyzed with the approach proposed in this chapter since, in addition to a large block size, the number of nonzero blocks is small and the blocks have a structure that can be exploited to perform the vector-matrix multiplications efficiently.

FS-ALOHA++

The FS-ALOHA++ algorithm is a contention resolution algorithm used for dynamic bandwidth allocation [12]. This algorithm operates on a time-division multiple access (TDMA) channel that consists of fixed-length frames. Each frame contains, among other things, $T = S + N$ minislots used to support the contention channel. When a user wants to transmit new data, it will send a request packet on the contention channel by selecting one of the first S minislots at random. If one user does not transmit in the same minislot as any other user, then that user is successful. All the users that were involved in a collision (in one of the first S minislots), on the other hand, form a transmission set (TS). Hence, in each frame either one or zero TSs are formed. If a TS is formed, then it joins the back of a (distributed) FIFO queue and is called a backlogged TS.

Backlogged TSs are served, in groups of $K \geq 1$, using ALOHA on the last N minislots, that is, all the users that are part of the first K backlogged TSs select one of the last N minislots at random. Users that are successful leave the contention channel; those involved in a collision retransmit in the next frame in one of the last N minislots. This procedure is repeated until the last N minislots are collision free. As soon as this occurs, the next set of K TSs can make use of the last N minislots (if the queue contains $i < K$ TSs, then only i TSs are served simultaneously).

Under the assumption that new requests form a Poisson process, one can analyze the FS-ALOHA++ algorithm (with parameters S, N, and K) by means of a GI/M/1-type Markov chain with a generalized boundary condition. This is achieved by keeping track of the number of backlogged TSs $N(t)$ at the start of frame t and the number of users $S(t)$ that will make use of the last N minislots in frame t (see [12] for more details). As at most one TS can be added to the back of the queue during a frame and K TSs may start service, one finds that the number of TSs may increase by one, remain fixed, decrease by $K-1$, or decrease by K (provided that at least K TSs are backlogged). Thus, if $N(t)$ represents the level of the Markov chain and $S(t)$ the phase, then one finds that only the matrices A_0, A_1, A_K, and A_{K+1} differ from zero (we do not discuss the boundary matrices here). Further, as the probability of having i users in a TS decreases quickly with i (due to the Poisson arrivals), we can easily truncate the value of $S(t)$, the number of users that are part of K TSs, by some $s_{\max}$.

It is not hard to see that the time needed to serve a group of K TSs can be represented by a PH distribution with an order $s_{\max} - 1$ representation (α_N, T_N). Further, if we denote by p_S the probability that a TS is formed in the first S minislots of a frame, we find that

$$A_0 = p_S T_N, \ A_1 = (1 - p_S) T_N, \ A_K = p_S T_N^* \alpha_N \ A_{K+1} = (1 - p_S) T_N^* \alpha_N,$$

where $T_N^* = e - T_N e$. For details on how to compute p_S, α_N, and T_N we refer the reader to [12]. As values for $s_{\max}$ equal to 20 typically guarantee a very small truncation error, one can easily compute the R matrix of this chain using traditional methods.

However, suppose we wish to modify FS-ALOHA such that the last N minislots are partitioned into M subsets of each N' slots (with $N = MN'$) such that up to M groups of K TSs can be served simultaneously. More specifically, for each of the M subsets that becomes collision free during a frame, we take a group of K TSs from the queue and serve this group using ALOHA on the N' slots of the subset. In other words, we replace the single-server queue with batch service and service time (α_N, T_N) with M batch servers with service time distribution $(\alpha_{N'}, T_{N'})$, where the order of the representation $(\alpha_{N'}, T_{N'})$ is also $s_{\max} - 1$.

In this case, we can still obtain a GI/M/1-type Markov chain in a similar manner, but the phase must maintain the state of *each* of the M servers, which implies that the block size grows very quickly with M and exceeds a few thousand even for $M = 3$ or 4. Further, as several TSs may become collision free in a frame, up to M groups of K TSs may be removed from the queue. This implies that the block matrices A_{iK} and A_{iK+1} will differ from zero for $i = 0, \ldots, M$. Hence, in this case the traditional approach of computing R to obtain the steady state is no longer feasible, but the approach taken in this chapter still applies as the block matrices have a Kronecker product form.

Acknowledgements The first author was supported by the FWO G.0333.10N project entitled "The study of a two-echelon integrated production/inventory system solved by means of matrix analytic models."

References

1. Bini, D.A., Meini, B., Steffé, S., Van Houdt, B.: Structured Markov chains solver: algorithms. In: SMCtools Workshop. ACM Press, Pisa (2006)
2. Bobbio, A., Horváth, A., Telek, M.: The scale factor: a new degree of freedom in phase type approximation. Perform. Eval. **56**, 121–144 (2004)
3. Boute, R., Lambrecht, M., Van Houdt, B.: Performance evaluation of a production/inventory system with periodic review and endogeneous lead times. Nav. Res. Logist. **54**, 462–473 (2007)
4. Boute, R.N., Disney, S.M., Lambrecht, M.R., Van Houdt, B.: An integrated production and inventory model to dampen upstream demand variability in the supply chain. Eur. J. Oper. Res. **178**, 121–142 (2007)

5. Fernandes, P., Plateau, B., Stewart, W.: Efficient descriptor-vector multiplications in stochastic automata networks. J. ACM **45**, 381–414 (1998)
6. Golub, G.H., Van Loan, C.: Matrix Computations. Johns Hopkins University Press, Baltimore (1996)
7. Latouche, G., Ramaswami, V.: Introduction to Matrix Analytic Methods in Stochastic Modeling. ASA-SIAM Series on Statistics and Applied Probability. SIAM, Philadelphia (1999)
8. Neuts, M.F.: Matrix-Geometric Solutions in Stochastic Models. John Hopkins University Press, Baltimore (1981)
9. Philippe, B., Saad, Y., Stewart, W.J.: Numerical methods in Markov chain modeling. Oper. Res. **40**, 1156–1179 (1992)
10. Saad, Y., Schultz, M.: GMRES: a generalized minimal residual algorithm for solving nonsymmetric linear systems. SIAM J. Sci. Stat. Comput. **7**, 856–869 (1986)
11. Stewart, W.: Introduction to the Numerical Solution of Markov Chains. Princeton University Press, Princeton (1994)
12. Vázquez Cortizo, D., García, J., Blondia, C.: FS-ALOHA++, a collision resolution algorithm with QoS support for the contention channel in multiservice wireless LANs. In: Proceedings of IEEE Globecom (1999)

Index

G. Latouche et al. (eds.), *Matrix-Analytic Methods in Stochastic Models*, Springer Proceedings in Mathematics & Statistics 27, DOI 10.1007/978-1-4614-4909-6,